Two Manuals Combined into One Book

FM 7-85 Ranger Unit Operations

And

Technical Manual for 5.56MM, M16A2, M4 and M4A1 Carbine

Department of Defense

Pentagon
Publishing

*FM 7-85

FIELD MANUAL
No. 7-85

HEADQUARTERS
DEPARTMENT OF THE ARMY
Washington, DC, 9 June 1987

FM 7-85
RANGER UNIT OPERATIONS

IMPORTANT

U.S. Army Infantry School Statement on
U.S. NATIONAL POLICY CONCERNING ANTIPERSONNEL LAND MINES

TABLE OF CONTENTS

Preface

CHAPTER 1. Introduction

1-1.	Mission	1-1
1-2.	Purpose and Function	1-2
1-3.	Fundamentals of Ranger Operations	1-2
1-4.	Capabilities and Limitations	1-4
1-5.	Employment Considerations	1-5
1-6.	Command and Control	1-6
1-7.	Communications	1-12

CHAPTER 2. Organization and Equipment

2-1.	Headquarters and Headquarters Company	2-1
2-2.	Regimental Headquarters	2-1

i DISTRIBUTION RESTRICTION: Approved for public release; distribution is unlimited.

2-3.	Regimental Headquarters Company	2-2
2-4.	Ranger Battalion	2-3
2-5.	Special Teams and Elements	2-5

CHAPTER 3. Deployment

3-1.	Readiness	3-1
3-2.	Ranger Ready Force	3-1
3-3.	N-Hour Sequence	3-2
3-4.	Intelligence and Security	3-2
3-5.	Operations	3-3
3-6.	Logistics	3-4
3-7.	Communications-Electronics	3-7
3-8.	Remote Marshalling Base and Intermediate Staging Base	3-7

CHAPTER 4. Insertion, Extraction, Escape, and Evasion

4-1.	Intelligence	4-1
4-2.	Deception	4-1
4-3.	Speed and Mobility	4-2
4-4.	Stealth	4-3
4-5.	Suppression	4-3
4-6.	Security	4-3
4-7.	Night Vision and Electronics Devices	4-3
4-8.	Rehearsals	4-4
4-9.	Sand Tables and Terrain Models	4-4
4-10.	Communications	4-4
4-11.	Planning Insertions	4-4
4-12.	Air Insertion	4-5
4-13.	Airborne Insertion	4-7
4-14.	Air-Land Insertion (Fixed-Wing)	4-12
4-15.	Air Assault Insertion	4-13
4-16.	Amphibious Insertion	4-16
4-17.	Land Infiltration	4-18
4-18.	Stay-Behind Techniques	4-21
4-19.	Extraction	4-21
4-20.	Escape and Evasion	4-22

CHAPTER 5. Strike Operations

5-1.	Application	5-1
5-2.	Raid Operations	5-1
5-3.	Interdiction Operations	5-7

5-4. Personnel and Equipment Recovery Operations 5-10

CHAPTER 6. Special Light Infantry Operations

6-1. Movement 6-1
6-2. Ambush Operations 6-4
6-3. Antiarmor Operations 6-14
6-4. Military Operations on Urbanized Terrain 6-14
6-5. Perimeter Defense 6-15
6-6. Breakout from Encirclement 6-17
6-7. Linkup Operations 6-18
6-8. Passage of Lines 6-19
6-9. Reconnaissance Platoon Operations 6-22
6-10. Reconnaissance Planning 6-23
6-11. Execution of Insertion 6-27

CHAPTER 7. Combat Support

7-1. Intelligence Support 7-1
7-2. Electronic Warfare Support 7-3
7-3. Communications Support 7-4
7-4. Fire Support 7-5
7-5. Aviation Support 7-14
7-6. Engineer Support 7-18

CHAPTER 8. Combat Service Support

8-1. Ranger Support Element 8-1
8-2. Logistics and Personnel 8-3
8-3. Resupply 8-8
8-4. Military Police 8-17
8-5. Counterintelligence 8-18
8-6. Personnel Service Support 8-18

APPENDIX A. Organization and Equipment of the Ranger Regiment A-1

APPENDIX B. Composition and Equipment of Ranger Regiment Liaison Teams B-1

APPENDIX C. Equipment Density C-1

C-1.	Communication Equipment	C-1
C-2.	Weapons and Equipment Each Battalion	C-2
APPENDIX D.	**Standing Orders, Rogers' Rangers**	D-1
APPENDIX E.	**The Ranger Creed**	E-1
APPENDIX F.	**Ranger History**	F-1
APPENDIX G.	**Training**	G-1
G-1.	Application and Guidance	G-1
G-2.	Philosophy	G-2
G-3.	Mission Analysis	G-4
G-4.	Unit Training	G-4
G-5.	Individual Training	G-4
G-6.	Collective Training	G-5
G-7.	Performance-Oriented Training	G-6
G-8.	Evaluation	G-6
G-9.	Field Training	G-7
G-10.	Professional Development	G-9
G-11.	Physical Training	G-9
G-12.	Environmental and Specialized Training	G-10
G-13.	Ranger Indoctrination Program	G-10
Glossary		Glossary-1
References		References-1
Authorization		Authorization Letter

PREFACE

This publication contains doctrinal guidelines for the employment and training of ranger units. The intended audience is intermediate and higher level commanders, staffs of those headquarters, and newly assigned officers and NCOs of the ranger regiment. It is to be used with all other published US doctrine concerning NBC operations and the climatic or geographic influences on combat. The fundamentals apply to the employment of ranger units on worldwide operations and across the spectrum of conflict from low to high intensity warfare. This includes all aspects of the integrated battlefield: nuclear, biological, and chemical. It is not intended to apply to the employment of long-range reconnaissance patrol (LRRP) companies, even though such units carry the designation "ranger." These LRRP companies, as well as the long-range surveillance company (LRSC) (corps level) and the long-range surveillance detachment (LRSD) (division level), are organized, trained, and equipped for long-range reconnaissance in enemy territory.

The scope of this publication addresses the mission, organization, equipment, capabilities, limitations, planning, training, operations, and logistical support of ranger units. It describes how the ranger regiment and its subordinate battalions train, receive taskings, plan missions, task-organize, deploy, and integrate supporting elements to perform the mission. Once in combat, the ranger regiment maneuvers to accomplish the mission in the manner of other light infantry units. Ranger units make full use of published light infantry doctrinal guidance. This manual will repeat procedures or techniques contained in other publications only if necessary for understanding or continuity.

This publication does not contain everything the ranger needs to know to successfully execute a unit mission. Rather, it supplements unit training and the ranger unit commander's guidance; it does not supersede it.

The proponent of this publication is HQ, TRADOC. Submit changes for improving this publication on DA Form 2028 (Recommended Changes to Publications and Blank Forms) and forward it to Commandant, US Army Infantry School, ATTN: ATSH-B-ID, Fort Benning, GA 31905.

Unless otherwise stated, whenever the masculine gender is used, both men and women are included.

CHAPTER 1
Introduction

The ranger regiment is a major component of the US Army's special operations forces. It is a unique light infantry unit tasked to conduct special military operations in support of national policies and objectives. These operations require highly trained, well-disciplined units capable of employment in any environment, either alone or in concert with other military forces. Within this publication, the term *ranger force* describes any size force consisting mainly of members of the ranger regiment and led by a member of the ranger regiment's chain of command. A ranger force may be a TOE unit or it may be a specially organized task force for a specific mission.

1-1. Mission.

a. The mission of the ranger regiment is to plan and conduct special military operations. These operations are conducted by specially trained, equipped, and organized forces against strategic or tactical targets in pursuit of national military, political, economic, or psychological objectives. They may support conventional military operations or they may be performed independently when conventional forces cannot be used. (See Figure 1-1.)

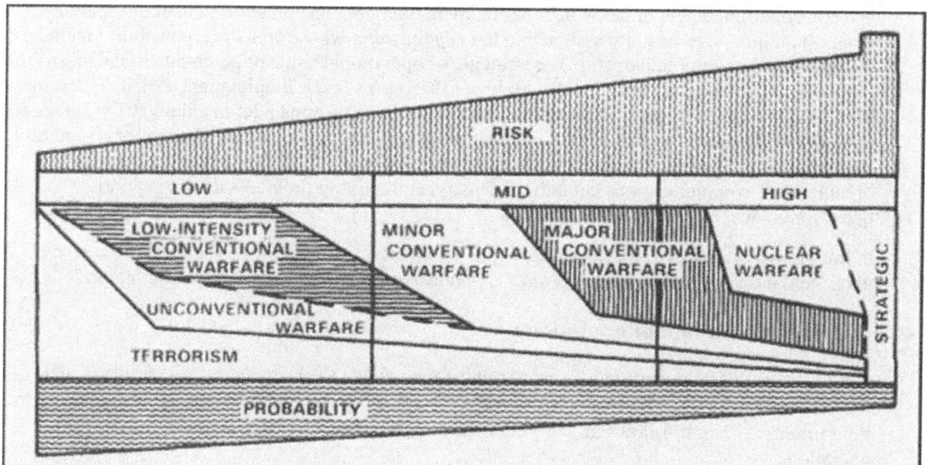

Figure 1-1. Spectrum of conflict.

b. Special military operations conducted by the ranger regiment include strike operations, usually deep penetration, and special light infantry operations. Strike operations include raids, interdiction, and recovery operations. Special light infantry operations include many of the light infantry missions assigned to airborne, air assault, or light infantry battalions and brigades. These operations are conducted in support of the AirLand Battle at all levels of intensity.

1-2. Purpose and function.

The ranger regiment provides the national command authority (NCA) the ability to move a credible military force quickly to any region in the world. The regiment uses the entire spectrum of intelligence support, from national systems to organic assets. Ranger units maintain a readiness posture that supports their immediate commitment to battle once deployed. They are often tailored for specific missions and may require augmentation from external sources. Tactical mobility may be augmented by USAF or Army special operations aviation (SOA) aircraft.

a. The ranger regiment conducts light infantry operations on the integrated battlefield as well as in low intensity conflicts. Special light infantry operations are conducted to accomplish certain contingency missions during peacetime or as part of the echelons above corps (EAC) battle.

b. The regiment conducts both strategic and operational missions. Its efforts are combined into an overall plan to destroy, delay, and disorganize the enemy, or to cause him to divert his attention and combat forces to rear area security. The limited number of ranger units and the diverse targets dictate a careful assignment of missions.

c. The mission, enemy, terrain, troops and time available (METT-T) factors govern the command and control arrangement under which the regiment operates. The ranger regiment or separate ranger battalions are normally assigned to the headquarters whose area of responsibility includes the regiment's operational area(s). The strategic or operational value of potential targets means that ranger units are normally employed at no lower than corps level. Employment at EAC is routine. The EAC headquarters employing ranger units could be the commander in chief's (CINC's) special operations command or field army. The controlling headquarters could also be that of a joint task force (JTF) or the Army component command (ARFOR) of a JTF. The ranger regiment's organization, communications equipment density, and training programs do not support operational level reconnaissance missions.

d. Ranger battalions are not oriented to a specific theater. Current force structure, contingency plans, and training needs (see Appendix G) preclude committing battalions to one region.

1-3. Fundamentals of ranger operations.

a. The success of an operation by a ranger unit depends on the observance of the fundamentals of the US Army's AirLand Battle doctrine (see FM 100-5). The ranger regiment fights the enemy as a light infantry force. It follows infantry doctrine and observes all the basic rules of infantry operations.

b. Along with initiative, depth, agility, and synchronization, ranger operations require--

(1) Detailed planning and coordination that allow the ranger unit to discern and exploit the enemy's weaknesses while avoiding his strength. Both permit the evaluation and use of information gathered from all sources. Detailed coordination integrates all supporting units and services. During planning, the ranger unit conducts pre-mission training, briefings, and rehearsals for all personnel. The use of special mission equipment is practiced and perfected.

(2) Decentralized execution IAW mission orders and the commander's intent. Special operations forces must use individual and unit initiative. Mission-type orders give the ranger force commander the flexibility to take advantage of opportunities on the battlefield.

(3) Surprise, achieved through the ranger unit's ability to move by uncommon means, along unexpected routes, and over rough terrain. Ranger units normally conduct operations during poor weather and reduced visibility, aided by night vision devices. This adds to the attainment of surprise.

(4) Survivability, achieved by using the classic infantry combat techniques of stealth and concealment. The ranger unit engages the enemy at the time and place of its own choosing. It takes full advantage of terrain, and destroys or suppresses enemy weapons. The ranger unit seeks to destroy or neutralize the enemy's command and control systems, his surveillance assets, and his mobility assets. Survivability is enhanced by rapid mission accomplishment and a prompt departure from the objective area.

(5) Mobility, speed, and violence of execution that allow the ranger unit it to close quickly on the objective area and complete the mission before the enemy can react. The speed at which events take place confuses and deceives the enemy as to the intent of the ranger unit. This forces the enemy to react rather than to take the initiative. Tactical mobility lets the ranger unit break contact and withdraw from the objective area.

(6) Shock effect, which is a psychological advantage achieved by the combining of speed and violence with the precision of the ranger attack. The ranger unit strives to apply its full combat power at the decisive time and place, and at the point of the greatest enemy weakness. The ranger unit strives to achieve maximum physical and psychological effect on the enemy by exhibiting aggressiveness and reasoned audacity.

(7) Multiple methods of insertion and attack, trying not to repeat operations thus decreasing the chance the enemy will detect a pattern. This is achieved through imaginative training and planning.

(8) Deception, practiced by the ranger unit during all phases of its operations, from deployment through the insertion phase, to the actions in the objective area and extraction. The enemy is kept unaware of the ranger unit's intentions, is fed conflicting and wrong information, and is kept from knowing the size or mission of the ranger force. The ranger force commander makes full use of ruses, feints, false insertions, electronic countermeasures, and dummy transmissions. Concurrent military operations by US or allied forces may be used as cover for ranger operations. Ranger units are careful to hide every possible aspect of an operation and to disguise those that cannot be hidden. While in the operational area, or when departing any location, ranger units ensure that no material is left that could provide the enemy with information concerning the ranger force.

(9) Audacity, achieved by a willingness to accept risk. The ranger force commander considers what the enemy expects the unit to do and the actual ranger abilities, He then chooses a course of action that may confuse the enemy, while remaining within the capability of the ranger unit. This fundamental is combined with deception and surprise to disrupt the enemy command and control.

c. Detailed planning and coordination finds the enemy. Surprise, mobility and speed, variety, deception, and audacity combine to shock and disorient the enemy, fixing him in place. The violence and precision of the ranger attack finishes the enemy while ensuring the ranger force survives.

1-4. Capabilities and limitations.

a. The ranger regiment has the following capabilities:

(1) Deploying quickly to conduct operations on all types of terrain and in all kinds of weather.

(2) Establishing a credible American presence in any part of the world to show US interest or resolve.

(3) Infiltrating and exfiltrating an area of operations, and assaulting an objective by land, sea, or air.

(4) Conducting strike operations to include raids, personnel and equipment recovery operations, and interdiction of key areas.

(5) Conducting special light infantry operations to include seizing and securing airfields, communications centers, command and control facilities, and key bridges; and other special light infantry operations.

(6) Performing short-duration reconnaissance of assigned ranger objectives for the ranger force commander.

(7) Operating for up to three days without resupply, and for longer periods when provided with accompanying or airdropped supplies.

(8) Providing liaison, communication, and coordination personnel and equipment to integrate the deployed ranger force into the logistical, intelligence, and operational system of the theater or joint task force (JTF) commander.

(9) Assuming operational control, for a limited time, of other US military forces such as engineers or infantry, airborne, or air assault battalions.

(10) Providing the focal point for all-source intelligence support to attached and assigned units of the regimental task force.

(11) Conducting limited combat operations under conditions of chemical, nuclear, or biological contamination.

b. The ranger regiment has the following limitations:

(1) Limited capability against armored or motorized units in open terrain.

(2) No organic transportation.

(3) Limited sustained combat capability due to the shortage of organic combat support and combat service support elements.

(4) Limited organic air defense weapons.

(5) Limited organic indirect fire support.

(6) No casualty evacuation capability.

(7) Reconstitution and retraining needed to replace combat losses.

1-5. Employment considerations.

a. Ranger units are characterized by the quality, motivation, training, and individual skill of their members. This produces units with superb collective abilities, able to adapt well to changing, complex situations.

b. Ranger units can conduct either deliberate or quick-response operations.

(1) Deliberate operations rely on careful planning, reconnaissance and surveillance of the target area, deception, secrecy, thorough preparation and rehearsals, and violent execution. A deliberate operation aims to complete the mission even though the enemy may have heavy forces on or near the objective area. Deliberate operations allow for detailed planning, evaluation, rehearsal, and coordination before insertion. A deliberate operation is likely to succeed against targets that the enemy has protected in depth, that have strong natural defenses, or that need a detailed and long insertion process.

(2) Quick-response operations rely on the high level of training and readiness of the ranger regiment to execute a mission before the enemy can react. These operations are conducted when there is little time for long, detailed planning. They rely on set procedures set forth by the ranger regiment and its supporting elements. A quick-response operation aims to complete the mission before an enemy can react. This type of operation may be chosen due to the time-sensitive nature of the target, political or military goals, the time frame of other operations, or the increased chance of enemy detection.

(3) Whether an operation is to be deliberate or quick response is often a difficult and time-sensitive decision by a high-level command authority. Decision-makers must consider the enemy's strength in the area, his intentions, his ability either to reinforce or to alter the target area, and the consequences of success or failure of diplomatic or military initiatives in related areas. The ranger unit commander tries to use the existing time, manpower, and resources to complete a detailed and coordinated plan. He refines that plan up to the insertion into the objective area.

c. Ranger units train to operate in any environment or weather condition. They regularly perform operations during periods of limited visibility. Ranger units maintain a high state of physical fitness and often train in close quarters combatives. In addition to completing advanced marksmanship training with standard US weapons, each member of a ranger unit trains with many foreign weapons. Ranger units are trained to operate on urbanized terrain, becoming specialists in entry and clearing techniques and quick-fire methods, especially during periods of limited visibility.

d. Because ranger units have limited vehicles, logistics operations capability, indirect fire support, and heavy weapons systems, they are not designed for continuous operations. During all phases of operations and training, ranger units need responsive external support.

e. Ranger units are normally employed against targets and under conditions that need their unique skills. Although targeting priority is set by the overall commander, ranger units are not normally assigned missions that can be done by conventional aerial bombardment or by other units.

f. Ranger units are oriented toward offensive operations. They are not normally employed as a rear area protection force. Although the ranger regiment has a small reconnaissance unit, ranger units do not normally conduct long-range reconnaissance missions. The structure, communications, and training of the ranger unit do not prepare it for LRRP missions. Ranger units engaged in strike and special light infantry operations have a secondary mission to collect and report combat information.

g. Ranger units can be deployed worldwide when US military presence or participation with a host national military activity would serve US interests. This deployment shows a readiness to commit forces into a threatened area or proves US national resolve. After the deployment, other activities include staging operations, rehearsing combat operations, securing base areas for use and deployment of other forces, and so on that provide a clear signal of US intent. Ranger units are not trained or organized to provide mobile training teams (MTT) to train indigenous forces. The US Special Forces or other special operation forces are trained to conduct such security assistance operations.

h. Ranger units can serve as an example to a host country and provide limited military advice and training. They will normally be augmented with linguists and technicians to increase their abilities. The regiment would normally still function as a unit. Most of its assistance would be through short-term, high-impact, unit-oriented operations. They would not be long-term individual efforts associated with advisory-type activities.

i. Ranger units may be deployed to engage in combined training exercises with allies. This enhances US national image by demonstrating the outstanding abilities of the American ranger. These activities may include ranger, light infantry, airborne, air assault, or amphibious operations.

j. Ranger units use standard US nuclear, biological, chemical (NBC) warning; detection; protection; and decontamination equipment and doctrine when operating on the integrated battlefield.

1-6. Command and control.

a. **External.**

(1) The ranger regiment is a key component of the US Army's special operations forces. The other elements of special operations forces are special forces (SF), psychological operations (PSYOP), civil affairs (CA), and special operations aviation.

(2) The 1st Special Operations Command (1st SOCOM) is responsible for command of all active component special operations forces elements, to include the ranger regiment, in peacetime. The 1st SOCOM has command and control of all assigned special operations forces, less operational command (OPCOM) of special operations forces units forward deployed. (See Figure 1-2.) The ranger regiment, while based in the continental United States (CONUS), is under the control of 1st SOCOM. During peacetime, the OPCOM of deployed ranger units is through channels chosen by the NCA through the Joint Chiefs of Staff (JCS) to the unified command.

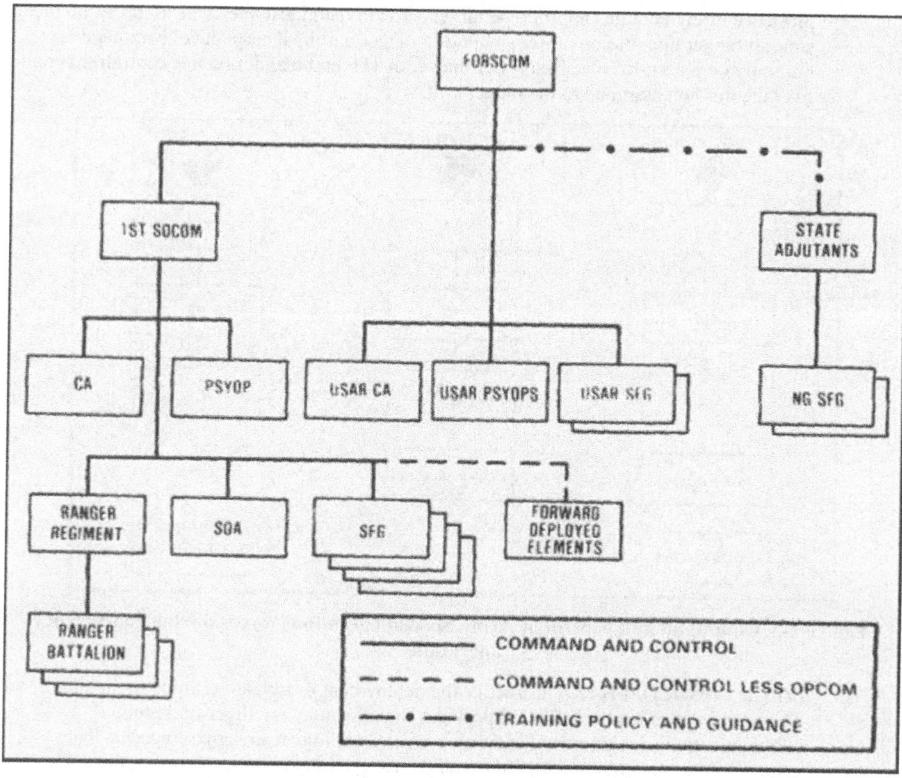

Figure 1-2. Command and control of Army Special operations forces during peacetime.

(3) In support of national policy, the NCA may direct contingency operations involving US

forces. These are politically sensitive military operations normally characterized by the short-term rapid projection or employment of forces in conditions short of conventional war. One aim of the NCA is to conclude them without going to war. The flexibility and effectiveness of the ranger regiment make it possible to use military force quickly to complete the mission and limit the spread of conflict. Command and control of special operations forces in a contingency operation is usually by a special operations task force (SOTF). (See Figure 1-3.) The SOTF and designated special operations forces are under the OPCOM of the highest level of command responsible for the operation. This command may be the unified command responsible for the contingency area (Option A), or a JTF designated by the NCA through the JCS (Option B). If no SOTF is formed, the ranger regiment could be OPCOM to the Army component commander (Option C). The ranger regiment or its battalions are not normally placed OPCOM to a division. A division's area of influence rarely contains appropriate targets. The division also lacks the assets to adequately support ranger operations. Employment of a ranger unit by a corps-level command is normally on a case-by-case basis. The theater or JTF commander retains centralized control over the mission assigned to the ranger unit.

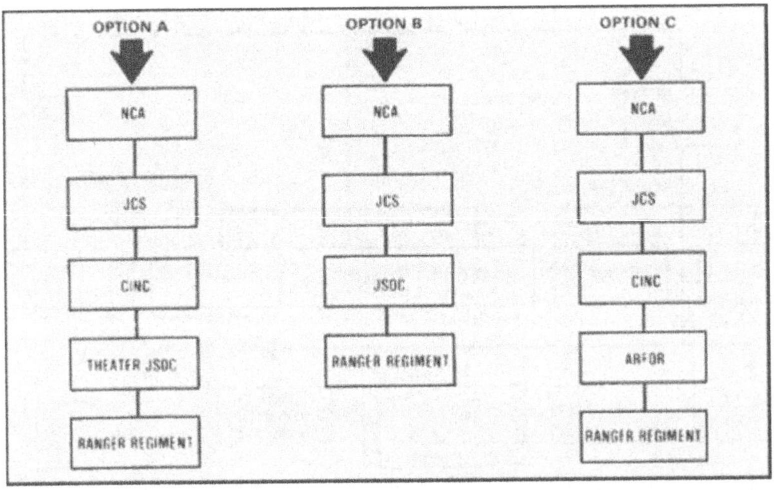

Figure 1-3. Command and control of Army Special operations forces during contingency operations.

(4) The 1st SOCOM is responsible for the deployment of the ranger force to the theater area. The headquarters having OPCOM provides the all-source intelligence, secure communications, insertion and extraction assets, and logistical support needed. The responsible commander in chief, through the Army component commander, provides support during the contingency operation.

(5) During wartime, ranger units are deployed as strategic assets to the theater of operations.

Strategic mission support is provided to the unified commander (CINC) from the NCA through the Joint Chiefs of Staff. It may involve the execution of sensitive operations. The employment of ranger units is controlled at the highest level of command directly responsible for the mission. Ranger units are assigned to a REMAB based on mission, force protection, and OPSEC considerations. The employing command and other elements provide timely and accurate intelligence, communications, logistical, and administrative support. Command and control of these units conforms to the provisions of JCS Pub 2. (See Figure 1-4.)

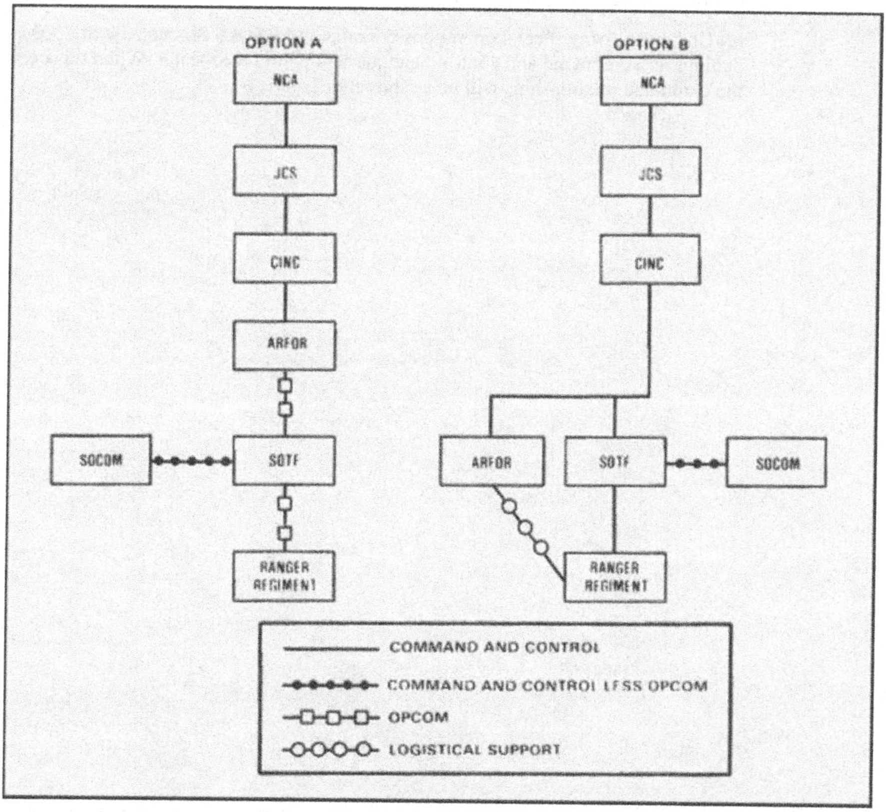

Figure 1-4. Command and control of Army Special operations forces during wartime.

(a) A theater joint special operations command (JSOC) is subordinate to the unified command to supervise the command and control, employment, and support of all US and allied special operations forces. The JSOC is a key element in the command and control of ranger units performing strike missions or special light infantry operations (Option A).

(b) The JSOC may be placed on a lateral line with other in-theater service components (Option B). This can occur when ranger units are -employed far from conventional forces and against strategic targets. It can also occur when the JCS or a CINC forms a JTF to conduct a limited-duration operation in an area where ranger units and conventional forces may operate near each other, or where operations must be coordinated on a regional level.

(c) Under the Army special operations command (ARSOC) concept, 1st SOCOM deploys as a command and control headquarters within the theater. When this occurs, the command relationships will be as shown in Figure 1-5.

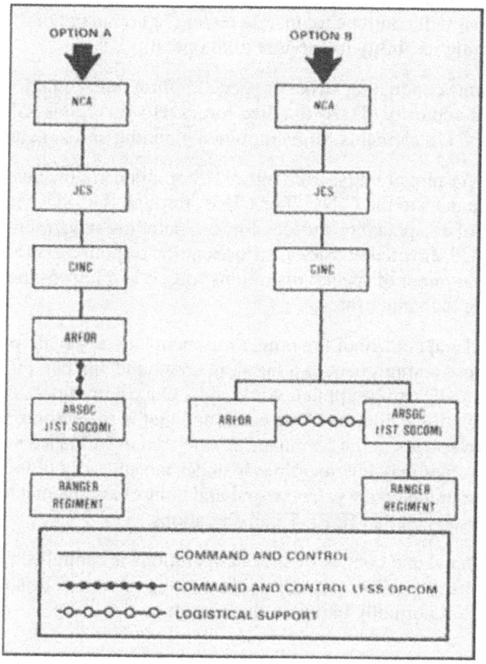

Figure 1-5. Command and control of Army Special operations forces under the ARSOC concept.

(d) Recently passed legislation will modify existing high level SOF command and control arrangements. All US Army, USN, and USAF special operations forces will be consolidated into joint commands at both national and theater level. The ranger regiment will continue to operate under the command of the 1st SOCOM, while increasing its ability to conduct joint operations.

(6) Ranger units conducting strike or special light infantry missions are not normally offered for transfer of authority (TOA) to allied forces. However, this will not preclude allied commands and US elements from combined planning and targeting for ranger missions.

(7) Requests for ranger units, from either US or allied commands, go through normal command channels to the CINC. The CINC, through the JSOC, is responsible for the employment of all special operations forces, including ranger units. This provides flexible response to JCS-directed missions, ensures quick response to the CINC's needs, enhances the joint employment of special operations forces, and lessens the chance of overcommitting and degrading the ranger force.

(8) Command and control of the ranger regiment and its battalions is IAW AirLand Battle doctrine. Before commitment to a mission, command and control is kept at a level where the unit's unique skills can be applied worldwide. Once committed, the ranger unit is placed under the operational control of the command that is responsible for the mission. This normally corresponds to the command whose area of influence includes the target area. Ranger task forces may operate directly under the authority of the senior American representative in the country. Command and control arrangements may vary according to the theater involved and METT-T considerations.

(9) The command and control of special operations is centralized and begins at the highest level of the military and civilian decision-making process. The specific command relationships are normally based on the mission.

b. **Internal.**

(1) The ranger regimental headquarters is organized to operate like that of a brigade headquarters. In addition to commanding and controlling all three ranger battalions, the regimental headquarters can assume operational control (OPCON) of conventional combat and combat support units, and other special operations forces for limited periods. It provides the regimental commander with a well-balanced responsive staff. It is structured as an operational headquarters that can deploy to the mission area and act as the ground tactical headquarters.

(a) The regimental headquarters prepares for combat by forming two small command and control groups. Each of these can deploy to an objective area and control combat operations. Other elements of the staff operate the tactical operations center (TOC) and the logistical operations center (LOC). These may be collocated depending on the situation. The TOC and LOC are normally located well behind the FEBA, at the REMAB or at the intermediate staging base (ISB). The ranger regiment also has two small liaison teams consisting of LNOS, staff representatives, and communications elements. (See Appendix B.)

(b) When a single ranger battalion is committed to an operation, the regimental commander normally exercises control from the TOC. A liaison and communications team is detached from the regimental headquarters to augment the ranger battalion. This team is attached to the deployed battalion until the end of the mission.

(c) If two or more ranger battalions are used in a single operation, the regimental commander normally deploys both command and control groups. He commands from the objective area, The use of two small, mobile control elements makes up for the possible loss of the regimental commander and the primary control element.

(d) Whenever the regimental commander is controlling an operation from the REMAB or the objective area, he normally provides a liaison team with a communications element to the higher headquarters exercising operational control. This cell can provide ranger representation to the controlling staff, operations interface, logistics interface, fire support planning and coordination, intelligence analysis and dissemination, and communications between the higher headquarters and the ranger battalion.

(2) The ranger regimental headquarters can serve as a ranger or special operations force task force headquarters. With augmentation, it can function as the Army special operations force command element. The regimental headquarters can also function as the Army component command of a JTF.

1-7. Communications.

a. Ranger missions need unique and dedicated communication support assets. Along with the ranger regiment's organic communications means, communications support assets are organized within the 1st SOCOM that-can deploy in support of the ranger regiment. This ensures that there are reliable communications to all elements of the SOTF. A special operations communications battalion (SOCB), deployed by the 1st SOCOM, provides more communications systems to support the special operations plan. The SOCB does not replace or duplicate existing in-theater communications. It augments the CINC's communications with the US Army's portion of the critical special operations force command, control, and communications link unique to the SOTF.

b. The ranger regiment's communication platoon can provide secure, long-range communication links from the CINC to a deployed ranger force. It can also enter the SOCB and the joint communications support element (JCSE) nets to provide a direct link with the NCA. Its equipment consists of four light vehicles equipped with various multiple communications systems. The two regimental liaison teams use one of these vehicles each. The TOC and LOC each use one.

c. Secure satellite communications (SATCOM) and radios with AM or FM capability are the main means of communication within the ranger regiment. These are also used for communications within the ranger battalions. The regimental commander has two light vehicles equipped with multiple communications systems. One of these is used to support each of the two mobile command and control elements of the regimental headquarters. Some of the radio systems can be dismounted and man-packed.

d. The ranger regimental headquarters has both man-packed and vehicular tactical satellite communication systems. Each of the ranger battalions has one base station and three man-packed systems.

(1) The man-packed radio is battery-operated. It provides two-way communications in both line-of-sight (mobile) and satellite (at-halt) modes. The digital message device group (DMDG) is the input and output device for data transmission. Secure voice communications is achieved using various encryption devices.

(2) The vehicular net control station uses the basic receiver/ transmitter unit installed on a vehicle. Each can serve up to 15 terminals in a communications net.

e. The reconnaissance platoon is equipped with multiple, man-packed communications systems. These let its teams operate in most of the communications nets within the regiment. Each five-man reconnaissance team has this communications ability, as does the platoon headquarters.

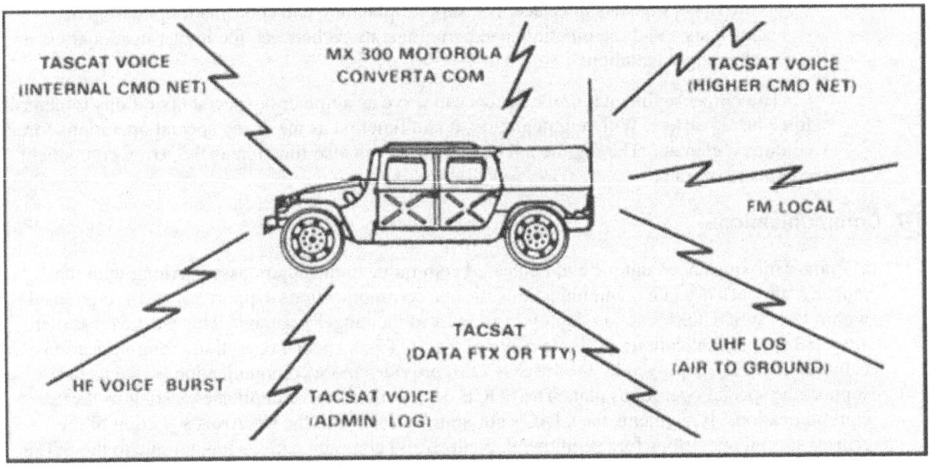

Figure 1-6. Communications capabilities within the regimental headquarters.

CHAPTER 2

Organization and Equipment

The ranger regiment is a specially organized and trained light infantry force consisting of about 2,000 personnel. It includes a regimental headquarters and headquarters company and three ranger battalions. (See Appendix A.) The regiment can operate as a single unit, separate battalions, or separate companies on specific missions. Independent missions are not normally assigned to ranger platoons.

2-1. Headquarters and headquarters company.

The HHC of the ranger regiment consists of 130 personnel, the majority being senior noncommissioned officers, commissioned officers, or soldiers with special skills to support the operations of the ranger battalions. (See Appendix A.) The HHC consists of two elements: the regimental headquarters and the headquarters company (minus).

2-2. Regimental headquarters.

a. The regimental headquarters consists of the regimental commander and his staff. The primary staff is composed of five sections: personnel (S1), intelligence (S2), operations and training (S3), supply (S4), and civil affairs (S5). These staff sections perform the standard functions and are supervised by the executive officer (XO). The regimental commander has a deputy commander and a special staff consisting of a communications officer, fire support officer, surgeon, and staff judge advocate. The S2 officer has a large intelligence analysis and counterintelligence section. The S3 section has extra liaison officers and assistant operations officers to coordinate the widespread and complex ranger missions. A United States Air Force (USAF) staff weather officer and a tactical air control party (TACP) are permanently attached.

b. The regimental headquarters:

- Analyzes assigned missions.
- Plans, coordinates, and controls training and operations.
- Coordinates the infiltration, resupply, exfiltration, and debriefing of the reconnaissance teams.
- Provides liaison support.
- Provides administrative, medical, NBC, and logistical planning and coordination support to the committed ranger battalions.
- Coordinates special support requirements (rigger, transportation, communications security [COMSEC]) for the ranger battalions.

2-3. Regimental headquarters company.

The regimental headquarters company (minus) consists of the company commander and his administrative personnel, plus the limited combat support (CS) and combat service support (CSS) elements assigned to the regiment. These are the fire support element, the communications platoon, the reconnaissance platoon, and the medical treatment team.

a. The company headquarters:

(1) Provides command and control of assigned elements.

(2) Provides administrative, personnel, logistical, and maintenance support to the assigned elements.

b. The fire support element:

(1) Plans fires to support the commander's ground tactical plan.

(2) Coordinates the integration of all available fire support means (mortars, artillery, armed helicopters, close air support, naval gunfire) to support mission execution.

c. The communications platoon (see Appendix A):

(1) Provides secure SATCOM teletype (TTY) and facsimile transmissions (FAX) abilities at the remote marshalling base or intermediate staging base (REMAB or ISB).

(2) Provides secure SATCOM, TTY, and FAX to the regimental headquarters and ranger battalions in the objective area.

(3) Provides cryptographic support to the regimental headquarters.

(4) Provides limited repair of radio and cryptographic equipment.

d. The reconnaissance platoon (see Appendix A):

(1) Trains, equips, and controls the employment of three reconnaissance teams.

(2) Infiltrates the objective area by parachute (high-altitude, low-opening [HALO]; high-altitude, high-opening [HAHO]; static line), helicopter, fixed-wing aircraft, self-contained underwater breathing apparatus (SCUBA), small boat, or other means.

(3) Remains undetected in the objective area for up to five days.

(4) Performs reconnaissance.

(5) Performs demolition target analysis.

(6) Uses small boats and inflatable rafts.

(7) Emplaces and monitors unattended ground sensors.

(8) Uses night image enhancement devices.

(9) Uses photographic equipment.

(10) Emplaces electronic target acquisition and designation devices.

(11) Collects specific combat information to satisfy priority intelligence requirements (PIR).

(12) Reports information using secure, long-range communication equipment.

Figure 2-1. Reconnaissance platoon communications.

(13) Performs drop zone (DZ) and landing zone (LZ) selection, marking, and reception duties.

(14) Escapes and evades enemy areas.

(15) Performs limited attacks or ambushes when tasked by the regimental commander.

(16) Reports weather in the objective area.

e. The medical treatment team:

(1) Provides medical support to the regimental headquarters.

(2) Coordinates required medical support and evacuation.

2-4. Ranger battalion.

a. The ranger battalion is organized and equipped to conduct both special and conventional combat operations. It contains organic combat, CS, and limited CSS elements. It consists of a battalion headquarters, a headquarters company, and three ranger rifle companies. (See Appendix A.)

b. The ranger battalion is the combat element of the ranger regiment. It can conduct independent combat operations or subdivide into its separate ranger rifle companies and their subordinate elements who can conduct semi-independent operations.

c. The ranger battalion headquarters and headquarters company (see Appendix A) includes:

(1) **Battalion headquarters.** The battalion headquarters consists of the commander; the executive officer; the command sergeant major; the S1, S2, S3, S4, and S5 officers; the chaplain; the communications officer; the fire support officer; and two radio operators.

(2) **Battalion headquarters section.** The battalion headquarters section consists of the S3 air and assistant S8, an assistant S2, an intelligence sergeant and a senior intelligence analyst, two operations sergeants and an operations assistant, a personnel staff noncommissioned officer (PSNCO), the chaplain's assistant, two clerk typists, two radio operators, a chemical officer and NCO, a personnel administration center (PAC) NCO, three PAC clerks, and a legal specialist. There is a USAF tactical air control party permanently attached.

(3) **Communications section.** The communications section consists of one telecommunications technician, one tactical communications chief, two combat signal radio team chiefs, one tactical communications systems supervisor, three radio operators, one switchboard operator, one single channel radio operator, and two equipment mechanics. The section provides communications support and associated maintenance.

(4) **Battalion medical section.** The battalion medical section consists of one general medical officer, one physician's assistant, two medical specialists, and twelve aidmen. The section provides unit-level medical support and trains selected members of the battalion in advanced medical procedures.

(5) **Battalion support platoon.** The battalion support platoon consists of a battalion support platoon leader, an ammunition NCO, and a food service section. The platoon provides limited CSS for assigned and attached units. This platoon does not have organic transportation. The battalion food service section consists of a food service sergeant (E7), a first cook (E6), and eight cooks.

(6) **Fire support team headquarters.** The fire support team (FIST) headquarters is supervised by the battalion fire support officer. It consists of the fire support sergeant and two fire support specialists. The headquarters controls three 9-man fire support teams. These teams are normally with the ranger rifle companies in a habitual support relationship. When operations are conducted near water, the fire support team headquarters may be augmented by spotter teams from the naval shore fire control party.

(7) **Company headquarters.** The HHC company headquarters consists of a commanding officer, an executive officer, a first sergeant, a supply sergeant, and an armorer.

d. The ranger rifle company (see Appendix A) includes a company headquarters, three rifle platoons, and a weapons platoon. Each company can conduct semi-independent operations for a short time when augmented by assets from the ranger battalion headquarters company.

(1) **Ranger rifle company headquarters.** The ranger rifle company headquarters consists of a commanding officer, an executive officer, a first sergeant, a supply sergeant, a tactical communications chief, two radiotelephone operators, an armorer, and an NBC NCO. It also has a two-man sniper team equipped with a sniper weapon. It normally has one of the 9-man FISTs attached from the ranger battalion headquarters company. Elements of this FIST may be further attached to the ranger rifle platoons.

(2) **Ranger rifle platoon.** The ranger rifle platoon consists of a headquarters section, three rifle squads, and a machine gun squad. The headquarters section personnel include a platoon leader, a platoon sergeant, and a radiotelephone operator. Each rifle squad consists of a squad leader and two balanced fire teams of four men each. The machine gun squad consists of a squad leader and three M60 machine gun crews of three men each. The machine gun squad is normally employed as a single element to provide quick, accurate, long-range supporting fire. One or more crews may be attached to other squads within the platoon.

(3) **Ranger weapons platoon.** The ranger weapons platoon consists of a headquarters section, a mortar section, and an antiarmor section. The headquarters section consists of a platoon leader, a platoon sergeant, and a radiotelephone operator. The mortar section has two 3-man 60-mm mortar squads. The mortar squads are normally employed as a section. They may be attached by section or squad to other platoons. The antitank section normally consists of a [illegible]ergeant and three 3-man 90-mm recoilless rifle or Dragon missile teams. They are normally employed as a section. The antiarmor section may also be attached by section or team to other platoons.

2-5. Special teams and elements.

The ranger regiment forms special teams to do critical tasks. These teams are made up of members of the unit and are used when needed. Examples of these special teams and elements are--

a. SCUBA teams whose members are graduates of the Special Forces Underwater Operations School. They conduct special insertion and reconnaissance duties.

b. HALO teams whose members are graduates of the Special Forces Freefall School and who conduct high-altitude, low-opening parachute operations.

c. HAHO teams whose members are also graduates of the Special Forces Freefall School and who conduct high-altitude, high-opening parachute operations, allowing offset parachute insertions.

d. Scout swimmer teams whose members are specially trained by the Landing Force Training Command (United States Marine Corps [USMC]) to conduct operations along coastlines, waterways, and inland waters. These teams conduct special insertion, reconnaissance, and security tasks often in support of amphibious operations.

e. Sniper teams, whose members are authorized by the table of organization and equipment (TOE), are used for both conventional and special operations. Along with the two-man sniper teams in the rifle company headquarters, each rifle platoon has two trained snipers.

f Demolition teams whose members are trained in advanced demolition and incendiary techniques and the use of special explosives,

g. Air defense teams trained in the use of man-portable air defense system (MANPADS) weapons. Such teams give the ranger battalion limited self-defense against air attack.

h. Liaison teams that are used by the ranger regimental commander to augment the existing command and control links from the employing unit headquarters to the ranger regiment. These teams are placed where they are most needed and may assist the higher headquarters staff in planning and target selection, as well as performing other liaison duties. See Appendix B for a detailed listing of the personnel and equipment included in the ranger liaison teams.

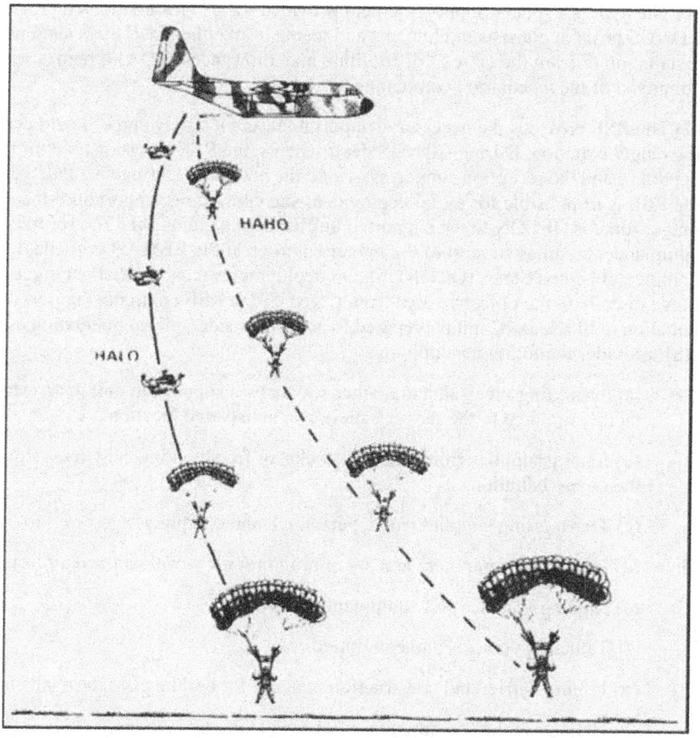

Figure 2-2. HALO and HALO parachute operations.

Figure 2-2

i. The ranger support element (RSE).

(1) The RSE is a special support element provided by host installation commanders at CONUS posts. It consists of elements and teams from either TOE units stationed at the installation or from the table of distribution and allowances (TDA) organizations under the command of the installation commander.

(2) The RSE provides the necessary support to ensure a timely and efficient deployment of the ranger battalion. During CONUS deployments, the RSE is responsible for both the departure and the reception support provided the battalion. During OCONUS deployments, the RSE is responsible for initial deployment and continuing support until the deploying ranger force is OPCON to the supported unified command or JTF. The responsible theater commander assumes support of the ranger battalion at the REMAB or the ISB if outside continental United States (OCONUS). On deployments that require the ranger battalion to move directly to the objective area from CONUS, the RSE continues support of the ranger battalion until released, and is prepared to accept considerable augmentation assets. The RSE provides the following support:

(a) Furnishing meals that are either served by a supporting unit in an established facility or taken to the ranger battalion at an isolated location.

(b) Issuing supplies from existing stocks, or from contingency items stored for use by the ranger battalion.

(c) Transporting supplies, food, personnel, and equipment.

(d) Palletizing equipment and ammunition for air movement and airdrop.

(e) Rigging vehicles and equipment for airdrop.

(f) Refueling vehicles and equipment.

(g) Issuing barrier and construction material for building of rehearsal sites.

(h) Helping in the construction of rehearsal sites and targets.

(i) Securing, storing, moving, and issuing ammunition and other Class V items.

(j) Maintaining ranger weapons and equipment, to include communications equipment. This is normally done by maintenance contact teams.

(k) Setting up and operating the RSE emergency operations center.

(l) Securing the REMAB if it is in CONUS.

CHAPTER 3
Deployment

Ranger units can quickly deploy by any means of transportation to any location in the world.

3-1. Readiness.

One ranger battalion is always in an advanced readiness condition as the Ranger Ready Force (RRF) available for immediate worldwide deployment. A second battalion is prepared to deploy later and a third battalion to follow. The regimental headquarters maintains command and control, liaison, communications, and reconnaissance elements immediately available for deployment. Higher status of readiness in response to specific world situations can be achieved.

3-2. Ranger ready force.

In order to maintain advanced readiness conditions, ranger battalions are rotated as the RRF.

a. Before taking over the duties of the RRF, battalions must do the following:

(1) Complete preparation for oversea movement (POM) qualification, to include an individual records check and required inoculations.

(2) Test fire all weapons, confirm battlesight settings, and repair or replace faulty weapons.

(3) Assemble and load on pallets the unit basic load (Classes I and V) and other supplies and equipment.

(4) Check unit-derived packing list components for accountability and serviceability.

(5) Prepare special operations troop listings.

b. During an RRF period, ranger units:

(1) Have a positive recall system for all personnel.

(2) Store enough air items for a battalion-size parachute assault.

(3) Conduct no off-post exercises, unless approved by the regimental commander.

c. Each unit keeps an alert roster and a telephone notification system for recalling personnel. The rosters ensure that the duty officer and the unit charge of quarters make a minimum number of calls. Each commander briefs personnel assigned to his unit on the recall system and tests it often.

d. Ranger units can maintain an increased readiness condition for long periods, but they should not be kept in that condition longer than needed. The effort to maintain this level of readiness will degrade the unit's training and effectiveness. If the needs of the operations dictate long periods at the highest readiness condition, provisions must be made for sustainment training. Also, added support will be required from the RSE.

e. The ranger regiment maintains a strong family support system when the unit is training or deployed on combat operations. Each member of the chain of command is responsible for family support.

3-3. N-hour sequence.

The N-hour sequence is a departure time sequence designed to help the ranger force complete all required deployment actions within a certain time. The length of the N-hour sequence depends on the readiness condition of the battalion before deployment. The N-hour sequence begins when the battalion is alerted and ends when the first elements depart. The N-hour sequence does not include travel time to the objective area.

3-4. Intelligence and security.

a. **Essential elements of friendly information.** Items of sensitive information and subject areas, if not secured, can be collectively analyzed by a hostile force. This analysis can reveal the intentions of a unit or compromise a mission. These items of information are designated as essential elements of friendly information (EEFI). They are never transmitted without encryption. The EEFI associated with a ranger unit alert normally are the mission, the use of staging bases, nature of the target, special terrain or weather conditions, and time limits. The ranger force commander must ensure that EEFI are not discussed outside of secure areas or made available to personnel without a security clearance and a need to know.

b. **Security plans.** Security plans for billets, marshalling areas, and out-loading areas must remain flexible to adapt to any situation. Certain items are always addressed in security plans. These are access control and physical security.

(1) **Access control.** Ranger units control access to their respective areas by allowing only authorized personnel to enter or leave. All entrances are secured and an armed guard with the access roster is posted. Access to telephones is restricted.

(2) **Physical security.**

(a) Equipment is open to theft and sabotage during the early phases of deployment. Units must provide guards to secure all unit and personal equipment.

(b) Weapons, ammunition, and sensitive items must be stored in a restricted area and guarded.

(c) Classified documents are open to compromise during deployment. Steps must be set up for maintenance, transportation, turn-in, and destruction of classified material. The most secure facility available must be used.

c. **Rear detachment security.** Security after deployment (maintained by the rear detachment) is as important as security before deployment. No information should be given to the news media from other than official sources. The rear detachment may continue to play a role in the unit cover and deception plan.

d. **Tactical operations center security.** Only personnel listed on TOC access rosters are allowed free entry into the TOC area. Personnel wanting to enter the TOC, who are not listed on the access rosters, are detained until their identity and purpose are proven. Such personnel are escorted at all times while within the TOC area.

e. **Marshalling and out-loading area security.** The ranger unit normally marshalls its personnel and equipment in the unit area, providing its own security force. The USAF security police provide this support at the departure airfield. On Army airfields, Army military police provide security. En route security is provided by the RSE.

f. **Personnel and unit equipment sanitization.**

(1) Before deployment, personnel are relieved of non-mission-essential material having intelligence value. This reduces the intelligence available to hostile forces if they capture a ranger force member.

(2) The ranger force commander sets the degree of sanitization IAW mission requirements.

(3) If strict sanitization is imposed, guidelines are issued as follows:

(a) The only personal items retained are identification card and tags.

(b) Unit insignia and markings are removed from uniforms and equipment.

(c) Personal items, such as wallets and photographs, are secured by the rear detachment commander.

(d) Personnel are briefed on how to compose and retain combat diaries. No personal diaries are allowed.

(4) The commander decides how much US currency is to be taken with the unit. '

g. **Area sanitization.** To avoid the compromise of classified or sensitive mission-related material, searches are conducted before and after leaving the host installation, REMAB, or ISB. These searches are held to locate and remove any material, indicators, litter, or other items that could compromise the ranger unit's mission. Unit commanders ensure that a thorough sanitization of their areas is conducted, to include dismantling rehearsal areas, destroying terrain models, erasing chalkboards, and emptying all classified and nonclassified waste containers. During their searches, commanders must pay close attention to emergency operations centers, tactical operations centers, and marshalling and loading areas. The ranger unit's intelligence officer coordinates support from local counterintelligence elements to search the area again after the ranger unit leaves.

3-5. Operations.

a. **Actions upon receipt of the OPORD.**

(1) Task-organize and equip units for special mission(s) IAW the warning order.

(2) Allocate/reallocate mission-essential equipment.

(3) Distribute augmentee personnel.

(4) Redistribute key personnel,

(5) Prepare a deployment time sequence.

(6) Prepare a movement plan.

(7) Prepare and maintain a current operation estimate.

b. **Advance elements.** A planning element is normally deployed to the REMAB/ISB or JTF headquarters in advance of the main body if the ranger force does not deploy directly to the objective area.

c. **Emergency operations center.** An EOC is setup by the ranger battalion headquarters. The EOC is responsible for relaying all reports to higher headquarters. It remains operational until ordered to cease.

d. **Air movement.** The ranger force always deploys by USAF aircraft. This is normally an administrative air movement to a REMAB or ISB. It may be followed by tactical air deployment from there. For some missions, tactical deployment begins at the unit's home station. Even if the movement is administrative, key personnel are cross-loaded. This is done under the direction of the operations section air officer based on aircraft allocations and load plans.

3-6. Logistics.

The following describes logistics readiness needs and steps for deployment.

a. **Supply.**

(1) **Class I.**

(a) Combat rations are normally palletized when the unit is RRF. These rations are to be used during the tactical operation. If the mission calls for deployment directly into the objective area, the palletized stock of rations is off-loaded in the unit area and issued as needed.

(b) Each ranger deploys with emergency combat rations.

(2) **Classes II and IV.**

(a) **Individual clothing and equipment.**

- Individual A bags are packed IAW a predetermined packing list, tagged, and marked. They are palletized, moved to the departure airfield by the RSE, and loaded IAW the air movement plan.
- Individual B bags are packed, assembled, and forwarded on order by the rear detachment.

- Individual environmental equipment is issued. Equipment shortages are replenished out of stocks maintained by the RSE.

(b) **Air items.**

- While the unit is on RRF, a contingency load of main and reserve parachutes is maintained by the RSE.
- Individual and unit air items are palletized by the RSE and shipped in bulk to the REMAB or ISB.
- For direct-deployment missions, air items and packing material are taken to the unit by the RSE.

(c) **General supplies.** The RSE maintains pallets of common contingency items and special equipment. Pallets are picked up by the out-loading support unit and moved to the unit area for issue or moved with the other pallets to the departure airfield.

(d) **Bulk supplies.** Barrier material is identified, relocated, palletized, and out-loaded IAW the air movement plan.

(3) **Class V.**

(a) The unit basic load of ammunition is secured, palletized, and loaded by the RSE IAW the air movement plan.

(b) For direct deployment missions, selected pallets are taken to the unit area and ammunition is issued. The convoy route is coordinated through the post EOC, and secured by the RSE.

(c) Designated items are rigged by the RSE for airdrop, if enough time exists before movement to the REMAB or ISB begins.

(4) **Class VI.** Personnel deploy with comfort items as needed.

(5) **Class VII.**

(a) All weapons and sensitive items not deployed are assembled and secured by the rear detachment personnel.

(b) Unit vehicles are prepared for airland or airdrop.

(c) Inoperable mission-essential equipment is repaired or replaced.

(d) Special equipment requests are forwarded to the logistics section as they are identified.

(6) **Class VIII.**

(a) Individuals deploy with one set of individual Class VIII items.

(b) Unit medics deploy with modified aidbags.

(c) Controlled medical items (less drugs) are palletized and deployed with section supplies. Controlled drugs are secured and issued to unit aidmen on order.

(d) Medical supply chests are maintained in the unit aid station.

(7) **Class IX.**

(a) Units deploy with initial prescribed load list (PLL) items.

(b) Critical PLL shortages are identified to the RSE for follow-on issue.

b. **Maintenance.**

(1) Upon alert, priority maintenance or replacement of key items of equipment is coordinated by the logistics officer. Only minimum mission-essential equipment is replaced.

(2) Items to be returned from supporting maintenance activities are identified, and pickup is coordinated by the property book officer.

c. **Out-loading coordination and support.**

(1) Upon alert, the RSE provides a liaison officer to the ranger logistics section.

(2) The support platoon leader (SPI,) attends the air movement conference and coordinates the following:

(a) Movement times (for both troops and equipment) to the departure airfield.

(b) Steps for weighing and joint inspection of the pallets and vehicles upon arrival at the departure airfield control group (DACG).

(c) Limitations on the pallet height and weight due to aircraft type, allowable cargo load (ACL), and availability.

(d) Changes in the planned number of pallets due to the mission requirements or airframe availability.

(e) Rigging and loading of the heavy drop and supply bundles.

(3) Empty pallets are placed in position for loading by the RSE in coordination with the SPL.

(a) Companies and staff sections prepare pallets.

(b) For airdrop operations, vehicles are prerigged at the unit location. On order, vehicles are driven to the heavy drop rigging site. The SPL and the officer in charge of the rigger detachment coordinate for a rigging site and the delivery of the required air items to that location. Rigging is done by the RSE, and supervised by a member of the ranger unit. Loading aboard the aircraft is IAW the air movement plan and supervised by RSE and USAF aerial port personnel.

(4) The DACG is provided by the host installation. Upon arrival at the DACG area, unit personnel and equipment are under the control of the DACG. As the deployment begins, a

member of the ranger unit is present at the DACG to ensure an orderly loading of personnel and equipment.

(5) The DACG plans for troop-holding areas in the event of delayed or aborted aircraft.

(6) Unit movement data are maintained by the air officer in the operations section.

3-7. Communications-electronics.

a. **Alert notification communications.** Alert notification of the ranger battalions by the regiment is done by secure communications means. Units in local training areas are informed by radio or through their assigned range field telephone. Units operating out of normal FM radio range during RRF periods set up a high frequency net, or other communications, to their higher headquarters.

b. **Telephone service.** Strict control is maintained over key telephones being used during unit alert. Care is taken to ensure that COMSEC and OPSEC violations do not occur. All telephones are assumed to be monitored; classified, sensitive, or official information is not discussed.

c. **Radio communications.**

(1) **Voice communications.**

(a) A secure radio net is set up early in the deployment. The ranger unit EOC, the installation EOC, and the RSE EOC operate in this net. A liaison officer from the ranger unit is placed at the installation EOC.

(b) The ranger support element liaison officer, attached to the logistics section, operates a secure voice station on a separate logistics net to the RSE EOC and the RSE commander.

(2) **Cryptographic material.** During RRF, COMSEC codes and enabling equipment are kept available to set up the secure voice net.

3-8. Remote marshalling base and intermediate staging base.

a. Ranger units can deploy from their CONUS base directly to the objective area. A more common method would be for the ranger unit to deploy first to either a REMAB or an ISB before insertion into the area of operations. (See Figure 3-1.)

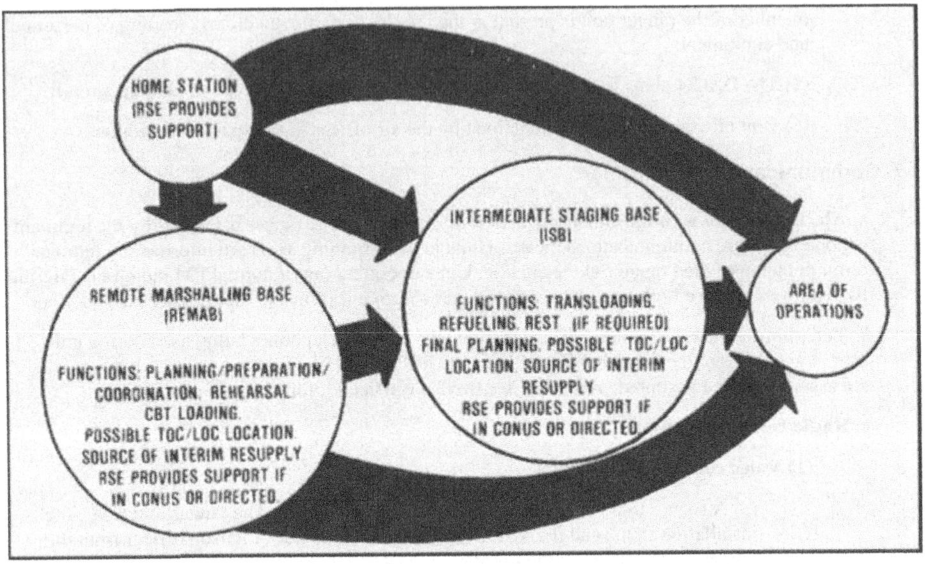

Figure 3-1. Deploying ranger units.

b. The REMAB is a secure base to which the entire ranger unit, including organic and attached support elements, deploys and where mission planning continues. The REMAB is normally within the geographical area encompassed by the command authority of the theater or JTF commander. This ensures that the CSS elements providing support to the ranger unit are operating within their normal area. It prevents or lessens out-of-sector support requirements for CSS elements.

c. The REMAB also provides a secure location for the ranger unit to conduct detailed planning and coordination with the controlling headquarters staff. In the REMAB, the ranger force commander conducts rehearsals, refines and modifies plans, determines PIR, and coordinates with the proper intelligence source to receive that information. It is in the REMAB that specialist augmentees to the ranger force are integrated into the unit, if they have not already joined. Specially trained supporting units, such as aviation and communications elements, also join the ranger force at the REMAB. The REMAB must provide--

(1) Access to the controlling headquarters staff.

(2) Physical security of billeting, planning, maintenance, and communications areas.

(3) Mess, billeting, latrine, and shower facilities for the ranger force and its supporting elements.

(4) Access to a C-141 or C-130 capable airfield, possibly with all-weather operations.

(5) Access to secure communications and processed mission intelligence.

(6) Access to rehearsal areas where sites can be built and live-fire rehearsals can be conducted.

(7) Access to the unit locations of major supporting elements, such as naval landing craft or Army aviation units.

(8) An external security force and an active counterintelligence agency.

(9) Vehicular transport for troop lift, equipment transfer, and administrative use.

(10) Access to maintenance support facilities.

(11) Medical support facilities to augment the ranger force medical personnel.

(12) Covered area to allow for packing of parachutes and rigging of airdrop loads.

d. The ISB is a more austere location. Elements of the ranger force deploy to an ISB to perform final planning, coordination, and task organization. The unit's organization and composition are finalized for movement to the objective area. The ISB is normally not occupied for long periods. Ranger units may deploy to the ISB from the REMAB or CONUS. Deployment to the ISB is more common when terrain or distance precludes insertion to the objective area directly from the REMAB or CONUS. Quick-reaction missions often involve the use of an ISB. Facilities needed at the ISB are limited to the following:

(1) Austere messing arrangements for the combat elements of the ranger force and its attachments.

(2) Austere billets or rest areas.

(3) Water point.

(4) Fuel for aircraft and vehicles.

(5) Areas for test firing of weapons.

(6) Covered and concealed areas for assembly of the ranger force and rigging of parachutes and door bundles.

(7) Austere airfield support facilities, possibly capable of all-weather operations.

(8) Remote location far away from civilians or traffic routes.

(9) Security and counterintelligence elements.

(10) Secure communications.

e. The REMAB and ISB should be located in an area similar in terrain and climate to the objective area. Time spent at the REMAB or ISB lets the ranger force begin acclimatization. Ranger units train yearly in all climates and emphasize physical conditioning. This allows acclimatization quicker than units not as well trained or conditioned. However, ranger units still need an acclimatization period if the terrain and climate of the objective area are different from that of their CONUS base. Not enough acclimatization means that the ranger force could be less efficient, especially when the objective area is hot, dry, or at a great altitude.

CHAPTER 4

Insertion, Extraction, Escape, and Evasion

Insertion is the entrance of a ranger force into a hostile area to execute a particular mission. It is the first critical phase of a ranger operation. These operations often involve deep penetrations of enemy territory by air, water, or land. This chapter emphasizes fundamentals and techniques employed by ranger units when their mission requires insertion and extraction. Extraction is conducted quickly after the mission to avoid casualties.

4-1. Intelligence.

a. Insertion plans are based on timely and accurate intelligence. The headquarters directing the operation must provide current intelligence information to the ranger force. The ranger regiment normally takes action to solicit required information from the controlling headquarters and from other agencies, if security permits. The type of intelligence needed varies depending on the specific mission, but normally includes the following:

(1) Current photographs from aerial or overhead reconnaissance.

(2) Detailed maps.

(3) Data on the population, terrain, and weather from area studies.

(4) Current enemy order of battle, enemy troop locations, and enemy capabilities.

(5) Details on the target area and infiltration routes.

b. To meet its need for current intelligence, the ranger force also relies on aerial reconnaissance and surveillance of the terrain over which it will move. Emphasis is on getting information on the enemy's ability to detect and engage the inserting forces. The location and abilities of air defense radar and weapons systems are critical. Terrain analysis is detailed and focuses on areas suitable for the insertion. Weather information must be detailed and current.

c. In getting information and intelligence data, OPSEC measures are stringent and must be enforced. Coordination with other agencies is kept within security constraints approved by the directing headquarters.

4-2. Deception.

There are always plans to deny the enemy knowledge of the ranger unit's insertion, or to deceive him as to the location or intent of the operation. Feints, false insertions, and other deception operations all add to ranger deception plans. Selection of unexpected means, time, place, and routes of insertion, as well as use of speed and mobility, give less reaction time to the enemy. Planning may include the use of diversionary fires to direct the enemy's attention away from the inserting unit. Deception techniques that may be used include--

a. Multiple airdrops, water landings, or both. This precludes the entire force from becoming engaged at the same time and place, if detected.

b. Dispersion of insertion craft (air or water) both in time and location.

c, Landing a force in an area closer to other potential targets than to the actual targets.

d. Leaks of false information and false messages.

e. False insertions or radio traffic to disguise actual insertions.

f. Diversionary actions such as airstrikes in other than the target areas.

g. Increased reconnaissance flights, either by manned aircraft or remotely piloted vehicles, over false targets.

4-3. Speed and Mobility.

a. Speed of action must be emphasized when the enemy can react quickly to the ranger force presence or when there are limited means of withdrawal. The operation must be conducted swiftly, allowing the ranger force to withdraw before the enemy can react.

b. Major problems facing ranger units are the need for speed and mobility, and the heavy loads carried by soldiers for sustained operations. Ranger units are required to be flexible and to be able to react to any problem within the objective area. However, if equipment is carried to cover every encounter, the load will become too heavy. This will reduce the speed and mobility of the ranger unit. The ranger force commander must always have a balance between the equipment carried and the mobility of his unit. He does this by making three different types of loads: fighting, mission, and existence. He decides what items will make up these loads, where they are carried, and what will be done with them upon contact with the enemy.

(1) The fighting load consists of those items of equipment, weapons, ammunition, food, and water that are common to all soldiers and are necessary for immediate use in combat. These items are normally carried and are retained at all times while in contact with the enemy.

(2) The mission load consists of those items of equipment or ammunition that are required by the mission, but are not normally carried by all rangers in a unit. The load may include spare batteries, demolitions, radios, antitank weapons, or mines. This load is normally spread out among members of the unit and is rotated often. It may be dropped upon enemy contact, but is normally carried to the objective and either used or emplaced there.

(3) The existence load consists of any items designed to provide protection from the elements: sleeping gear, changes of clothes, spare rations, or tentage. This load may or may not be brought into the objective area, depending on the tactical situation. If it is brought, it is normally dropped upon enemy contact or left in an assembly area and retrieved later.

4-4. Stealth.

Methods and steps chosen for insertion must emphasize stealth and try to avoid detection by the enemy. Multiple routes or methods of insertion may be used to preserve the ranger force if some soldiers or units are seen.

4-5. Suppression.

Effort must be made to suppress enemy detection devices, weapons systems, and command and control facilities. This may include electronic countermeasures or direct attack. This helps prevent the enemy from finding and engaging the ranger force en route to the objective area. Deception tactics add to suppression activities. If the ranger force is inserting by infiltration, suppression must not disclose friendly activities.

4-6. Security.

Emphasis on OPSEC prevents compromise of an impending operation. The operation must not be disclosed by overt rehearsals or training activities, the open use or procurement of special items of equipment, or the location of the marshalling area. Several measures can help in maintaining security:

a. Guarded and secure facilities for the headquarters and staff during planning.

b. Issue of operation orders only after units are moved to secure marshalling areas.

c. Alerts, rehearsals, and training to get the local people used to conditions that will exist during actual marshalling.

d. A secure facility set up to isolate units selected for an impending operation.

4-7. Night vision and electronic devices.

a. Rangers use night vision devices to detect and avoid the enemy's forces and his detection devices. These devices also assist in controlling and speeding movement and allow units to traverse almost impassable terrain.

b. The use of electronic homing devices permits precise navigation during reduced visibility. These devices may also be used with visual marking systems.

c. Passive night vision devices are used to assemble rapidly and to reorganize.

d. The use of active night vision devices in assembling, and the degree to which drop zones or assembly areas are "sterilized," depends on the following:

(1) Mission success may depend on not being seen during insertion. When a clandestine insertion is made, the time needed to get to the objective will normally be greater, the use of passive devices will prevail, and the sterilization of the area will be vital. Each ranger must be briefed on sterilization plans and techniques for erasing signs of the insertion.

(2) When detection is likely and the mission depends on getting to the objective before the

enemy can react, speed becomes crucial to the success of the insertion. Insertion may be near or on the objective. Rangers must get in and out quickly. They do minimum sterilization of the area and can use active RSTA devices.

4-8. Rehearsals.

Rehearsals should be close to the actual conditions of insertions and extractions. Security of rehearsals and rehearsal sites is important to the success of the operation. Rehearsals are held on terrain and structures like those in the objective area. If such structures are not available, they are built by the ranger force and the RSE with its supporting engineers.

4-9. Sand tables and terrain models.

Sand tables are useful to acquaint personnel with insertion sites and surrounding terrain. The use of sand tables and terrain models during the issuance of prejump orders and briefings enhances an orderly and speedy tactical assembly.

4-10. Communications.

Reporting during insertion and extraction missions is normally by exception, using an operations schedule.

4-11. Planning insertions.

a. **Centralized planning.** Planning for insertion of ranger units is centralized at the lowest level responsible for the overall operation. This is normally at the joint task force level, with detailed coordination required among all supporting forces. These forces may include other Army elements, USAF, United States Navy (USN), United States Marine Corps (USMC), other US agencies, and allies. Preparation must include detailed contingency planning for use of firepower, shock effect, speed to break contact (if detected), control measures, procedures, and planning. When planning takes a long time, intelligence must be continually updated through all available sources. This includes overhead photography and ground sources to ensure that the initial planning data are still current. Planning stresses the use of deception. Escape and evasion plans, external communication nets, and the location of caches must be known to leaders at all levels.

b. **Reverse planning sequence.** Insertion planning is done in the reverse planning sequence with the ground tactical plan being prepared first. Actions at the objective serve as the basis for the landing and assault plan, the plan for insertion and movement, the loading plan, and the plan for marshalling. All plans are interrelated. Some may be developed at the same time. Consideration of the factors of METT-T guides all planning. Special considerations that apply to these operations are:

(1) **Mission.** The nature of actions at the objective and requirements for special items of equipment and nonorganic personnel impacts on how the unit inserts. When the mission requires rapid deployment, the fastest method of insertion may be needed. In other operations, where success of the mission depends on maintaining secrecy, speed may be less important.

(2) **Enemy situation.** Enemy disposition, strength, and security measures along the route(s) near the objective area affect the method of insertion to be used. Planners must take into consideration whether the enemy:

(a) Uses RSTA detection devices or not.

(b) Has a strong air defense system or possesses air superiority.

(c) Has gaps between his defensive positions.

(3) **Terrain and weather.**

(a) **Topography.** Terrain such as mountains, swamps, or heavy forests favors land infiltration. Deserts, arctic regions, and open prairies favor air or water insertion. Areas not inhabited or with a friendly but scattered population favor land infiltration. A heavily populated area, or one which contains large enemy forces, favors air or water insertion. Hydrography--tides, currents, surf, reefs, and sandbars--must be considered when planning water insertion.

(b) **Weather.** Bad weather with reduced visibility and high winds favors land infiltration, and adversely affects air or water insertion. Limited visibility favors land infiltration; good visibility does not. Adverse weather aerial delivery systems (AWADS) lessens the impact of reduced visibility as a limiting factor on air infiltration. High winds may preclude airborne insertion.

(4) **Troops.** The number of men to be inserted and their level of training may be limiting factors. A need for special skills may call for use of nonranger unit personnel whose physical stamina and ability to perform many infiltration techniques may not equal that of the rangers. Whether supporting forces and special items of equipment are available must also be considered.

(5) **Time.** The main concern is whether the ranger operation is quick-response or deliberate-response. During a quick response mission, planning, preparation, and rehearsal time is short. The timing of the insertion is crucial with respect to weather, the enemy, and if an insertion craft is available.

4-12. Air insertion.

The most rapid form of insertion is by air. Rangers and equipment can be delivered by parachute (static-line or freefall technique), fixed-wing aircraft (airlanded), or helicopter (airlanded, parachute, or rappeling).

a. Suppression of enemy air defense along the insertion corridor may be necessary. The USAF will normally assume the major responsibility for suppressing enemy air defense artillery (ADA) capabilities that may interfere with the inserting forces. This is done by a variety of sophisticated electronic countermeasures applied against enemy equipment and by strikes against known enemy positions. Army fire support elements, special electronic mission aircraft (SEMA), and ground jammers may participate, as well as Navy aircraft, jammers, and gunfire.

(1) A main danger area is the forward edge of the battle area (FEBA) or frontier areas where

the enemy employs most of his sophisticated weapons systems and air defenses. Other dangerous areas are population or troop concentration areas behind enemy lines protected by point air defense systems.

(2) Since the majority of the enemy's detection devices and air defense weapons may be located near the point of entry, fire support, smoke screens (even at night), and suppressive measures may be critical. The use of special items of equipment may be needed to counter the enemy's RSTA effort.

(3) Long-range fires (artillery and naval gunfire) should be planned on known and suspected enemy antiaircraft locations and on prominent terrain features along the route(,s). Once beyond the FEBA, inserting units may be beyond the range of conventional indirect fire. More reliance will then be placed on air assets for fire support and electronic countermeasures.

b. There must be close coordination between the ranger unit and the transporting unit. What the transporting unit does during the insertion has a great impact on the success of the mission. To decrease the chance of detection, maximum use is made of periods of reduced visibility and tactical cover and deception. Drop zones and landing zones are chosen behind tree lines, in small clearings in forests, or on other types of terrain not easily seen.

c. All flight routes over enemy territory should be planned in order to avoid occupied areas and enemy air defenses. Routes should complement the cover and deception plan. They should avoid giving away final ranger objectives.

d. In-flight emergencies, mainly in deep-pedetration flights, should be considered, The ranger force receives a preflight briefing on the route to be flown and is informed of checkpoints along the route. Simple ground assembly plans for contingencies are set up before enplaning. In an emergency, the leader of the unit involved decides whether to continue the mission or abort. This is based on METT-T factors, contingency plans, and distance to the objective as compared to the distance to friendly territory. Contingency provisions for air and water rescue are made.

e. Deciding the DZ location is a joint effort. After receiving input from the air and ground commanders, the joint task force commander makes the final decision. During air assault operations, the ground commander is responsible for the overall process. After reviewing MEYF-T factors, it may be determined that surprise is more important than speed, or vice versa. This may determine the number, size, and locations of DZs and LZs.

f. Planning and preparation continue during the marshalling phase of any operation using air delivery for insertion. A review must be made of items of equipment that are crucial to the ranger force reaching its target area, accomplishing the mission, and returning safely. The need for special supplies and equipment, their transport, the distance to be traversed, the delivery means, and external support are all factors that enhance or reduce the chance for success. Once the means of insertion has been decided, logistics planning continues with emphasis on requesting and coordinating external support. The following must be considered:

(1) Detailed planning for mission logistical needs.

(2) Extra sets of key items of equipment to ensure redundancy and self-sufficiency of

inserting units.

(3) Plans for accompanying, planned, and emergency resupply. The ranger force should prepackage loads to meet specific contingencies. It should consider the use of supplies and equipment already in the operational area.

(4) In quick-response operations, only manportable items of equipment needed to accomplish the mission are carried. Plans and signals must be established for resupply of key items.

(5) Specific plans to evacuate casualties.

(6) All attached personnel and augmentation equipment must be properly prepared for movement.

(7) Provisions must be made for evacuating captured enemy equipment and personnel.

(8) All inserting units should be cross-loaded. Items to be considered in a cross-loading plan are the ground tactical plan, the assembly plan, the configuration of the DZ, the light route and formation of the aircraft, and the number of aircraft available. Cross-loading ensures that key leaders and equipment--for example, door bundles--are evenly distributed throughout the assaulting force. If one or more aircraft abort or are shot down, some key leaders and equipment will still arrive at the DZ. This permits mission accomplishment. Cross-loading aids rapid assembly on the DZ in support of the ground tactical plan. If a ranger is to assemble near the leading edge of a DZ, he is loaded on the aircraft to exit near the front of his stick.

(9) A bump plan is used if an aircraft should abort while still on the ground and there are no spare aircraft. The bump plan ensures that key leaders and equipment are not left behind when the main body takes off. If an aircraft aborts on the ground, key personnel move to another aircraft and replace nonessential personnel. This concept promotes success on the objective.

4-13. Airborne insertion.

a. During the planning phase of an airborne operation, there is no room for error in judgment. It can be planned and conducted well only if there is a constant exchange of information. All personnel involved must be informed of all changes in plans, times, and locations.

b. The reverse planning sequence is the key to the success of an airborne insertion. Most of the planning is based on the ground tactical plan, followed by the air movement plan to support it.

c. The USAF is responsible for flight planning, initial point (IP) selection, and crew procedures throughout the flight. The senior ranger in the aircraft must remain oriented and keep abreast of any last-minute changes. He must coordinate with the aircraft commander.

d. Most airborne insertions are made at either very high altitudes (HALO/HAHO) or very low. For combat operations, drop altitudes may be less than 500 feet above ground level (AGL), depending on the type of parachute used.

e. In airborne insertions, emphasis is on the use of special delivery or navigational techniques since normal insertion is during periods of limited visibility.

(1) **Adverse weather aerial delivery system.**

(a) A self-contained navigational system in some USAF C-130 aircraft is AWADS. The system allows the aircraft crew to fly a specific route to a DZ during darkness or bad weather. The system works with a Doppler radar that determines the ground speed of the aircraft, and an AWADS computer programmed with special flight route information (start point, route, checkpoints, wind speed and direction, and air speed). These two instruments show where the aircraft is in relation to the programmed checkpoints. This eliminates the need for airdrops under visual weather conditions.

(b) Only certain aircraft in formation are equipped with the actual AWADS device. Other aircraft are equipped with station-keeping equipment (SKE). This SKE consists of a computer that tells the aircrew where their aircraft is in relation to other aircraft in the formation.

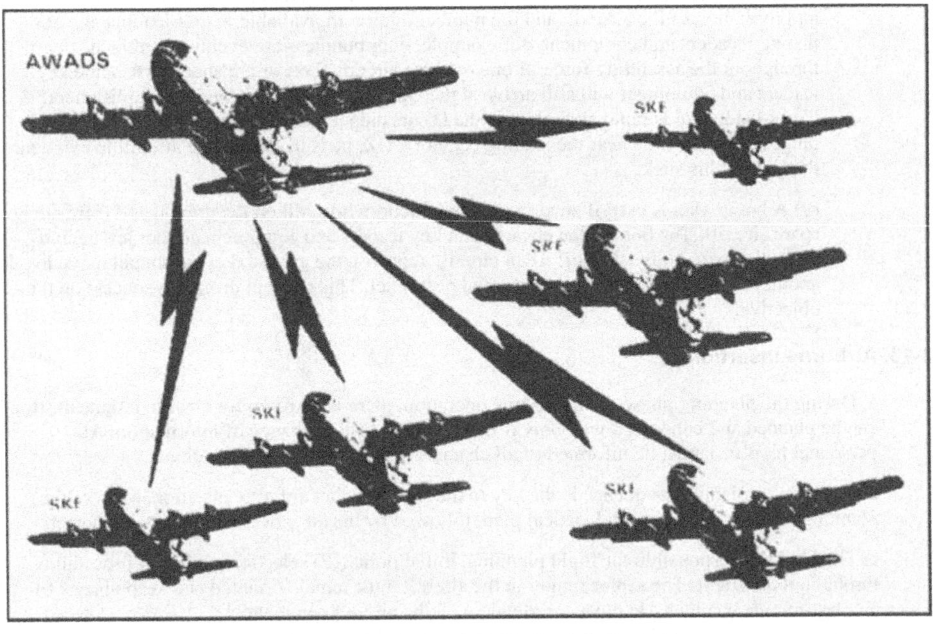

Figure 4-1. AWADS abd SKE aircraft formation.

(c) Airborne insertions using AWADS give the ranger commander the ability to airdrop personnel and equipment during low or zero visibility. This increases the safety of USAF aircraft. It protects them from visually tracked enemy antiaircraft guns and non-all-weather fighters. It also uses adverse weather as a combat force multiplier. Using AWADS, personnel and equipment may be inserted day or night without a pre-positioned USAF combat control team (CCT) or an Army assault team. Time length of the air train is shortened. The air corridor has to be cleared only once. This enhances the element of surprise. The CCTs are brought in with the assault elements. They are able to direct other low altitude parachute extraction systems (LAPES), container delivery systems (CDS), and air-land or heavy drop missions. Use of AWADS enables the ranger commander to conduct rapid, vertical reinforcement of units threatened by an enemy attack during bad weather.

(d) The use of AWADS requires detailed planning, rehearsal, and close coordination among the forces involved. Intense training of both USAF and Army personnel is required.

(e) Personnel must be briefed on terrain through use of photographs, maps, and terrain models. Limited visibility relies mainly on compasses and audible and visual assembly aids.

(f) During reduced visibility, essential heavy-drop serials may precede personnel aircraft. This reduces the time lag between personnel and heavy-drop chalks due to the need to move friendly troops off the DZ. Remaining heavy-drop serials or LAPES can be scheduled following P-hour.

(2) **Reconnaissance teams.**

(a) The operations of the main ranger force may be enhanced by use of reconnaissance teams. These teams are inserted ahead of the main body to provide early warning or to conduct reconnaissance of key objectives. If reconnaissance teams are required at a point along the flight path or near the main DZ, they may be cross-loaded on board any aircraft within the formation. They exit at planned times so as to land as close to their objective as possible. Should such teams be required at a point perpendicular to the axis of the main formation, one of the following is used:

- Another flank aircraft delivers teams to designated points before or with arrival of the main body.
- Selected aircraft deliver the main body, turn, and then deliver the reconnaissance teams.
- The teams are cross-loaded as complete elements on aircraft delivering other units whose flight path crosses the desired locations.

(b) Missions assigned to reconnaissance teams include:

- Providing local security for a USAF CCT, in which case, by joint doctrine, they form the Army assault team (AAT).
- Establishing surveillance or blocking positions on likely avenues of approach into the objective area or DZs.
- Locating and marking targets or assembly areas for the main assault force.

(c) Reconnaissance teams may be inserted by static line or HALO/HAHO operations.

(d) If the reconnaissance element is to locate a specific target or objective, it may be inserted up to 24 hours early. The longer the reconnaissance team remains near the objective area, the greater the danger of it being seen.

(e) In order for the information to be used by the ranger force commander, secure communications and a linkup plan must be prepared and rehearsed.

(3) **Assembly.**

(a) The ranger airborne force can conduct a quick assembly of combat power (weapons, personnel, and command and control elements) on the DZ. Quick assembly and reorganization is crucial because the unit is open to attack. Assembly must be fast and precise. Assembly areas and plans are set forth after considering all the factors of METT-T, mainly the location of the enemy, visibility, terrain, drop information, dispersion pattern, and cross-loading.

(b) The number of assembly areas depends on the location of the DZ, size of the assembly areas, and enemy's detection ability. Unit members move directly to the assembly areas without breaching security. Dissemination of information to all jumpers is the key to a successful assembly plan.

(c) For battalion-size or larger operations, general assembly areas are prescribed in the ranger force OPORD. Company commanders select and designate the exact location of their assembly areas. Although assembly area locations are terrain-dependent, they are generally on the DZ if the drop is at night and off the DZ if the drop is during daylight.

(d) Personnel chosen to carry assembly aids are cross-loaded in the center of their unit sticks. Strong, aggressive leaders should jump first and last in each stick. Upon landing, the leaders roll up their sticks toward the center, locate and move to the proper assembly aid. They may also carry or wear assembly aids. Individual rangers then follow these key personnel. During daylight operations, radiotelephone operators (RATELOS) can display color-coded pennants on their radio antennas to facilitate linkup with leaders. Key personnel may carry and jump with their own radios, and also display color-coded pennants on the radio antennas. Key personnel and RATELOs put their radios into operation before moving to assembly areas. For drops occurring near dawn or dusk, both daylight and night assembly plans are required.

(e) When tactical vehicles or door bundles are dropped, they are normally marked with panels, chemical lights, or strobe lights. The delivery parachutes are often color-coded.

(f) A ranger force should be able to assemble on an unfamiliar DZ, at night, within 45 minutes of the final aircraft pass. Because most ranger operations are time-sensitive, the unit normally moves from the assembly area to the objective when enough of the force has assembled to accomplish the mission. A straggler control point remains at the assembly area under the control of a designated leader. When everyone assembles, or at a specific time, the leader moves to link up with the unit. If available, a radio is left with the straggler control point.

(g) Assembly area security plans are prepared and disseminated. These plans must address hasty defense, sectors of responsibility, observation posts, fire support, and chain of command.

(4) **Contact on the DZ.** During the conduct of parachute insertion, rangers must be ready for enemy engagement at all times. Rangers are open to attack on the DZ before assembling. Just as contingency plans and chance immediate-action drills are developed, briefed, and rehearsed for other chance enemy contacts, rangers must be prepared to execute immediate-action drills for enemy contact on the DZ. The following actions are taken when there is enemy contact on the DZ:

(a) Rangers immediately assemble in fire team-size groups and fight off the DZ to designated assembly areas. Elements attempt to break contact and assemble.

(b) Assembly areas designated by commanders in the mission OPORD serve as primary rally points.

(c) Alternate assembly areas (normally 2 to 5 kilometers from the DZ) are designated in the mission operation order (OPORD). They are used by elements that cannot link up or assemble in the primary assembly area. These elements continue with the mission once assembled.

(d) During limited visibility, personnel should move directly and quickly off the DZ. They must avoid mistaking other friendly elements for enemy, which could cause firefights among friendly forces.

(5) **Reorganization.**

(a) The unit reorganizes according to plans, using designated assembly areas, aids, and identification markings for personnel and equipment. Replacements are named for key leaders or personnel who are lost or injured.

(b) Security teams assemble and move to their positions. Remaining units move quickly to their assembly areas, carrying supplies and equipment.

(c) Subordinate leaders keep the commander informed during assembly and reorganization.

(d) Runners may be used to report assembly status. This maintains radio silence. Designated personnel quickly recover supplies and equipment and move to the assembly areas.

(e) Reorganization is complete when all units have assembled or have been accounted for, and control has been established.

(6) **Pop-up technique.** This involves close terrain flying until the DZ is reached, then the aircraft quickly climbs to jump altitude. Special navigational equipment and aircrew training

are needed but may not always be available.

(7) **HALO/HAHO.** This involves high-altitude, low-opening and high-altitude, high-opening jumps with high performance parachutes designed to maneuver the jumpers to a specific point on the ground. During these operations, midair assembly procedures may be used.

(8) **Arctic.** Airborne insertion in extremely cold conditions is difficult. Times allocated for all parachute and equipment rigging must be increased. Rigging of individual equipment must be IAW FM 57-220. Ahkios can be rigged, packed inside a double A22 container, and dropped in platoon bundles. Because of the increased amount of individual equipment needed, allowable cargo loads for USAF aircraft are reduced.

(9) **Water.** Because of a lack of suitable land DZs, or to avoid detection, rangers may conduct airborne operations onto water DZs close to land. Procedures and techniques are similar to land DZ operations.

(10) **In-flight rigging.** On long-range operations involving extended flight time to the objective area, personnel may rig in flight. In-flight rigging requires detailed planning, coordination, of all jumpers. A rehearsal of these procedures should be conducted before deployment. Aircraft loads must be reduced to carry out in-flight rigging procedures. (See FM 57-220.)

4-14. Air-land insertion (fixed-wing).

a. **Planning factors.** Most of the planning factors that apply to airborne insertion operations also apply to air-land insertion operations. Most USAF aircraft cannot conduct air-land operations in total instrument weather conditions. The AWADS equipment is accurate enough for airdrop navigation, not air-land. Air-land operations are conducted, however, under conditions of extremely reduced visibility. Specially trained USAF crews, using sophisticated navigational techniques, can locate and land on small airfields in bad weather. As in airborne operations, ranger reconnaissance teams, as well as specially trained CCT elements, can jump before the air-land aircraft arrive. They emplace navigational and location markers to assist the pilots. Rangers may rely solely on air-land operations for insertion. This method does not provide for delivery of the force in the shortest time possible. Air-land insertions are normally in concert with an insertion of airborne elements.

b. **Time.** Since turnaround time is crucial, loads in aircraft should be unchained as the aircraft taxies. This also applies to personnel and seat belts. Aircraft should stop as close as possible to the assembly areas where personnel and equipment are off-loaded.

c. **Method.** Air-land insertions normally begin after a ranger Army assault team (AAT) and a USAF CCT have parachuted into the initial objective area and conducted reconnaissance, clearing, marking, and security operations. The ranger ground force commander is the arrival airfield commander. He has control over all arriving, taxiing, and departing aircraft (fixed-wing or rotary) through the CCT, which locates near him. The ranger force command and control element is usually on the first aircraft in and the last aircraft out.

d. **Security.** Air-land operations concentrate large numbers of men and equipment, which presents an open target to the enemy. The ranger ground force commander must ensure rapid off-loading, assembly, dispersion, and movement toward the objective.

e. **Communications.** The ranger ground force commander must have secure radio communications with the ground force, CCT, fire support elements, and higher headquarters. Wire should be laid to connect ground command and control locations.

f. **Operation duration.** Depending on the factors of METT-T, the aircraft may remain on the ground for extraction while the mission is being accomplished or take off at once. They may then orbit, pending recall, or return at a prearranged time.

g. **Extraction.** Extraction of the air-landed ranger force is normally planned and accomplished like an insertion, only in reverse order. A simple, yet effective, way must be found to see that no ranger personnel are left in the objective area.

4-15. Air assault insertion.

Ranger air assault (air-land, rappel, or combination) insertions normally take place during limited visibility. They quickly place the ranger force on, or close to, the objective. The ranger force (depending on the mission) normally plans on fighting when it hits the ground, completing the mission, and then being extracted. Air assault insertions need detailed planning, surprise, flexibility, speed, shock effect, concentration of combat power, and precise timing.

a. **Ground tactical plan.** The reverse planning process is important. The ground tactical plan, developed from the mission assessment, is the first planning area. All other planning begins from this point. The ranger commander and his staff decide stated and implied tasks, and request the proper helicopters for the mission.

b. **Fire support plan.** The fire support available to the ranger commander may be artillery, naval gunfire, mortars, attack helicopters, and USAF or USN tactical aircraft. Because of the timing of ranger air assault operations, preparatory fires are often not used. If they are used, enemy ADA sites normally have priority.

c. **Landing plan.** This plan puts the ranger force into the objective area at the time, place, and proper sequence to support the ground tactical plan. The ranger commander selects primary and alternate LZs and the priority for landing. The exact location of an LZ is determined after a detailed analysis of METT-T. Rehearsals are important. The ranger force must quickly assemble, reorganize, and accomplish the mission. The senior ranger on each aircraft ensures that the personnel on his aircraft offload at the correct LZ.

d. **Air movement plan.** The air movement plan provides for the coordinated movement of the ranger force into the zone of action in support of the 'landing plan. Flight routes, air movement tables, flight formation, altitude, and air speed should be considered. The ranger commander must have secure in-flight communications with his subordinate commanders.

e. **Loading plan.** This plan supports the air movement plan and provides for the pickup zone (PZ) setup and control. It sets priorities for the commander for loading personnel, equipment, supplies, and cross-loading. Air-loading table, aircraft loading formations, bump plan, and rehearsals should be considered.

f. **Staging plan.**

(1) This plan includes PZ selection and operation. The selection of PZs and LZs needs planning and coordination for best use of air assault assets. Site selection in or near the objective must be coordinated in person between the supported ranger commander and the aviation commander. The tactical situation is the key factor in PZ and LZ selection. Other factors are size of landing zones, surface conditions, ground slopes, approach and departure directions, prevailing winds, obstacles, communications, cross-loading, aircraft command and control, PZ and LZ markings, and rehearsals.

(2) The PZs are set up and secured about 15 minutes before touchdown of first aircraft. The fire support plan must support the other plans. All supporting fires must be coordinated with the air mission commander (AMC). Because of the depth of most objective locations, USAF tactical aircraft will be the main fire support method. Orbit locations for fire support aircraft must be planned.

g. **Other planning considerations.** These include escape and evasion, actions at the last LZ, assembly plan, downed aircraft procedures, control measures, forward area rearm and refuel point (FARP), weather delays, and deception plans.

4-16. Amphibious insertion.

Insertion by water may be by SCUBA, surface swimming, small boats, submarines, surface craft, helocasting, or a combination of these methods.

a. **Planning.** Planning for waterborne operations must be thorough. The plan is decided upon between the ranger commander, the transporting unit, and the joint headquarters directing the operation. Plans must be made for all types of enemy action and weather while on board the transporting craft. The transporting unit is given information on a need-to-know basis. Information that could endanger the operation may be withheld until the mission is under way. Initial planning includes:

(1) Time schedule of all events, from the beginning until the end of the joint operation, to be used as a planning guide. Timing for each event must be exact; the success of the operation depends on it. (See Figure 4-2)

Figure 4-2. Waterborne insertion.

(2) Embarkation site, where troops and equipment are loaded on the transporting craft.

(3) Drop site, where the ranger force will depart the primary craft and load into smaller boats.

(4) Handing site, where the unit will beach its boats, or land directly from amphibious craft.

(5) Loading, when loads are lashed and waterproofed IAW unit standing operating procedures (SOPs). Inspections are important. A bump plan for key personnel and equipment must be developed.

(6) Cross-loading, which will help preserve the ranger force. location of leaders, crew-served weapons, and needed equipment are stressed.

b. **Landing site.**

(1) **Beaches.** Beaches are checked to find one that allows an undetected approach. Landing sites that cannot be approached from different directions are avoided. Stretches of defended or patrolled coastline should be avoided. If sand beaches are used, tracks and other signs of

operation that may compromise the mission should be erased. Rural, isolated areas are best. The area behind the landing site should provide an assembly area and concealed routes for exit.

(2) **Barriers.** Reconnaissance may find that good landing sites are not suitable because of barriers between the site and the objective. Such barriers may be patrolled roads, large areas giving little or no concealment, enemy obstacles or defensive positions, areas populated with enemy sympathizers, or an area used for enemy troop maneuvers and training. Unless a means of avoiding such barriers is devised, landing sites should be limited to those without barriers.

(3) **Other factors.** Beach selection should also consider:

(a) The scheme of maneuver.

(b) Surprise.

(c) Enemy dispositions.

(d) Distance to the target.

(e) Landing and exit sites.

(f) Trafficability of the beach.

(g) Cover and concealment.

c. **Routes.** The route the transporting craft takes to the drop site should be planned to deceive the enemy. A major change in the route immediately after the unit's debarkation could compromise the mission. The route should be the same as a route that would be used in some other naval operation--for example, minelaying or sweeping, or a naval patrol. An alternate route must be planned.

d. **Tactical deception.** There should be other plans to deny the enemy knowledge of the insertion besides the water approach route plan. Plans may include electronic countermeasures (ECM), diversions, feints, ruses, or demonstrations.

e. **Actions at the drop site.** Some operations may permit landing directly from the transporting craft. If not, primary and alternate drop sites must be set. The drop site should be at least 1,500 meters offshore to cover any noise in the loading and launching. If the enemy has a surface radar capability, a drop site should be chosen to avoid detection. The drop site may have to be several miles offshore or may need the use of ECM. The manner in which men, small boats, and equipment are loaded must promote fast action at the drop site.

f. **Navigation.**

(1) Ship-to-shore navigation (to the landing site) may be done by dead reckoning. The course may also be maintained by compass navigation, reference to a shoreline silhouette, radar, or prearranged signals from the shore.

(2) Boat handlers, navigators, and other experts are relied on in various phases of the

transport operation.

(3) Lead reconnaissance teams or scout swimmer teams may be used to plant radar, radio, or infrared buoys, or distinct, shielded lights. Boats may be equipped with radar reflectors or beacons to permit accurate tracking.

g. **Actions on the beach.**

(1) Boats hold offshore while ranger scout swimmers move to a covered and concealed position. After reaching the beach, they make a short listening halt. They check the beach and the area immediately beyond the beach for enemy activity. Swimmers signal the force "all clear," mark the landing site, and provide local security.

(2) When the boats land, chosen personnel move directly into covered and concealed security positions. They prepare to defend the landing site.

(3) If the ranger unit is not carrying equipment, boats may be moved off the beach and buried or camouflaged. If heavy equipment makes it hard to move the boats, chosen personnel start unloading at once. The remainder of the force secures the landing site. Each boatload moves into position under the control of a designated leader. Teams are positioned with the best cover, concealment, and fields of fire available.

(4) The ranger force leader sets up a command post from which he can control the operation. He may establish secure communication with his headquarters.

(5) Boats may be deflated and buried or camouflaged near the landing site or away from it. If the boats are to be hidden near the landing site, a team may be chosen to dig holes or cut brush for camouflage along with the unloading of the boats. A small team may be chosen to sweep the beach to erase tracks and drag marks.

(6) The ranger force should be assembled and prepared to move within minutes after the last boat lands. Every ranger must be briefed on the unit's plan should enemy contact be made upon landing.

h. **Insertion by air from ship.**

(1) Helicopters from a ship can launch the ranger force from over the horizon and move at a low level to the LZ.

(2) Helicopters may be vectored from ships to a landing zone. The aircraft may return to the ship while the ranger force conducts the mission, they may orbit offshore, or they may stay at the LZ. Other aspects of landing and assembling are the same as discussed for helicopter insertions.

i. **Helocasting.** This form of insertion combines helicopters and small boats in the same operation. It is planned and conducted the same as air assault operations, except that the LZ is in the water. While helicopters are moving at low levels and low speeds, rangers drop small boats and jump into the water. They then assemble, climb into the boats, and move to shore. Ranger elements are usually assembled and loaded into the small boats within 20 minutes of the last ranger entering the water. This type of operation is normally conducted during limited visibility.

j. **Fire support.** Fire support may be ship-to-air, ship-to-shore, air-to-ground, or ground-to-ground. Specific missions include:

(1) Suppressing enemy air defense.

(2) Neutralizing enemy field artillery within range of boat or helicopter routes.

(3) Employing on-call fires against shoreline targets.

(4) Destroying enemy aircraft.

k. **Contingency planning.** The following must be planned for:

(1) Actions on enemy contact.

(a) En route.

(b) Helocast site.

(c) Flares.

(d) Aerial attack.

(e) Small arms fire.

(f) Indirect fire.

(2) Bump plan with priority to key leaders, crew-served weapons, and radios.

(3) Downed craft procedures.

(4) Escape and evasion.

(5) High surf.

(6) Adverse weather.

(7) Separation.

l. **Rehearsals.** The ranger force must rehearse all aspects of the amphibious insertion. This includes boat launching, paddling, beaching and assembly, boat commands, and capsize drills.

4-17. Land infiltration.

Land infiltration to the objective may be the best way to complete a mission. This is normal when the enemy has total air control or effective air defenses in the operational area to be penetrated, or when the objective is close to friendly territory. Rangers can infiltrate over any type of terrain, in any climate. However, thick forests, swamps, and broken or steep terrain offer the best chance for success.

a. **Planning.** There are many methods of infiltrating by land, and each has its good and bad points. Plans are designed to get the infiltrating force to the objective area without being seen. Plans should include:

(1) Selecting concealed routes for surprise and protection. Primary and alternate routes are based on detailed map and aerial photo study, ground reconnaissance, and data from agent reports.

(2) Avoiding obstacles, populated areas, silhouetting (when forced to cross obstacles or ridge lines), enemy positions, main avenues of approach, and movement along often used routes and trails.

(3) Selecting the time of infiltration to make use of reduced visibility, such as darkness, rain, snow, and fog, and periods of reduced enemy alertness, such as early morning, holidays, and times following protracted movement or combat. The time of the infiltration is important during critical phases (crossing borders or passing through enemy troop concentrations or populated areas).

(4) Using speed and stealth. All rangers must know routes, linkup points, alternate linkup points, time schedules, danger areas, and actions on enemy contact and at linkup points.

(5) Centralizing planning and decentralizing execution. Centralized coordination must ensure that units are acting IAW cover and deception plans. Control should be given to the leaders of the infiltrating groups. Within their designated zones, these leaders select their own routes and formations. Leaders control the rate of advance. When the enemy has electronic detection devices, radio communications must be kept to a minimum. Key leaders normally go with the group that is using the fastest lane or the lane least likely to be detected.

(6) Seeding infiltration routes to the assembly area with unattended ground sensors. These sensors can help the infiltrating force by reducing the chance of surprise contact during the infiltration, and by helping in controlling the forward progress of the units.

(7) Planning for fire support, which must include coverage of routes, danger areas, assembly areas, objectives, and PZs.

(8) Rehearsing actions at the objective, actions on enemy contact, actions at danger areas, and movement techniques.

b. **Movement.** Land infiltration is conducted using one of three techniques:

(1) **Movement in small groups along one axis.** All members of the force use the best route. Small groups are harder to detect, easier to control, and do not compromise the total force if detected. This technique may take too long and it needs an assembly area or linkup point. If the lead group is detected, other groups may be ambushed.

(2) **Movement in one group.** This technique has no reassembly problem. The same route is used by all, making navigation easier. A large force can fight out of a dangerous situation easier than a small one. It has no coordination problems between other infiltrating units. This form of movement is easier for the enemy to detect. If the force is detected, the mission may be endangered.

(3) **Movement in small groups along several routes at the same time.** This technique avoids putting the total force in danger, is less likely to be seen, forces the enemy to react in

many locations, and makes it harder for the enemy to determine the size of the force or its mission. Some groups travel over poorer routes and have problems of assembly, control, and support; the capture of personnel may endanger the mission. Regardless of which technique is used, ranger units normally move to reduce the chance of enemy contact.

c. **Assembly area.** Following movement into the general area of operations, infiltrating units converge on the assembly area or linkup point. Assembly may begin at once after the passage of lines, or it may take place at any time before reaching the objective. The infiltrating force must remain undetected before assembly. If an infiltrating unit is seen at this time, the enemy may learn the location of the assembly area, and the objective. It may take action to destroy the infiltrating force.

d. **Linkup point(s).**

(1) Selection characteristics include terrain that--

(a) Is free of enemy troops.

(b) Offers cover and concealment from enemy observation and patrols.

(c) Does not restrict movement of friendly units.

(d) Accommodates the unit.

(e) Is easily identified.

(f) Offers escape routes to alternate linkup points.

(g) Offers concealment from air attack.

(2) Action sequence for occupying the linkup point.

(a) Halt short of the point.

(b) Establish local security.

(c) Conduct a reconnaissance of the linkup point.

(d) Occupy the linkup point.

(e) Emplace observation posts and listening posts.

(f) Link up with other groups.

(g) Conduct final preparations.

(h) Continue the mission.

(3) Security may be enhanced with the aid of devices such as night vision goggles and night sights. Unattended ground sensors may be placed on likely avenues of approach into the linkup point. Normally, only passive observation devices are used to assist in avoiding detection.

(4) Plans must set a time schedule for actions at the linkup point. All unit members must move in quickly, consistent with security, so that the unit may proceed to the objective rally point (ORP) and on to the objective.

4-18. Stay-behind technique.

A ranger force can be positioned behind enemy lines by employing the stay-behind technique. In this method, a ranger force allows itself to be bypassed as the enemy advances. This is done to allow the ranger force to perform a specific mission. Stay-behind operations may require the concealment, or cache, of supplies before the enemy bypass. Rangers may need to use captured enemy supplies. The location of the assembly area and the plan for extraction are important parts of stay-behind plans.

4-19. Extraction.

The ranger force is normally extracted after the mission is completed. The enemy may also force the unit to abort its mission and to call for extraction. Contingency plans must be made to extract a force under enemy pressure, as well as one that is not in contact. Primary and alternate extraction sites are planned and coordinated with all ,porting forces. Ranger operations are normally conducted against deep targets. Extraction may begin with land movement but normally it includes the use of boats or aircraft. Plans may include primary I alternate means of extraction.

a. **Terrain.**

(1) The terrain is important in choosing the point of extraction - for example, whether or not the extraction site offers good cover from enemy direct-fire weapons and positions from which the extracting force can suppress the enemy. Tidal conditions and PZ size must be considered.

(2) Unlikely locations such as swamps, jungles, or mountain areas are often used for extraction. Specially trained air and naval crews, using the latest extraction devices, make such terrain useable.

b. **Enemy.**

(1) Extraction is best conducted without enemy pressure. However, such pressure could develop during the operation. Detailed plans are made for emergency or contingency extractions forced by the enemy as well as clandestine extractions not under enemy pressure. If under pressure, units will have to use planned fire support to suppress enemy weapons. At the beginning of the extraction with security posted to provide early warning, the unit may start to thin out the same as with a night withdrawal. Ground security may be relied on until the last unit withdraws under cover of suppressive fires.

(2) If enemy contact is made during an extraction, the commander must decide whether to reinsert units or break contact and move to another extraction point. Fire support and suppression are important and include suppressing enemy air defense by deception, ruses, or decoys.

(3) The ranger unit may have special weapons systems or equipment to aid in the extraction. Shoulder-fired antiaircraft missiles may be used to defend against enemy aircraft threatening

the extraction.

c. **Extraction by air or water.** Extraction by air or water is preferred when the resources are available and their use will not endanger the mission. Other factors that favor their use are when--

(1) Long distances must be covered.

(2) Time of return is essential.

(3) The enemy does not have air and naval superiority.

(4) Heavily populated hostile areas block land movement.

(5) Ranger forces are burdened with prisoners of war,casualties, or critical items of equipment.

d. **Extraction by land exfiltration.** Stealth and evasiveness are key elements of exfiltration. This method of extraction is favored when--

(1) The enemy has air superiority.

(2) The enemy is able to prohibit air or water extraction.

(3) The distance to friendly lines is short.

(4) The terrain provides cover and concealment (for movement on foot) and limits enemy mobile units.

(a) Multiple routes of exfiltration may be used to prevent destruction of the force if soldiers or units are detected. Captured enemy vehicles and equipment may be used to assist in the exfiltration.

(b) The ranger force may exfiltrate in one body or in small groups. Exfiltrating in small groups avoids the time delay of assembly.

(5) The exfiltrating force is lightly equipped and not burdened with captured personnel or material.

(6) The exfiltration route passes through an area occupied by friendly civilians or guerrilla forces who can assist the movement.

(7) Areas along exfiltration routes are not inhabited.

(8) The enemy force is dispersed or is not able to concentrate against the exfiltrating force.

4-20. Escape and evasion.

a. A vital part of all premission planning is the development of en-route and post mission E&E plans. Such plans enhance survival of the ranger force, and the transporting air or boat crews. The ranger force commander is responsible for--

(1) Helping in the development of an E&E plan in coordination with all supporting agencies.

(2) Ensuring that all members of the ranger force and the supporting boat or air crew are briefed on the E&E plan.

b. Each E&E plan is unique. Each situation has unique problems associated with E&E. The E&E plan devised by the ranger force commander must address these problems, while profiting from the individual abilities and training of the rangers and their supporting air and boat crews. The following considerations apply to E&E plans devised during ranger operations:

(1) The plans enhance survival of personnel who can no longer accomplish their assigned mission. The senior ranger in a group must decide if there are any remaining missions that the group can accomplish. If not, then he must try to evade and escape enemy capture if unable to link up and be extracted with the rest of the ranger force.

(2) Because of the depth of penetration behind enemy lines, most successful E&E plans may involve either air or water movement away from enemy-held territory. The air and boat crews supporting ranger operations are highly skilled and dedicated. If there is any chance of a successful rescue, these forces continue to try to retrieve rangers escaping from an enemy-held area.

(3) Rangers possess stamina and endurance. They can move a great distance (especially at night) and over rugged terrain to reach an area where they can rendezvous with search-and-rescue (SAR) aircraft or boats. Rangers are also resourceful and highly trained in communication, stealth, and camouflage. This, coupled with their self-discipline, means that successful E&E can be conducted over a great distance, through heavy concentrations of enemy, and over a long period.

(4) Escape and evasion plans for ranger elements should include avoiding contact with the local civilians. However, the aid of friendly insurgent forces may be enlisted. The ranger E&E plan may also include the use of other special operations forces E&E networks that are in place behind enemy lines. These nets must not be compromised by the volume of evaders.

c. Escape and evasion plans should be based on three phases:

PHASE ONE: That portion of the insertion following entry into the hostile zone where personnel survival and recovery is the only thing considered. Should an aircraft or boat carrying rangers be disabled during phase one, the ranking survivor assumes command. He must consider E&E planning and the experience and skills of other survivors. He should be concerned with the security of the mission, caring for wounded, and moving away from enemy activity while contacting SAR elements for recovery.

PHASE TWO: That portion of the insertion that is near the objective and could permit the ranger element to successfully pursue its mission. Although part of the E&E plan, actions by the ranger force during this phase become a forced land infiltration rather than an E&E. The senior surviving ranger must decide if enough personnel and equipment can be moved to the objective area in time to accomplish the mission. If so, he must proceed, taking along any surviving air or boat crew members. If not, he may proceed directly to the extraction point and wait, or he may contact SAR elements for recovery.

PHASE THREE: That portion of the operation after mission accomplishment. It is a hard time to escape and evade since the enemy is alert and reacting to the ranger forces. The confusion caused by the violence of the attack, coupled with the rangers' ability to move quickly cross-country away from the objective area, offers the best chance of a successful escape and evasion. The surviving ranger elements that cannot be extracted can do one of the following:

(1) Move over land at once to a PZ or beach for extraction by SAR elements.

(2) Move to an area where reentry into friendly lines can be made.

(3) Move to a neutral or nonaligned country where either internment or return to friendly control can be effected.

d. Any plan for E&E should provide for the issue to each ranger of an area map, a covert means of recognition, and a small compass. An emergency ration may also be included and any items needed for the climate in that objective area. The ranger force commander must balance the need for other survival equipment against the added weight and bulk involved.

CHAPTER 5

Strike Operations

The ranger regiment conducts strike operations in support of the Air Land Battle. Ranger strike operations are integrated into an overall plan designed to destroy, delay, and disorganize the enemy. They may also cause him to divert a major part of his combat forces and power to rear area security. Strike operations also help create an environment in which other special operations forces can operate. Strike operations can complement air power in disrupting enemy forces in depth. They allow the theater commander to regain the initiative and lead to decisive action.

5-1. Application.

a. The ranger regiment can conduct strike operations either as deliberate-response or quick-reaction missions. It can be inserted into the area of operations by land, sea, or air.

b. Strike operations can be conducted throughout the depth of the battlefield in support of conventional or unconventional warfare. Multiple strike operations can have a synergistic effect and can create situations that help friendly forces seize and maintain the initiative.

c. Strike operations include raids, personnel and equipment recovery, and interdiction of lines of communication. They are best accomplished by ranger units who use conventional techniques and special skills. Strike operations vary in size. When well augmented and supported, the ranger regiment can conduct strike operations deep in enemy territory using up to three battalions. Strike operations normally are limited in time and may be conducted unilaterally or in conjunction with other special operations forces or allied forces. They are normally ended by the extraction or exfiltration of the ranger force.

5-2. Raid operations.

a. A raid is a strike operation conducted behind enemy lines against strategic objectives, targets of high tactical value, time-sensitive targets, or key personnel and facilities in enemy rear areas. It normally involves a swift penetration of hostile territory to confuse the enemy or destroy his installations. Ranger units can conduct deep penetration raids when given the appropriate assets for insertion.

b. A strategic raid by the ranger regiment would be directed by the NCA and conducted under the OPCOM of a unified or JTF commander. An operational raid would be directed by the unified or JTF commander and conducted by that command directly or under the OPCOM of a corps commander. Typical targets would include:

(1) Command, control, communications, and intelligence centers at front level or combined arms army level.

(2) Nuclear and chemical weapons storage sites and delivery means.

(3) Key logistical centers, warehouses, ammunition complexes, or fuel pumping centers supplying logistical support to a specific front.

(4) Air defense and air traffic control integrating centers and air defense weapons sites.

(5) Key power generating and transmitting facilities, hydroelectric dams, and irrigation systems.

(6) Key ports or rail complexes.

(7) Key installations or facilities, such as airstrips, buildings, bridges, dams, or tunnels.

c. The size of the ranger raid force is determined by a detailed analysis of METT-T. The unit committed to the ground phase of the raid mission is kept as small as possible. It may be as small as a ranger company or it could include all the battalions of the ranger regiment. The force must be tailored to complete the mission quickly, violently, and with few casualties. The size of the force may also be affected by the augmentation of specially trained personnel needed for that mission.

d. Raids are normally conducted in five phases. (See Figure 5-1).

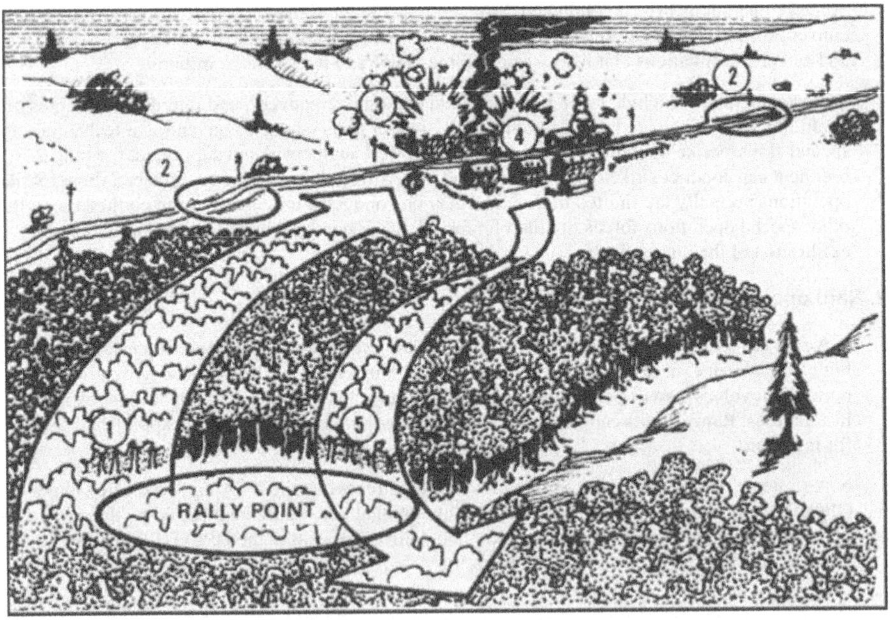

Figure 5-1. Five phases of a raid.

PHASE 1. The ranger force inserts or infiltrates into the objective area.

PHASE 2. The objective area is then sealed off from outside support or reinforcement, to include the enemy air threat.

PHASE 3. Any enemy force at or near the objective is overcome by surprise and violent attack, using all available firepower for shock effect.

PHASE 4. The mission is accomplished quickly before any surviving enemy can recover or be reinforced.

PHASE 5. The ranger force quickly withdraws from the objective area and is extracted.

e. This paragraph discusses the raid after insertion/infiltration. All considerations discussed in Chapter 4 must be applied in order for the unit to get deep behind enemy lines. Upon receipt of a raid mission, the following basics are emphasized:

(1) **Maximum use of intelligence information.** The gathering and dissemination of information must be continuous and provided to the raid force even while en route to the target area. To ensure mission accomplishment, the ranger unit must be kept informed of the latest enemy developments in the objective area to prevent being surprised. Only then can the full combat power of the rangers be concentrated at the decisive time and place.

(2) **Plan development.** The reverse planning sequence and the planning steps set forth in FM 101-5 are followed in the plan development.

(3) **Coordination.** Coordination is through the operational headquarters with friendly units (such as higher and supporting Army or joint headquarters, intelligence agencies) for fire, RSTA, special equipment, and personnel and logistical support.

(4) **Rehearsals.** Rehearsals help validate all aspects of planning for the raid and ensure precision in execution-They allow changes to be made in the plan before it is carried out.

f. A successful raid is ensured by--

(1) Launching the raid at an unexpected time or place by taking advantage of darkness and other periods of limited visibility, and moving over terrain that the enemy may consider impassable.

(2) Avoiding detection through proper movement techniques and skillful camouflage and concealment to include taking advantage of natural cover of the terrain, and using sophisticated equipment to detect and avoid enemy forces.

(3) Timing the operation as close as possible.

(4) Using all available support, both organic and nonorganic, to include use of special weapons such as Air Force "smart bombs" and artillery cannon-launched guided projectiles, with the ranger unit using laser target designators.

(5) Performing quick, violent, precise, and audacious actions that focus full combat power at the decisive time and place.

(6) Disengaging quickly upon mission completion.

(7) Withdrawing swiftly using planned routes and including a deception plan.

g. Four functions are normally performed by ranger tactical formations when conducting a raid. Each subelement is organized and equipped to do a specific part of the overall mission. Depending upon the specific mission, nature of the target, enemy situation, and terrain, these four functions are as follows:

(1) The command group controls movement to and actions at the objective. This unit normally consists of the ground commander, other subordinate leaders in the raid organization, and communications to support these leaders.

(2) The security element, whose organization is determined by the mission of the raid force, size and type of enemy force and its mobility and state of alert, terrain and avenues of approach into the area, and the time needed to seal off the objective area. The security element may perform the following:

(a) Secure the ORP.

(b) Give early warning of enemy approach.

(c) Block avenues of approach into the objective areas.

(d) Prevent enemy escape from the objective.

(e) Provide overmatch for the units at the objective and suppressive fires for their withdrawal.

(f) Provide short-range air defense.

(3) The support element provides the heavy volume of fire needed to neutralize the objective. Because fires from this unit are violent and devastating, they must be closely controlled to ensure the precision needed. On order or as planned, fires are lifted and shifted to cover the maneuver of the assault element by suppressing enemy fire from the objective or aerial fires. The support element may also be given specific locations to cover by fire in support of the security element if an enemy quick-reaction force moves toward the objective area. These may include routes into and out of the objective site, key terrain features, or installations adjacent to the main objective. Once the assault has been completed, or on order from the raid force commander, the support element displaces to the next planned position. Organization of the support element is determined by the following:

(a) Size of the objective, the geography of surrounding area, and the enemy threat (to include air) in the area. This element should be able to neutralize the objective (when supported by air or naval gunfire) and to lift or shift fires either when the assault is launched or when ordered by the raid force commander.

(b) Mission of the assault unit.

(e) Suitable firing positions.

(d) Size and nature of the enemy force in the objective area and those enemy forces capable of reinforcement at the objective.

(e) Fire support from other units (air strikes, naval gunfire, surface-to-surface missiles, and artillery fire).

(4) The assault element seizes and secures the objective and protects demolition teams, search teams, prisoner-of-way teams, or other teams.

(a) The organization of the assault element is always tailored to the mission. Each target area must be examined carefully. To assault, seize, and destroy an installation, position, or equipment, the assault element could be organized into one or more assault teams. The element's mission is to overcome resistance and secure the objective and to destroy the installation or equipment. Other specialized teams may also be needed. For example, sniper teams could be needed to remove key sentries. To capture prisoners, liberate personnel, and seize or destroy equipment, the assault element could be organized into assault teams, prisoner teams, search teams, medical teams, demolition teams, or breach teams.

(b) To destroy a point target or installation in a heavily defended area where the USAF cannot get close en to be effective, the assault element might be organized with one small team equipped with laser target designators. From covered and concealed positions, members of the team could then provide guidance for USAF delivery of laser-guided munitions from a safe distance.

h. The site chosen for the raid LZ or DZ must support the planned actions at the objective. There are two options when choosing sites:

(1) The ranger force can land on or near the objective and seize it before the enemy can react. (See Figure 5-2.) This avoids forced marches over land carrying heavy combat loads. If there is no suitable landing area near the objective, or the enemy has a strong reaction force nearby, this option is not favored.

Figure 5-2. Airborne landing on the objective.

(2) The ranger force can land unseen far from the objective. It then assembles, reorganizes, and moves into an ORP near the objective. (See Figure 5-3.) The objective is seized after security and support elements are in place. This option makes coordinated action easier by setting up control of small units before engaging the enemy.

Figure 5-3. Airborne landing distant from the objective.

5-3. Interdiction operations.

a. AirLand Battle doctrine set forth in FM 100-5 is a shift in emphasis toward keeping the initiative in battle. The deep battle seeks out enemy elements separated from the main battle area and attacks them before they can be deployed. Interdiction of selected enemy forces by attacking them or blocking chokepoints hinders, delays, or interrupts the enemy's use of his lines of communication; it denies him the use of certain areas; it destroys industrial facilities, military installations, and equipment; and it stops or slows his advance along a particular route of approach. The ranger regiment's ability to conduct interdiction is vital to execution of the AirLand Battle doctrine.

b. Interdiction by ranger forces includes blocking or channeling approaches, holding key terrain for a limited time. and destroying facilities.

c. Ranger forces conduct interdiction missions that supplement those conducted by long-range Army and USAF weapon systems. Large-size ranger forces are normally used only when other systems fail or are not available, and the target is of such priority that it must be attacked. Small ranger laser target designator teams may work in concert with USAF or USN strike aircraft to direct terminally guided weapons systems. Rangers should be directed against crucial interdiction targets that disrupt, confuse, or delay the enemy and allow friendly forces to gain the initiative at the forward line of own troops (FLOT).

d. Typical interdiction operations are--

(1) Destroying rail lines, switches, rail yards, and rolling stock.

(2) Destroying bridges, overpasses, tunnels, and truck parks.

(3) Closing mountain passes or routes in restricted terrain.

(4) Cutting oil or natural gas pipelines and destroying pumping and distribution systems.

(5) Destroying dams, locks, spillways, and flood control gates.

(6) Destroying rail transload points, switches, and yard engines.

(7) Destroying shipyards, dry docks, wharfs, lighters, cranes, and stevedore facilities.

(8) Destroying electrical generation facilities, power lines, transformers, and grid monitoring centers.

(9) Destroying water pumping stations and filtration systems.

(10) Destroying radio and TV stations, phone lines, microwave terminals, satellite receiving stations, and telephone exchanges.

(11) Ambushing enemy forces and personnel moving along lines of communication.

e. Considerations when using rangers for interdiction are:

(1) Insertion and extraction.

(2) Support resources available.

(3) Ranger force vulnerability.

(4) Target identification.

(5) Threat environment.

(6) Other or supporting means of interdiction.

(7) Command and control.

(8) Time and magnitude of the effect on the enemy forces at the forward edge of the battle area.

(9) Collateral effects of target destruction.

f. Interdiction targets are not identified and attacked at random. They are part of an overall deep attack plan to destroy or cripple an entire system. Interdiction is based on the assigned deep campaign plan that directs which targets are to be attacked, by whom, the results desired, and the priorities of attack for specific systems. Based on his plan, the ranger force commander selects those ranger elements best suited to conduct the attack. Target selection, which requires detailed intelligence and planning, is based on:

(1) **Target criticality.** A target is critical when its destruction or damage would have a major influence on the enemy's ability to conduct or support operations. Each target is considered in relation to other elements of the particular target system designated for interdiction. The criticality of a target changes with the situation. For example, when the enemy has few locomotives, a railroad bridge is less critical.

(2) **Target accessibility.** The ranger force must be able to insert into the target area. Terrain, security controls, location, and means of insertion are considered when studying a target for accessibility. systems. Based on his plan, the ranger force commander selects those ranger elements best suited to

(3) **Target restorability.** The enemy's ability to restore a damaged facility to normal operating capacity affects target selection. The enemy's repair and maintenance ability and his reserve stocks must be considered.

(4) **Target vulnerability.** A target is vulnerable when it is open to attack by means available to ranger forces. Vulnerability is affected by the nature of the target such as type, size, disposition, and security. It is also affected by the means available to attack it such as explosives, incendiaries, and special devices.

g. Key to the interdiction process is the ability to locate and evaluate target systems for attack and the timing of the attack. A target system is a series of related functions that serve a common purpose.

(1) A target is one element, installation, or activity identified for attack such as a locomotive, train, bridge, or prison.

(2) A target complex is several targets in the same general area, such as a railway marshalling yard, an airfield, or port and dock facilities.

(3) A target system consists of an industrial system and its sources of raw materials; the rail, highway, waterway, or airway systems over which these materials are transported; the source of power and method of transmission; the factory complex; and the means by which the finished product is transported to the user. The targeting process is usually accomplished in the special operations force cell at EAC and the battle coordination element (BCE) at corps level.

h. Interdiction techniques include ambush, emplacement of conventional demolitions, target designation for laser-guided munitions, and emplacement of beacons or sensors.

NOTE: The enemy surface-to-air threat normally precludes the use of an airborne platform for direct observation and designation. This may be accomplished by the on-ground ranger force.

i. Through interdiction, ranger units are an ideal force to contribute to the Army's AirLand Battle operational concept. This includes taking the initiative, moving quickly to strike deep in the rear of the enemy, and defeating the enemy force beyond the FEBA. Planners and commanders must ensure that the ranger units are used on missions that are synchronized to contribute to the overall deep battle campaign. Strikes against critical targets must combine with efforts of other assets to take the initiative from the enemy, limiting his flexibility, initiative, and momentum.

5-4. Personnel and equipment recovery operations.

Recovery operations are specialized raids to liberate imprisoned or detained personnel or to return certain items of equipment to friendly control. These operations include recovering and extracting downed or hijacked aircrews and political or military leaders. Rangers use surprise and combat power to overwhelm resistance before harm can come to the detainees or prisoners.

a. **Planning.** The planning of personnel and equipment recovery operations is the same as for the raid. The main difference is that the ranger force commander must plan for the extraction of the recovered personnel and the loading and extraction of sensitive equipment. Personnel and equipment recovery raids are often executed as quick-response missions. They normally require the ranger force commander to plan and execute the recovery quickly.

b. **Augmentations.** This type of strike operation often requires that the ranger force be augmented with nonranger personnel with special skills or abilities. Examples are medical personnel, technical experts, mechanics or crew members trained in repair and retrieval operations, linguists or translators, and other special operations force elements.

c. **Organization.** The organization of the recovery force is the same as for a raid. Sometimes special teams are needed to perform certain missions involved in the recovery. This mission is conducted aggressively, usually ending in close combat during darkness or other limited visibility conditions. Reconnaissance teams may be inserted ahead of the ranger force to reconnoiter the objective and to locate the detained personnel.

(1) All planning and execution takes place as described for the raid. The emphasis must be on detailed, timely intelligence.

(2) Medical teams must be available to care for the detainees.

(3) The ranger personnel must be adept at searching and clearing of buildings.

(4) The ranger force must--

(a) Locate the specific building or area where the friendly detainees are being held.

(b) Gain entry into buildings, use stun grenades, and avoid obstacles and booby traps.

(c) Engage all enemy personnel using selective-kill techniques.

(d) Locate, identify, and secure detained personnel.

(e) Move personnel to a secured extraction site.

(f) Leave no detainees or rangers behind in the objective area.

d. **Execution.** The key to success is surprise and speed. The enemy must be assaulted when least expected. The ranger force must ensure that friendly detainees are not harmed during the assault - either by rangers or enemy personnel.

e. **Extraction.** The ranger force commander should plan for the extraction point to be as close as possible to the point of recovery. This prevents the detainees from having to move cross-country or for the ranger force to transport bulky equipment long distances. For small groups of detainees or small items of equipment, the extraction point can be further away. Recovery operations may use any method or combination of methods of insertion and extraction (see Chapter 4). Close planning and coordination is normally required with US Army, USAF, or USN aviation for both delivery and evacuation of the target area.

CHAPTER 6

Special Light Infantry Operations

Ranger units are normally employed on strike operations. However, broad US strategic needs demand a wide range of other military abilities. The ranger regiment may conduct special light infantry operations when larger conventional forces could not be used. The regiment conducts these operations using light infantry doctrine, methods, and techniques.

These operations include, but are not limited to, movement on foot, military operations on urbanized terrain (MOUT), passage of lines, ambush, and reconnaissance. Ranger units do not normally conduct planned operations against enemy armored forces. Battlefield surprises or METT-T considerations may cause ranger units to engage armored forces for short periods. The ranger regiment can conduct a conventional light infantry defense for short periods. A defense is not normally conducted adjacent to other friendly forces. It is normally a perimeter defense from which the unit then breaks out from encirclement and links up with friendly forces.

6-1. Movement.

a. Most missions involve movement on foot in tactical operations. Ranger units infiltrating behind enemy lines seek to avoid chance enemy contact. They choose the movement that allows them to retain security and control. To avoid loss of surprise and initiative, casualties, and mission failure, ranger units normally--

(1) Avoid chance enemy contact, if possible.

(2) Move on covered and concealed routes.

(3) Avoid likely ambush sites and other danger areas.

(4) Practice camouflage, noise, and light discipline.

(5) Maintain all-round security.

(6) Make unavoidable enemy contact with the smallest element possible.

(7) Retain the initiative to attack at the time and place of the ranger unit's choice.

b. In planning the movement of his unit, the ranger commander must consider all the elements of METT-T and determine how they will affect the movement. He must determine the route, formations, security measures, methods of command and control, location of key leaders and weapons, and action to be taken on chance enemy contact.

(1) **Routes.** The ranger force commander conducts a careful route analysis, using the factors of observation, concealment, obstacles, key terrain, and avenues of approach (OCOKA). He chooses a route that maintains the security of the force while ensuring surprise. He tries to choose a route that will avoid enemy contact and speed movement. The commander makes a detailed route reconnaissance using maps, aerial photographs, and reports from persons familiar with the area. If possible, he reconnoiters the route himself or uses a reconnaissance team. He may have that team act as guides or mark the route using covert marking devices.

(2) **Formations.** Ranger units use the movement techniques found in current light infantry doctrinal publications. The traveling, traveling overmatch, or bounding overmatch technique is chosen based on the likelihood of enemy contact. The ranger force normally tries to avoid detection and contact with the enemy while moving. The commander chooses a movement technique that ensures that if unexpected contact were made, it would be with the smallest element possible and not the entire formation. The distance between moving ranger elements depends on the terrain, visibility, and enemy situation. March interval is normally great enough to allow each succeeding element either to deploy or change direction if enemy contact is made. Distances are close enough that each element could quickly assist the element to its front.

(3) **Security.** Security is to the front, rear, and flanks when enemy contact is likely. A front security element is placed well forward with adequate communications with the main body. The security element is used to detect the enemy and warn the main body. The main body then either changes direction or hides until the enemy is past. If the enemy does detect the main body, the security element may be used to attack the enemy from the flanks or rear. Flank security elements are placed next to the route of march. They move forward either by alternate or successive bounds. An alternative is to have the security element moving adjacent to the column along routes paralleling the direction of march. Rear security is handled the same as front security. Plans are made for the rear guard to support the main body during chance enemy contact. This is done either by maneuver or by furnishing supporting fire.

(4) **Command, control, and communications.** All available communication is used, consistent with OPSEC and movement security, to assist in maintaining control during movement. March objectives, checkpoints, and phase lines may be used to aid in control. The number of reports is reduced, normally only exception reports are needed. The ranger force commander should be well forward in the formation but may move throughout as the situation demands. Communications with security elements are mandatory. Operations security often precludes the use of radios. Therefore, connecting files, runners, and visual signals are normally used. Detailed planning, briefing, rehearsals, and control are valuable if there is enemy contact. Alternate plans are made to cover all possible situations.

(5) **Weapons.** Rangers are placed within the formation to allow a mix of their individual weapons to cover the front, flanks, and rear of the formation. Rifles, grenade launchers, and light automatic weapons cover all sectors. Crew-served weapons such as machine guns, antitank weapons, and mortars should be given designated locations within the formation. This allows the element leader to employ them quickly. They should be placed near the lead element, but not so close as to be engaged along with it. Crew-served weapons are used to

provide overmatch, especially when employing the bounding overmatch technique of movement.

(6) **Rehearsals.** The ranger force, expecting movement as part of an operation, conducts the following combat drill rehearsals:

(a) Conduct movement.

(b) React to enemy contact.

(c) Maneuver under fire (fire and movement).

(d) Cross danger areas.

(e) Break contact.

(f) Conduct a hasty ambush.

(g) Make a temporary halt.

(h) Establish a perimeter.

(i) Establish a rally point.

(j) Take actions upon consolidation.

(k) Use arm-and-hand signals.

(l) React to flares and indirect fire.

c. After a ranger force is inside enemy territory, it must be alert to avoid detection before reaching the objective.

(1) Should the ranger force security element become aware of enemy presence, it alerts the main force. The main force tries to move away without alerting the enemy and continues the mission.

(2) If the enemy does make contact, the security element then engages the enemy, trying to deceive him as to the size of the ranger force and its direction of movement. After breaking contact, the security force divides into small groups and moves to a rally point to rejoin the main force.

(3) The ranger force does not fight en route to the objective unless there is no alternative, then it quickly breaks contact and moves away.

(4) Ranger units may need to set up a temporary defensive position for resupply, evacuation of wounded, planning, or extraction. One technique that can be used is the temporary perimeter defense.

(5) If the moving ranger force comes under indirect fire, it moves quickly out of the impact area and continues the mission.

(6) The most effective means of combating an ambush is through the immediate return of a

high volume of fire by all weapons. This requires alertness, discipline, dynamic leadership, and rehearsed plans.

(a) Regardless of the method of movement, all weapons should be positioned for immediate use. A well-executed ambush is hard to counteract. However, the effectiveness of an ambush can be reduced by the immediate return of fire. Enemy weapons may be silenced, thus creating weak points in the ambush position. Smoke and high-explosive (HE) grenades are effective in executing counterambush actions. They create confusion, provide a screen for movement, and disrupt the ambush plan.

(b) When all elements of a ranger force are not trapped within the ambush, the elements that are free to maneuver should initiate an immediate flank or rear assault against the ambush force. A flank assault permits better coordination with personnel in the ambush. Supporting fires are provided by those weapons that cannot be hand carried and fired in the assault.

(c) Specific actions against a near and far ambush are as follows:

- **Near ambush.** Elements caught in the kill zone return fire at once and move out of the kill zone. Elements not in the kill zone help by maneuvering and conducting a flanking attack. The unit assaults the ambush position.
- **Far ambush.** Those ranger elements not in the kill zone set up an overmatch position and lay down a base of fire for the elements in contact. The elements in contact move out of the kill zone and prepare to assault the enemy, using individual movement techniques and fire and movement.

(7) The reorganization after enemy contact involves the use of rally points, plans for local security, reorganization of the force, evacuation of dead and wounded, and movement based on the unit mission.

6-2. Ambush operations.

a. **Definition.** An ambush is a surprise attack, from a concealed position. It is used on a moving or temporarily halted enemy to destroy or capture him and his equipment. An ambush is a useful tactic because--

(1) Small, well-trained, disciplined forces, with limited weapons and equipment, can destroy much larger enemy forces.

(2) It reduces the enemy's overall combat effectiveness by destruction and harassment of his forces.

(3) Enemy morale and effectiveness suffer heavily at little cost to the force executing the ambush.

b. **Execution.** A successful ambush must be executed with precision, violence, speed, and audacity of execution. For success, ambush operations must emphasize the following:

(1) **Surprise.** Surprise, more than any other single aspect, enhances the value of an ambush. Surprise increases the potential for inflicting damage on the enemy with less danger to the

ranger force.

(2) **Coordinated firepower and shock effect.**

(a) Coordinated firepower is used for maximum shock effect. Massive volumes of accurate fire, explosives, and mines, coupled with an aggressive attack, breaks the enemy's spirit to fight back. Surprise increases shock effect and the chances for success. Shock effect can cover unexpected defects in an ambush--for example, ambushing a much larger force than expected.

(b) All weapons must be sited with interlocking fires in the kill zone and along likely avenues of entrance or exit. Mortars should be used if the terrain permits. Tripods and traversing and elevating mechanisms are normally used with machine guns to lock in fires. All riflemen use firing stakes to mark left and right limits, and elevation stakes. There is a tendency to shoot high in an ambush--especially at night. The M203 grenade launchers are sited to cover the dead space and routes of escape. When the rules of engagement permit, plastic bags containing CS crystals may be taped to the front side of Claymore mines.

(3) **Control.** Control is essential.

(a) Leaders must have contact with all members of their unit to alert them to the oncoming enemy. Leaders should not move around the ambush site during this crucial period. A method used to alert members can be to tie strings or vines to rangers' legs or arms. By a series of light tugs, all members of the ambush can be alerted to enemy presence.

(b) The leader must initiate the ambush with a casualty-producing device. A bank of Claymore mines on a double-ring main is an excellent device to spring an ambush. Other good techniques are to use a 90-mm recoilless rifle firing antipersonnel (APERS) rounds or a machine gun. All must be under the direct control of the ranger leader. Whistles or pyrotechnics must not be used. They will allow the enemy time to react.

- As soon as the enemy is hit, he reacts. The ambush force has only a few seconds to destroy the enemy in the kill zone. He will quickly recover from the initial shock and leave the kill zone--either with a direct counterattack or withdrawal, Subsequent fires and other banks of Claymore mines must be planned.
- The leader always executes the ambush, except when a member of the ambush knows he has been discovered. He then has the authority to execute--with killing fire, not by yelling.
- The cease-fire must be controlled by the ranger leader. A whistle or other device may be used to get attention, and then cease-fire is signaled.

(4) **Security.** The flanks and rear of an ambush site are open to counterattack. Flank and rear security may be enhanced by -

(a) Echeloning in depth.

(b) Designating sectors of observation.

(c) Positioning of RSTA devices.

(d) Enforcing noise and light discipline.

(e) Having a good withdrawal plan.

(f) Securing routes of withdrawal.

(g) Executing with speed and violence.

(h) Positioning security forces to sea] off the ambush area.

(i) Having good camouflage.

(5) **Simplicity.** A simple, direct plan improves the chance of success. The ambush plan must be clear yet concise to offer the greatest likelihood of success. For example:

(a) Mission statements for security, support, and assault elements must be clear, concise, and direct.

(b) Tasks to be performed by the ambush elements should be easy to understand.

(c) Contingency plans should be simple.

(d) Routes into positions and withdrawal routes should not cross. They should be the shortest, most secure routes.

(6) **Training and self-discipline.** Precise and violent execution is how ranger units win against enemy forces. All advantages must be exploited. Discipline must be strict. There must be no sleeping, talking, eating, or smoking in the ambush site. If an ambush is to be set up for long periods, then the elements of the ambush must be pulled back at set times to the ORP for rest. Extended ambushes of 24, 36, or 48 hours require six- or eight-hour shifts. It may take a company to man an extended platoon ambush position. Tired troops cannot man an ambush well. Other vigorous operations cannot be performed all day and have an alert ambush all night.

(7) **Maximum use of specialized equipment.** This includes the most sophisticated weapons, communications equipment, and RSTA devices. The effectiveness of any mission is enhanced by the use of this equipment.

(8) **METT-T.** Every ambush mission must be evaluated by the ranger commander IAW METT-T to design an ambush that works.

(a) **Mission.** Clearly state the mission--for example, stop enemy resupply, ambush enemy armor columns, or capture personnel or equipment.

(b) **Enemy.** State the type of enemy units--for example, guerrillas, main force, logistics carrying parties, or armor units. What is the size of the force, its habits when attacked, and type of support? What is the enemy night vision ability and expected time of movement?

(c) **Terrain.** Is it wooded, mountain, jungle, or urban? Some good ambushes can be set in a city or suburb. Are the routes available to the enemy suitable to his activity?

(d) **Troops.** Squads and platoons are the basic ranger ambush forces. An ambushing force should be able to engage a force two to three times its size--but evaluate enemy and terrain, A platoon ambushing an enemy tank company in daylight in open terrain is risky. But engaging that same column at night or in a forest may give the ranger platoon an advantage.

(e) **Time.** How long will it take to emplace the ambush? A complex, lethal ambush with plenty of "dirty tricks" takes hours to set up; a hasty ambush on a trail takes 15 to 30 minutes. How long is the ambush to be in place? What hours of the day, or at nighttime only? How soon must it be set up? Do not set up an ambush in darkness--even if it is to be executed in darkness. Try to allow every ranger to '6see" and "lay" in his weapon in daylight. Time is important.

c. **Organization of the ambush force.** A ranger unit conducting an ambush must be task-organized to perform the following functions: assault, security, and support. The ambush forces should be task-organized according to the TOE--by platoons, squads, and fire teams. The TOE should not be changed to create smaller elements for an ambush. The TOE formations may be reinforced with machine gun, recoilless rifle teams, or a 60-mm mortar squad.

(1) **Assault.** The elements assigned the assault mission either move directly into their positions or through a release point. The mission may include any combination of the following:

(a) Conduct the main assault.

(b) Halt an enemy's motorized column or any moving target,

(e) Kill or capture personnel.

(d) Recover supplies and equipment.

(e) Destroy vehicles and supplies.

Search teams are not always used. The ranger leader must decide how and when to use search teams. When rangers leave the security of their well-chosen, concealed ambush position, they are subject to the fires of the enemy who may also be hidden and ready. If it is at night, do not throw tripflares or shoot illumination to light the search area as this will illuminate and expose rangers to the enemy. Always assume there is hidden enemy--the ambush will not kill them all. Night observation devices (NODs) or a red filtered flashlight taped to M16s should be used to quickly search in ranger buddy teams. If the return fire from the enemy is great or if the ambush missed the main body, then the leader may choose to break contact and leave without searching the kill zone.

(2) **Security.** The elements assigned the mission of security may move to their positions

directly or by way of a release point. Their missions may include any or all of the following:

(a) Secure flanks, rear, or ORP.

(b) Provide early warning.

(c) Seal off the kill zone to prevent the enemy from escaping or reinforcing.

(d) Assist in executing the ambush.

(e) Cover withdrawal of main ambush force.

(3) **Support.** The units assigned a support mission provide fires that may include employment of--

- Heavy automatic weapon fires.
- Antitank fires.
- Mortar fires.
- Mines.
- Flame munitions.

d. **Ambush site.**

(1) **Choose an ambush site.** When choosing an ambush site, all sources of information must be used to enhance surprise, exploit the enemy's weak points, and take advantage of the terrain. An analysis of the terrain is also based on the other factors of METT-T to include maps, aerial photos, and, if possible, personal reconnaissance. Emphasis is on--

(a) Natural cover and concealment for the ambush force.

(b) Routes of entry and withdrawal (at least two) that should be direct and easy to reach.

(c) Good observation and fields of fire.

(d) Harmless-looking terrain.

(e) Few enemy escape routes.

(f) Limited enemy reinforcement ability.

(g) Nearby assembly or rendezvous area.

(h) Terrain that will canalize enemy into kill zones, and natural obstacles to keep him there.

NOTE: Try to select a site covered by friendly supporting indirect fires.

(2) **Take advantage of the terrain.** Emphasis must be on exploiting all natural cover and concealment afforded by the terrain. Site the ambush and individual positions based on the terrain rather than trying to adapt the terrain to a fixed geometric design.

(3) **Restrict enemy movement.** Restricting enemy movement by natural or man-made

obstacles should also be planned.

e. **Types of ambushes.**

(1) Ambushes have two categories for ranger purposes--area ambush and point ambush.

(a) Area ambushes may be set up by ranger platoons, companies, or battalions. They are used to interdict enemy movement in a given area or inflict casualties on his forces. Area ambushes consist of a series of point ambushes. The size and location of the ambushes are dictated by the METT-T analysis.

(b) Point ambushes are set at the most ideal location to inflict damage on the enemy. Such ambushes must be able to accept the enemy force from more than one direction. The ranger force levels may enable it to execute an ambush in two or three main directions. The other directions must be covered by security that gives early warning of enemy attack.

(2) The basic ambush is linear (Figure 6-1). An ambush is set on an expected avenue of approach. This ambush can accept contact from three basic directions--left, right, and front. The rear is secured by a security team.

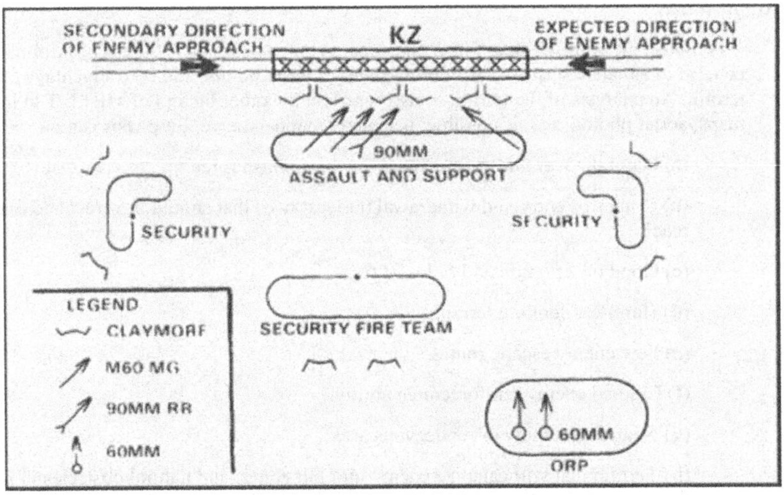

Figure 6-1. The linear ambush.

(3) A variation of the linear ambush is the L-shaped ambush (Figure 6-2). The L-shaped ambush is formed with the base (bottom) of the L perpendicular to the expected enemy direction of advance. This is a good ambush for a road, jungle trail, or an area where the enemy is canalized and his approach route is known. The L-shaped ambush can handle an enemy coming from the expected direction (toward the base of the L) from the front (stem

of the L). It is less effective against an enemy formation that comes from the opposite direction. The ambush must be executed when the enemy main body has cleared the base of the L. The flank security must protect the rear of the base of the L. In the L-shaped ambush, weapons must be sited to avoid direct or ricochet fire into friendly forces. The security elements must be considered as fire fans are being planned.

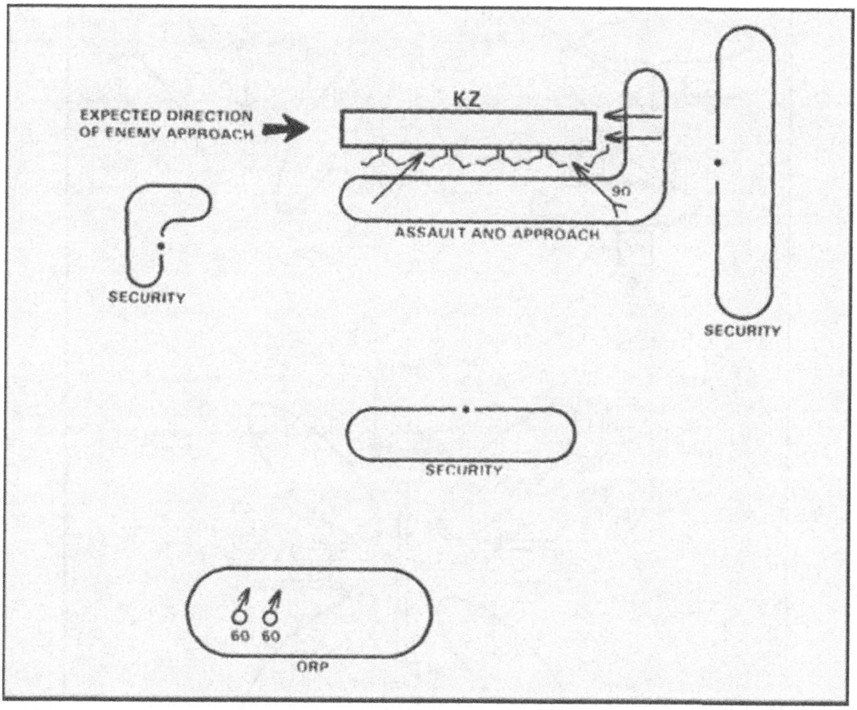

Figure 6-2. The L-shaped ambush.

(4) Other ambushes (the Z, the X, the V, and so on) are all variations of the linear and L formations (Figure 6-3). The more advanced ambush formations are difficult. Units must be expert in the basics before they are ready to employ such techniques.

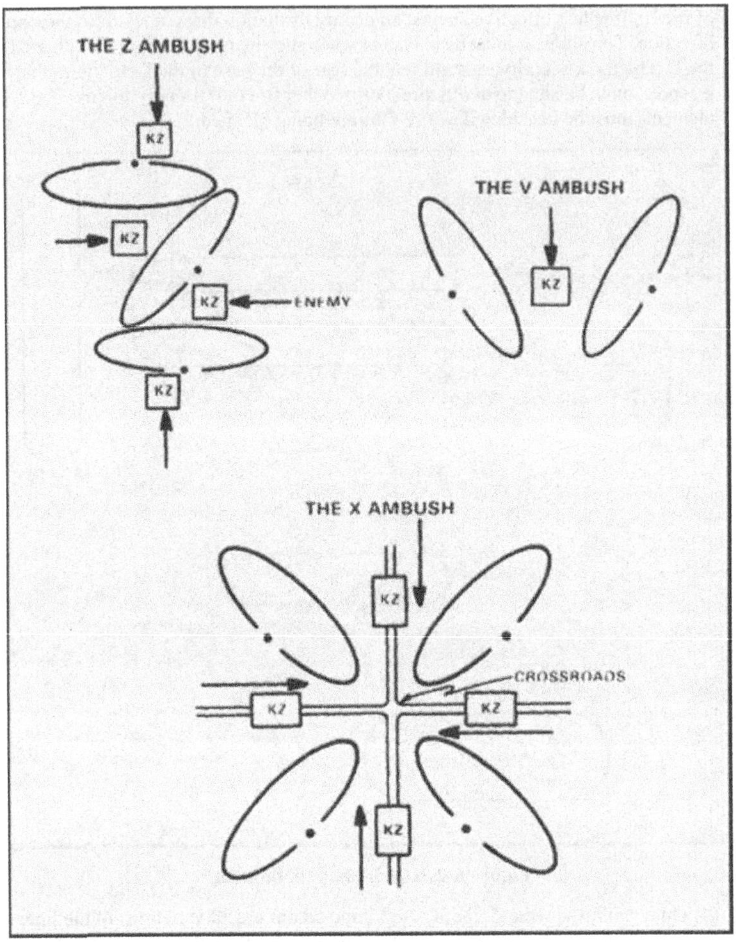

Figure 6-3. Other ambush variations.

(5) Another type of point ambush is the mechanical ambush (Figure 6-4). It consists of Claymore mines set in series with a double-ring main. It is command detonated or detonated by a triggering device activated by the enemy. Mechanical ambushes are normally manned. Rangers prepare to engage the enemy with direct fire if the mechanical ambush does not detonate or if it is wholly or partially ineffective. Mechanical ambushes are an effective way to interdict a large area using a small ranger force. If the mechanical ambush is effective and

rangers do not reveal their presence, the enemy is confused. This has a devastating effect on his morale and effectiveness.

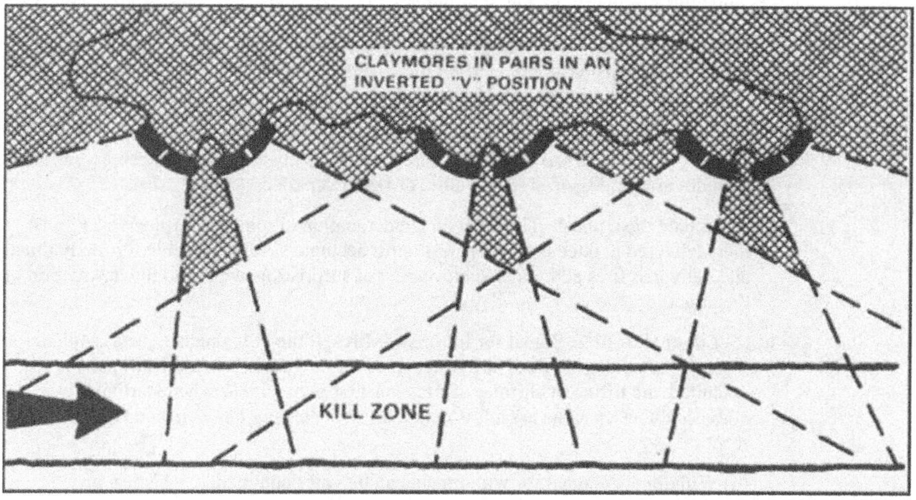

Figure 6-4. The mechanical ambush.

f. **Execution of the ambush.**

(1) **Occupation.** Stealth and security are important factors.

(a) Position security teams and early warning detection devices first.

(b) Use the rest route to main ambush position consistent with security.

(c) Quickly occupy the ambush position and set up communications and signaling devices.

(d) Position key weapons (automatic and antiarmor).

(e) Rig Claymore mines, tripflares, and booby traps.

(f) Ensure that all weapons are correctly positioned. Assign sectors of fire to provide mutual support and cover dead space.

(2) **Camouflage.** Each man must be hidden from the enemy, During preparation for the mission, each man camouflages himself and his equipment, and secures his equipment to prevent noise. At the ambush site, positions are prepared with minimum change in the natural appearance of the site. All debris resulting from preparation of positions is concealed to prevent evidence of occupation.

(3) **Movement.** Movement is kept to a minimum. The number of men moving at a time is closely controlled. Every man is as quiet as possible, especially at night. Light discipline is rigidly enforced at night. Smoking is forbidden.

(4) **Signals.** Audible and visual signals, such as whistles or pyrotechnics, must be changed often to avoid setting patterns and alerting the enemy. Three or four simple signals are needed to execute the ambush. Signals are used to--

(a) Provide early warning of an enemy approach. A signal by the security force to alert the patrol leader to the correct direction of enemy approach may be given. This includes arm-and-band signals, radio, or field telephone.

(b) Initiate the ambush. This may be the detonation of mines or explosives, Fire is then delivered at once in the heaviest, most accurate volume possible. Properly timed and delivered fires add to the achievement of surprise, as well as to the destruction of the target.

(c) Lift or shift fires, Signal for lift or shift fires, if the kill zone is to be assaulted. Voice commands, whistles, or pyrotechnics may be used. When the kill zone is assaulted, the lifting or shifting of fires must be as precise as when starting the ambush. Otherwise, the assault is delayed and the enemy has a chance to recover and react.

(d) Withdraw. A signal for withdrawal can be voice commands, whistles, or pyrotechnics.

(5) **Objective rally point.** The ORP is located far enough from the ambush site so that it will not be overrun if the enemy manages to attack thc ambush unit(s). Existence loads are normally left in the ORP. Routes of withdrawal to the ORP are scouted (when possible by each man), Withdrawal routes should provide cover and concealment for the unit and hinder enemy pursuit. Withdrawal routes are a main consideration in the selection of the ambush site. They may be the key to survival after execution of the ambush. On signal, the ambush force quickly (but quietly) withdraws to the ORP. If pursued, the withdrawal may be by bounds, with mines or hasty ambushes used to delay pursuing forces.

(6) **Ambush variety.** If one method of ambush is used predominantly, the enemy will develop an effective defense against it and will be affected less by the shock of the ambush since he knows what to expect. More than one method of ambush should be used. No single method will fit all combinations of terrain, equipment, weather, and enemy capabilities. Variety should also be used when signaling - for example, use multiple signals, both audible and visual. Use weapons fire, mines, and RSTA when possible and vary signals to avoid compromise.

(7) **Swift action.** Speed in the execution of the ambush and the withdrawal should prevent enemy reaction forces from engaging the ambush force. Speed is often a shield against casualties and failure. When there is contact with reaction forces, speed may enhance quick disengagement. Ranger units must not only perform better than the enemy but also move faster than the enemy can react.

g. **Successful ambush.** Emphasize the following:

(1) Intelligence to ensure the enemy is ambushed at a time and place when he least expects or is least prepared to fight. This is preferred during reduced visibility to achieve surprise.

(2) Detailed planning and thorough training and rehearsing of all elements in all phases of the ambush. This ensures maximum shock effect through swift, precise execution.

(3) All available RSTA devices to permit daytime effectiveness at night when moving, shooting, or detecting enemy movement. To avoid detection, active RSTA devices should not be used until after the ambush has been triggered.

(4) All available firepower with emphasis on antiarmor, area and automatic weapons, and precision-guided munitions.

(5) Speed to achieve surprise and enhance security of the force.

(6) Cover, concealment, and overall protection afforded by the terrain when moving or when occupying ambush positions.

6-3. Antiarmor operations.

a. Antiarmor operations are not primary missions for ranger forces. They lack antiarmor firepower and have limited antiarmor ability. Ranger units can, however, perform limited antiarmor missions on close terrain by interdiction or ambush. The antiarmor weapons they do have are man-portable, but the ammunition is too heavy to carry many rounds cross-country. A ranger battalion has ten Dragon missile launchers and nine 90-mm recoilless rifles. It is also equipped with light antitank weapons (LAWs) and laser target designators (LTDs) that can guide Copperhead rounds.

b. Every ranger must be skilled with the limited antiarmor systems available. Proper selection, fortification, and camouflage of antiarmor ambush sites, route selections, withdrawal procedures, and initiative are important. Armor kill zones (AKZs) are normally set up after infiltrating the ranger force behind enemy lines under limited visibility.

c. A ranger antiarmor force is organized into four elements: command and control, support, security, and armor-killer teams (Dragons, LAWs, 90-mm recoilless rifles). These elements function the same as a raid force. The exception is that the assault force is the armor-killer team with the mission of destroying enemy armor, and it will probably not enter the AKZ. After the ambush occurs, the ranger force withdraws to its ORP. Indirect fires (if available) are then brought to bear on the AKZ.

d. Ranger units may conduct antiarmor ambushes and interdiction operation by use of laser-guided Copperhead rounds or air-delivered precision-guided munitions. Air-delivered family of scatterable mines (FASCAM) can be used to provide the ranger force security against attack by armored forces.

6-4. Military operations on urbanized terrain.

a. The ranger regiment normally tries to avoid the manpower-intensive, high-casualty battles associated with MOUT. However, due to the nature of strike operations and the types of targets selected, the regiment may conduct MOUT for limited periods. As in all ranger operations, the ranger force commander tries to surprise the enemy, attack him from unexpected directions, and destroy him with a violent, precise attack. The maze of cellars, rooms, sewer systems, rooftops, and alleys common to urban areas presents many opportunities for the ranger force.

b. Offensive operations by a ranger force exploit the centralized planning and decentralized execution aspects of ranger unit training. Individual action and initiative are always channeled to accomplish the commander's intent. The ranger unit's offensive tactics are based on quick action and boldness. Attacks by the ranger force are short and sharp. Execution is marked by swiftness, daring, timing, and surprise. The enemy's sleeping habits, eating habits, and relief times are watched. launched almost always at night or under smoke screens, attacks are often made without mortars or artillery, relying on surprise for success. Snipers and sharpshooters are used to shift the balance of battle.

c. Deliberate defensive operations are not good ranger unit missions. However, if a ranger unit must defend for a short period, the close terrain of urban areas allows it to function well against a stronger force. The buildings in a city break up advancing enemy formations and force them to move along streets. The ranger force executes an active defense anchored on strongpoints, such as a building or a group of buildings. It launches surprise counterattacks against the enemy's flank and rear. The aim of ranger defense is to give the enemy no respite and to make him feel threatened by a sudden, deadly attack. Ranger units should be withdrawn from defensive positions in urbanized areas as soon as possible. Prolonged MOUT and the associated casualties quickly degrade a ranger unit's combat effectiveness.

6-5. Perimeter defense.

The Perimeter defense may be conducted to control key terrain, gain time or economize forces.

a. **Planning.**

(1) **Warning order.** The order must be issued soon after mission receipt. It should contain the mission, time schedule (including rehearsal and inspection times, mission time, and place for OPORD), uniform and equipment, and special instructions.

(2) **Operation order.** The leader normally gives the OPORD on the ground to be defended. Locations and sectors of fire for squads and crew-served weapons should be designated. If time permits, leaders walk the terrain with subordinates as the OPORD is issued. The OPORD covers security, communications, signals, target reference points (TRPs), armor kill zones, and priority of engagement rules for antiarmor weapons. The OPORD also states the priority of work.

(a) Set up security and communications.

(b) Position crew-served weapons and personnel.

(c) Clear fields of fire.

(d) Prepare fighting positions.

b. **Organizing.**

(1) Squads are positioned with overlapping and interlocking sectors of fire and observation.

(2) The M60 machine guns are positioned either with an assigned final protective line (FPL) of fire or with a primary direction of fire (PDF). The PI)Fs are employed on covered avenues of approach such as gullies. The M60 FPLs should interlock across the unit front.

(3) The 90-mm recoilless rifles and Dragons are positioned with frontal cover to engage attacking armored vehicles from the flank.

(4) Riflemen are positioned to support and protect crew-served weapons. Riflemen are assigned FPLs that interlock at a point just beyond hand-grenade range.

(5) Dead space is covered with Claymore mines, antipersonnel mines, and M203 fires.

(6) Antitank mines, positioned to provide close-in protection from armored vehicles, are covered by LAW fires.

(7) Leaders plan indirect fire for long-range suppression of armored vehicles and short-range protection from dismounted assault. Mortar or artillery final protective fires (FPF) are placed on the most dangerous dismounted avenue of approach.

(8) Range cards are prepared for all crew-served weapons. Squad and platoon leaders prepare sector sketches and fire plans.

(9) The ranger force camouflages everything from ground and aerial observation.

(10) Leaders designate alternate and supplementary positions.

c. **Executing.** Leaders control the distribution and concentration of fires in their sectors. They constantly consider the overlap of fires with units to their left and right.

(1) Squads and crew-served weapons engage targets in sectors assigned by platoon leaders.

(2) Each weapon is used in its best role:

(a) Rifle, against enemy infantry within 300 meters.

(b) Machine gun, against enemy infantry within 1,000 meters.

(c) Grenade launcher, against infantry within 350 meters and armored vehicles within 200 meters.

(d) Light antitank weapon, against armored vehicles within 200 meters.

(e) Recoilless rifle, against armored vehicles within 400 meters.

(f) Dragon, against armored vehicles within 1,000 meters.

(3) Most dangerous targets are engaged first.

(4) Squads shift fires as needed or as directed by platoon leaders.

(5) The platoon masses fire on selected targets while maintaining coverage of the entire platoon sector.

(6) Platoon leaders control fires.

(7) After repelling an enemy assault, the platoon consolidates and reorganizes the defense.

6-6. Breakout from encirclement.

A breakout from encirclement is conducted when units, operating behind enemy lines, find themselves cut off from friendly forces and surrounded by superior forces. Given ranger unit missions, the chance of operating as an isolated force behind enemy lines is great. The breakout is characterized by determining enemy weak points, deception, massing of combat power, and a direct attack for a violent and timely breakout.

a. **Characteristics and fundamentals.** The breakout from encirclement involves the following characteristics and fundamentals:

(1) Set up a hasty defense. Adjust the perimeter as needed and make contact between subunits. Assign each a specific sector.

(2) Plan fire support. Plan FPFs on likely dismounted avenues of approach. Mass antiarmor weapons on the most dangerous mounted avenue of approach.

(3) Conduct reconnaissance. Identify enemy weak points. Locate enemy armored forces and the routes they must use. Locate enemy automatic weapons.

b. **Planning.**

(1) Organize all support personnel into maneuver elements--for example, company headquarters, mortars, and antitank. Organize the ranger force into a rupture force, reserve, support force, and a rear guard.

(2) Issue a fragmentary order. It should contain current local enemy situation and cover all pertinent areas of the five-paragraph field order.

(3) Select rally points.

c. **Execution.**

(1) Reduce the defense perimeter. This is done slowly until the unit is massed for the breakout attempt.

(2) Create a diversion. This is done either by fire from the rear guard force or by a limited attack in a direction other than that of the main breakout attack.

(3) Break through the defense with the rupture force. Hold the shoulders of the breakthrough.

(4) Move the reserve through to exploit the breakthrough.

(5) Move the support troops through with casualties. Have them collect casualties en route.

(6) Have the rear guard follow the support troops. It may have to fight a delaying action and use fire and movement to break contact with the enemy.

(7) Consolidate and reorganize following the breakout.

(8) Have the rupture force break contact with the enemy once the rear guard joins.

6-7. Linkup operations.

During operations behind enemy lines, ranger units may be required to link up with other friendly ground forces. Linkup may occur in airborne, air assault, amphibious, or shore-to-shore operations. It may occur during the breakout of an encircled force or during the convergence of separate forces. It may involve conventional or unconventional US forces, allied forces, or friendly guerrillas. Rangers may be required to link up with other rangers, partisans, or friendly ground forces that are conducting offensive operations.

a. **Planning.** Ranger commanders plan detailed linkup operations to avoid friendly forces firing on each other. Coordination and planning of communications, recognition signals, fires, command relationships and responsibilities, and control measures are essential. Coordination should be done before the start of the operation.

(1) Site. The linkup site should be easy to recognize, have cover and concealment, be located away from prominent terrain the enemy might use, be defensible, and provide multiple access and escape routes. An alternate site should be planned in case of compromise.

(2) **Communications and recognition signals.** Frequencies, call signs, codes, visual signals, and alternates for each of these should be planned before departure of friendly lines. Radios may be used to report the location of each unit to the site, and occupation and securing of the site. A system of mutual recognition must be devised to keep friendly patrols from firing on each other. This may include visual signals such as arm bands, panels, colored lights, or RSTA devices.

(3) **Fires.** The headquarters directing the linkup sets up fire control measures and priority of fires. The most commonly used control measure is the restrictive fire line (RFL), beyond which one force may not fire without coordination with another force. The RFL may be adjusted as two units draw closer together. Such an adjustment needs close coordination and should be planned before the operation. Successive phase lines between the two units can serve as on-order RFLS, provided that neither unit comes too close to the RFL before it is shifted.

(4) **Command relationships and responsibilities.** These are set up by the headquarters directing the linkup. Ideally, liaison personnel are exchanged before the operation. The key to success is liaison and guides.

b. **Execution.** Each member of the ranger force must understand the plan. Rehearsals are necessary in order to avoid confusion during the linkup. One uninformed ranger can start firing and cause enough confusion that friendly forces could be identified as enemy. Actions must be quick since units are open to enemy attack while they try to link up.

(1) **Moving units.** A linkup between moving units is hard. As the units move closer to each other, the chance of them engaging one another increases. Therefore, the linkup units must adjust their movements to each other and continuously coordinate on a designated secure radio net. If possible, one or both units should come to a short halt before linkup.

(2) **Stationary units.** Moving units must know the positions of the stationary units. They must keep stationary units advised of locations. Stationary units guide the moving units to the linkup point by radio. The stationary units must be ready to accept and position the moving units smoothly and quickly.

6-8. Passage of lines.

A passage of lines is an operation in which one unit moves through another unit that is stationary and disposed in a tactical formation on a FEBA. Ranger forces may conduct a passage of lines to get behind the enemy. Movement in forward unit areas must be controlled, coordinated, and kept to a minimum. This avoids conflict with friendly troops or the activation of RSTA devices. Ranger units treat the positions of forward units as danger areas. They must be assumed to be under enemy surveillance in all weather or visibility. The ranger force is open to attack during the passage. Personnel and units may be more concentrated than is desired. Fires of the stationary unit may be masked for a limited time. The passing ranger unit may not be in a good position to react to enemy contact. Detailed reconnaissance and coordination are crucial to ensure that the passage is conducted quickly and smoothly.

a. **Principles.**

(1) **Coordination.** The ranger force commander, or his representative, must coordinate departure and reentry of friendly front lines (FFL). This is done either directly with the units through which the rangers pass or through designated staff agencies. Items to be coordinated include fire support, passage lanes, control measures, guides, communications, intelligence, CSS, and actions on enemy contact.

(2) **Reconnaissance.** The ranger force commander should make a ground reconnaissance of the area through which the unit will pass. He should pick the initial rally point (IRP) at this time. The reconnaissance should include observing the area just forward of the friendly unit for routes, danger areas, and obstacles. During reconnaissance or coordination, rangers should wear the same uniform as the troops in the forward unit. This will avoid attracting the enemy's attention.

(3) **Security.** Maintain security to avoid contact with the enemy while departing or reentering. It is important that the ranger force make no enemy contact at this time. Control is hard if a fire fight starts just forward of an FFL.

b. **Departure of friendly front lines.**

(1) Designate an IRP.

(2) Maintain security. Use an appropriate movement formation when departing the FFL.

(3) Do not move within the friendly unit's area without a guide.

(4) Make final coordination with the friendly unit commander.

(5) Count the ranger force members as they depart.

(6) Make a security and listening halt.

c. **Departure of friendly units.**

(1) The ranger force arrives at the forward unit and is met by a guide. The guide leads the rangers to an assembly area chosen by the ranger commander during initial coordination with the forward unit commander.

(2) No one should move, either singly or as a unit, anywhere in the FFL without a guide.

(3) The ranger commander makes final coordination with the forward unit commander. He determines if any changes have taken place since initial coordination and learns of any recent enemy activity.

(4) Before leaving to coordinate, the ranger commander gives instructions (called a contingency plan) for what should be done while he is gone. These instructions state where he is going, who is going with him, how long he will be gone, what to do if he does not return, and actions to be taken if there is enemy contact.

(5) On the commander's return from final coordination, he may issue a fragmentary order (FRAGO) to cover any changes.

(6) Three common threats and techniques for countering them during departure from the FFL are:

(a) **Ambush and chance contact.**

- The ranger force commander must avoid enemy contact while departing.
- The ranger force goes to the friendly side of the departure point where a point team is dispatched to see that the area forward of the barriers is clear.
- The point element should check the area out to the first covered position or an area large enough to allow the unit room to maneuver if it is hit while departing.
- The distance varies with the size of the ranger force and the terrain.
- The point team notifies the ranger commander when the area is clear. The rest of the unit is brought forward.

(b) **Indirect fire.** If enemy indirect fire is falling, the ranger commander should not halt his unit after final coordination at the forward command post. Instead, he should move it quickly through the FFL. The point element should be far enough ahead of the main body to provide security from ambush or chance contact.

(c) **RSTA.** If the enemy has radar, sensors, or night vision devices, they may be offset by the following countermeasures:

- Use a well-hidden departure point, such as a reverse slope or a dense woods.

- Infiltrate the unit through the departure area and have the rangers rendezvous in a defilade.
- Pass through during rain, fog, snow, or darkness to help conceal the passage.
- Employ electronic countermeasures.

(7) Have a security and listening halt after the ranger force has moved out of sight and sound of the FFL. This is a short halt to see if enemy is in the area, and to accustom the unit to the sights and sounds of the battlefield. Halt in a position that covers the unit from chance friendly small-arms fire. If the security and listening halt must be in an exposed position, have each man lie flat on the ground.

(8) During movement, have frequent halts to see that the unit is not being followed and that no enemy is in the area.

d. **Reentry of friendly front lines.**

(1) Set up and occupy a reentry rally point (RRP).

(2) Maintain security at the RRP and at the reentry point.

(3) Use as few personnel as possible to locate the reentry point while the bulk of the unit remains in the RRP.

(4) Meet a guide at the reentry point. Normally, a forward-of FFL password is used since it may be overheard by enemy.

(5) Count the ranger force members to ensure no infiltrators follow the unit, especially during reduced visibility.

(6) Give the friendly unit commander a spot report, providing combat information of immediate tactical value.

e. **Reentry of friendly units.**

(1) The unit is moved into a rally point near the reentry point. This rally point should be on a prominent terrain feature where the ranger commander can pinpoint his location, with respect to the reentry point.

(2) The forward unit is alerted by radio that the ranger unit is ready to reenter. A code word is used for security and brevity. The code word must be acknowledged by the forward unit before the ranger force reconnoiters the reentry point. This shows that a guide has been sent to the reentry point and is waiting for the unit.

(3) If the ranger commander is certain of the reentry point location, he moves the entire unit directly to the reentry point. He has the point element coordinate the reentry.

(4) An effective method of locating the reentry point is through the use of a thermal imagery device. It detects the body beat of the reentry guide, even though he may be well camouflaged or his location masked by smoke. If ground surveillance radar is available, it can be used to vector the ranger unit to the reentry point.

(5) The reconnaissance party should never move parallel to friendly barriers or move around the wire. If the reentry point cannot be found by the initial reconnaissance or if the reconnaissance party finds only the barrier wire, the unit should notify higher headquarters and move to another rally point to wait until daylight. The unit should not stay in a rally point from which a radio transmission has been made.

(6) When the reentry point is found, the ranger commander can bring the unit forward. If the reentry point is easy to find, he can call by radio for the unit to come forward.

(7) The guide leads the unit through the barriers to the security position coordinated with the forward unit commander. The unit halts in the security position, remaining in movement formation. The ranger commander then gives a spot report to the forward unit commander. The ranger commander tells him anything of intelligence value or of immediate tactical use to the forward unit.

6-9. Reconnaissance platoon operations.

a. The reconnaissance platoon organic to the ranger regiment consists of a platoon leader, a senior reconnaissance sergeant, and three five-man reconnaissance teams that include a communications specialist in each team (see Appendix A). This platoon is responsible for short-duration reconnaissance of objectives for the ranger force commander. This reconnaissance is routinely in support of future strike or special light infantry operations. The teams can also be used by the ranger force commander to conduct deep target acquisition and designation, or to confirm or disprove information reported by other sources. Rarely will teams be used to attack enemy targets.

b. The reconnaissance teams of the ranger regiment do not replace (or perform the same reconnaissance tasks associated with) long-range reconnaissance patrol (LRRP) units or long-range surveillance units (LRSUs) that may be organic to the corps or division. These units are organized, trained, and equipped for long-term passive surveillance and reconnaissance in enemy territory. The ranger reconnaissance teams are trained and organized to support the immediate intelligence-gathering needs of the ranger force commander. They are not to act as a human intelligence (HUMINT) collection asset for other operational headquarters.

c. The reconnaissance platoon headquarters trains, equips, and controls the employment of the three reconnaissance teams. It operates the communications stations needed to receive reports from deployed teams. In the absence of the S2, it debriefs team members upon extraction or exfiltration. It can provide one or more teams to support each ranger battalion on independent operations. The reconnaissance teams can--

(1) Infiltrate the objective area by parachute (HALO, HAHO, or static lines), helicopter, fixed-wing aircraft, SCUBA, small boat, foot, or other means.

(2) Remain undetected in the vicinity of the objective area up to five days.

(3) Perform reconnaissance operations employing a full range of night observation devices, infrared detection devices, unattended sensors, and photographic equipment.

(4) Perform demolition target analysis.

(5) Operate small watercraft and inflatable boats.

(6) Emplace unattended ground sensors, omni-directional navigational beacons, hand-emplaced expendable jammers, and electronic target designation devices.

(7) Collect combat information to satisfy priority information requirements and mission-essential elements of information. Teams report that information by use of long-range, secure, burst-transmission communications equipment.

(8) Perform DZ selection, marking and reception duties.

(9) Report objective area weather conditions.

(10) Perform highly selective, limited attacks or ambushes when so tasked.

(11) Link up with the main body of the ranger force in the objective area, or escape and evade the enemy in order to return to friendly lines.

(12) Act as part of the AAT during airborne operations.

d. There are many limits of the reconnaissance teams.

(1) Mobility is normally restricted to foot movement in the area of operations.

(2) Use of radio and other active electronic or optical devices makes the teams open to enemy detection.

(3) All supplies and equipment needed by the team must be carried on the initial insertion, since resupply increases the probability of detection by the enemy.

(4) Conduct of insertion and extraction operations deep in enemy-held territory is difficult due to the need for secrecy, timeliness, security, and accuracy of location.

(5) Medical treatment of team casualties is limited to individual first aid. Casualty evacuation is difficult unless linkup with the main ranger force has been achieved.

e. The training of reconnaissance team members parallels that given to all members of ranger units with emphasis on infiltration, communications, and reconnaissance skills. Extensive training is also given in survival, evasion, resistance, and escape (SERE) and advanced first aid. Training emphasis is placed on actions at the reconnaissance site. The reconnaissance teams must conduct their mission and remain unseen. Every team member is trained to consider the need for information balanced by the likelihood of detection. Detection by the enemy compromises future operations.

6-10. Reconnaissance planning.

a. Detailed planning ensures that the reconnaissance mission is successful. It is also vital to the survival of the reconnaissance team. This planning is conducted by the ranger force planning group assisted by elements of the JTF staff. The supporting air or naval commander, navigator, and electronic warfare officer may aid the ranger force staff in the planning.

b. Upon receipt of the mission, the team selected begins an intense preparatory phase at the REMAB or the ISB. There the team receives its initial mission briefing. The items covered allow the team to complete its detailed planning. The reconnaissance platoon leader reviews the team plan. He ensures compatibility and integration into the ranger force commander's plan for employment of the main force. Staff specialists at ranger regiment and battalion--such as those concerned with intelligence, weather, or communications--provide help.

c. Planning for the use of reconnaissance teams is conducted within the framework of the ranger planning sequence for the entire mission. There are five planning considerations for the employment of reconnaissance teams:

(1) Operations security must be considered in all staff efforts to provide maximum protection for the reconnaissance teams. This includes intelligence, CE, logistics, administration, and maintenance. Operations security must be integrated throughout every ranger force operation. It consists of four main categories of security measures: signal security, physical security, information security, and deception. All are related and must be considered at the same time for each operation.

(2) Some missions may require rapid deployment to the operational area, thereby dictating the quickest method of insertion. However, others may depend on maintaining total secrecy, regardless of time.

(3) The transportation means selected for the delivery and recovery of the teams depends on the mission. This selection is also based on the abilities, limits, and availability of mission support platforms.

(4) The distance to and from the objective area must be considered, as well as the distance from the insertion point to the actual target. Reconnaissance teams can move great distances over rugged terrain. However, this movement does require time and limits the team's ability to carry bulky or sensitive equipment.

(5) Consideration must also be given to intelligence information available concerning the objective area which includes, but is not limited to--

(a) **Enemy situation.** The enemy threat--his abilities, disposition, security measures, detection, and air defense systems - affects the selection of the means for delivery or recovery.

(b) **Terrain.** Land formations must be considered in selecting the method of infiltration and exfiltration. Terrain affects the selection of altitudes, approach and exit routes, landing areas, drop zones, and beach landing sites. Air infiltration routes that provide terrain masking are desirable in static line parachute operations.

(c) **Weather.** Seasonal weather conditions affect team delivery and recovery. Factors to be considered are temperature, precipitation, visibility, clouds, and wind. If para-SCUBA techniques are used, high surface winds and their effect upon surf conditions or period of reduced visibility may preclude the use of parachutes, inflatable boats, or swimming as entry or recovery techniques. These same conditions favor land infiltration and exfiltration, The AWADS reduces the impact of visibility as a limiting factor for air infiltrations.

(d) **Astronomical conditions.** Periods of sunrise and sunset, moon phase, moonrise and moonset, and twilight must be considered.

d. A reconnaissance team mission must be specific and support the mission. Information concerning the target may not be gathered because the risk of detection by the enemy is so great. Detection may preclude the conduct of the planned operation. If more than one mission is assigned, priorities are set, All team missions must be coordinated to avoid other friendly forces operating near the target. Ranger forces conduct operations at such depth behind enemy lines that no other US or allied forces are normally present. There is a chance that a friendly indigenous guerrilla force may be operating near the objective, but ranger reconnaissance teams do not normally make contact with them.

e. The reconnaissance platoon leader joins with the intelligence and operations sections of the ranger force headquarters in the initial planning for reconnaissance team operations. The method of operation while moving, communication procedures, reporting, surveillance procedures, and actions on enemy contact should be included in the platoon's SOPS.

f. The ranger force intelligence officer normally prepares the reconnaissance plan IAW the guidance provided by the controlling headquarters. He is helped by the operations officer and the reconnaissance platoon leader. Selected team members and a member of the unit providing the insertion and extraction means are briefed on the mission early in the planning phase. They should participate in the detailed planning to follow. The essential details of a reconnaissance plan normally include--

(1) Areas to be observed and recommended areas or positions from which this can be done. These positions are determined in advance of the insertions of teams. They are based on a study of terrain, road and rail nets, enemy order of battle, delivery means, operations plans of the controlling headquarters, and the desires of the ranger force commander.

(2) Load plans and procedures. This includes both insertion and extraction. Alternate sites are chosen for possible use. Primary and alternate DZs are selected, if applicable.

(3) The flight plan. This includes insertion and extraction flight routes and alternate routes.

(4) The movement plan to and from a patrol base. This includes the times, routes,. and destinations when individuals or small groups are required to operate away from the patrol base.

(5) The fire support plan. This includes the location and times of planned air strikes in support of the main body, and all planned fires by the main assault force. The reconnaissance teams should know the frequencies and call signs for tactical air support.

(6) The timing and ground tactical plan for execution of the main force operation. This includes disposition of the main force, special mission teams from the ranger force, or any other special operations force elements.

(7) The communications plan. This includes frequencies, reporting schedule, emergency reporting procedures, and alternate communications plans. It also includes the ground-to-air

communications plan and the friendly recognition signal to be used during linkup with the main assault force or other friendly forces.

(8) Plans for use of technical specialists or special equipment.

(9) The plan for the treatment and evacuation of sick or wounded team members from the objective area.

(10) The E&E plan. This includes routes out of the objective area, passwords, and alternate extraction zones or methods far from the objective area.

(11) Mission-specific elements of information desired by the ranger force commander, and any special or on-order missions the team must conduct.

g. The team leader uses specific steps in planning, preparing, and executing missions. These procedures are comprehensive, yet adaptable to any situation. Three of the most important steps are:

(1) **Issuance of warning order.** This action gives the team maximum time to prepare for the mission. It should be issued as soon as the team leader has been notified of a mission and updated after he has made his tentative plan.

(2) **Issuance of OPORD.** The team leader conveys the information and instructions needed to accomplish the mission. He requires a briefback from each team member.

(3) **Rehearsals.**

h. The specific steps used when planning missions are as follows:

Step 1. Study the mission.

Step 2. Plan use of time.

Step 3. Issue initial warning order.

Step 4. Study terrain and situation.

Step 5. Make tentative plan.

Step 6. Organize.

Step 7. Update warning order.

Step 8. Coordinate (continuous throughout).

Step 9. Consult aerial photographs and map updates.

Step 10. Complete detailed plan. (To be combined with paragraphs 3, 4, and 5 of the operation order.)

Step 11. Issue operation order.

Step 12. Supervise (throughout), inspect, rehearse.

Step 13. Execute mission.

i. A workable E&E plan must be developed. Such a plan enhances the survival of the reconnaissance team members. It ensures that they are psychologically prepared to conduct the mission, by knowing there is a system and plan for returning them to friendly lines. The E&E plan must include a section describing the actions of the reconnaissance team members should their insertion craft be disabled en route to the objective area. It should also include what to do if the team is unable to link up with the main ranger force, or if the main ranger force has not been inserted into the objective area.

j. Rehearsals and briefbacks are best for finding flaws in procedures or planning. Thorough coordination of all procedures is essential. The exact type of aircraft or other means that will be used for insertion should be used for training. Rehearsals should be conducted under terrain, astronomical, hydrographical, and meteorological conditions close to those to be met on the operation. The team leader conducts a briefback with the platoon leader. This lets the platoon leader decide if the team is ready for the mission. It also allows him to make changes.

6-11. Execution of insertion.

a. Reconnaissance teams normally conduct blind insertions. Blind insertions are amphibious, heliborne, or parachute drops onto unmarked DZs without help from a DZ-marking party or a CCT. Blind insertions often depend on favorable light and weather conditions in the objective area. Favorable conditions for the reconnaissance team often means low-lying cloud cover, fog, and darkness. The teams are able to operate during reduced visibility by the extensive use of night observation devices. The team's extensive training and high level of land navigation skills allow it to quickly traverse rugged terrain and not be detected.

b. Team insertion into the objective area often needs a cover and deception plan. Team members can conduct HALO and HAHO operations. Teams can be inserted into areas not normally acceptable as a DZ for conventionally trained parachutists. By the use of HAHO techniques, reconnaissance teams can assemble in the air and glide to a landing at a location far from the drop point.

c. If reconnaissance teams need special or bulky equipment that cannot be carried on the individual team member, it can be airdropped, hidden near the DZ, and retrieved later. Such container drops are often used during low-level static line insertions. The high-altitude airdrop resupply system (HAARS) can be used to deliver such items at the same time as reconnaissance teams are inserted by HALO.

CHAPTER 7
Combat Support

Combat support is used to enhance the combat power of the maneuver companies. Combat support elements may be organic or nonorganic. They can conduct a variety of missions in providing indirect and direct fire support, security, intelligence, and maneuver and fire control.

7-1. Intelligence support.

a. The success or failure of a ranger mission often depends on the accuracy, detail, and timeliness of intelligence information. Both deliberate and quick-response missions need specific information that can be obtained only by a multidiscipline collection and analysis agency. An active interface with all levels of the intelligence system helps get the needed information quickly.

b. The theater or JTF commander is responsible for providing intelligence support. The 1st SOCOM also provides information and analyzed intelligence to the ranger force commander. The main source of processed intelligence is the Echelons Above Corps Intelligence Center (EACIC), similar to the division level military intelligence (MI) all-source production center (ASPC). The EACIC provides intelligence, security, and electronic warfare (EW) interfaces (see Figure 7-1). It is through the EACIC that ranger commanders can request support from national, joint, combined, or CONUS intelligence analysis systems.

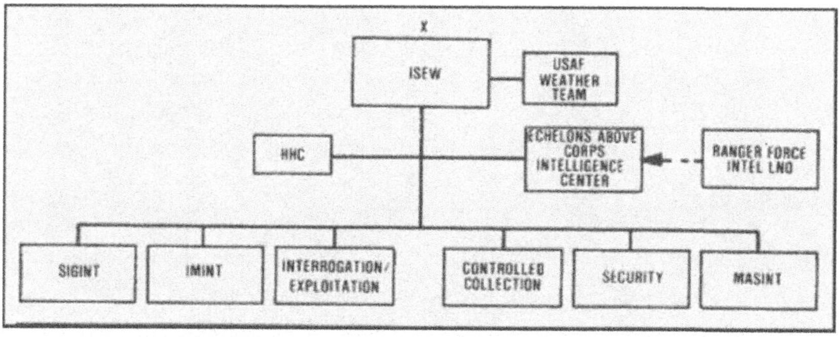

Figure 7-1. Intelligence at echelons above corps.

c. Intelligence requirements for ranger operations are unique, highly sensitive, and compartmented. The intelligence officer of the 1st SOCOM and the appropriate 1st SOCOM regional planning cell coordinate with theater intelligence agencies to provide needed information.

d. The ranger regimental staff and the ranger battalion staff have personnel in their intelligence sections to correlate and analyze information from many sources.

e. The reconnaissance platoon, organic to the ranger regiment, maintains surveillance of the objective area before the insertion of the main ranger force. It reports any last-minute developments that would affect the ground tactical plan. The platoon can also be tasked to determine any mission-specific items of essential information that the ranger force commander needs to complete his plan. This platoon is a highly specialized unit that is reserved for use by the ranger force commander. It is not a HUMINT collection agency for use by the theater or JTF commander. Reconnaissance teams normally link up with the ranger force in the objective area and are extracted with the main body. If this linkup cannot be made, they will exfiltrate enemy territory. They may use escape and evasion nets set up by other special operations forces or Department of Defense (DOD).

f. The ranger regiment normally places a liaison team at the headquarters of the command having OPCOM of a ranger force. This team provides command and control communications and liaison. It puts an intelligence liaison officer (LNO) at the EACIC or the appropriate ASPC (see Figure 7-2). This provides coordination to ensure that the intelligence needs of the ranger force are relayed to the intelligence processing center, and that the analysis is on the specific needs of the ranger force commander. This LNO coordinates with another located at the corps or EAC targeting center. Close coordination between the regiment, the intelligence analysis center, and the targeting center is essential to effective employment.

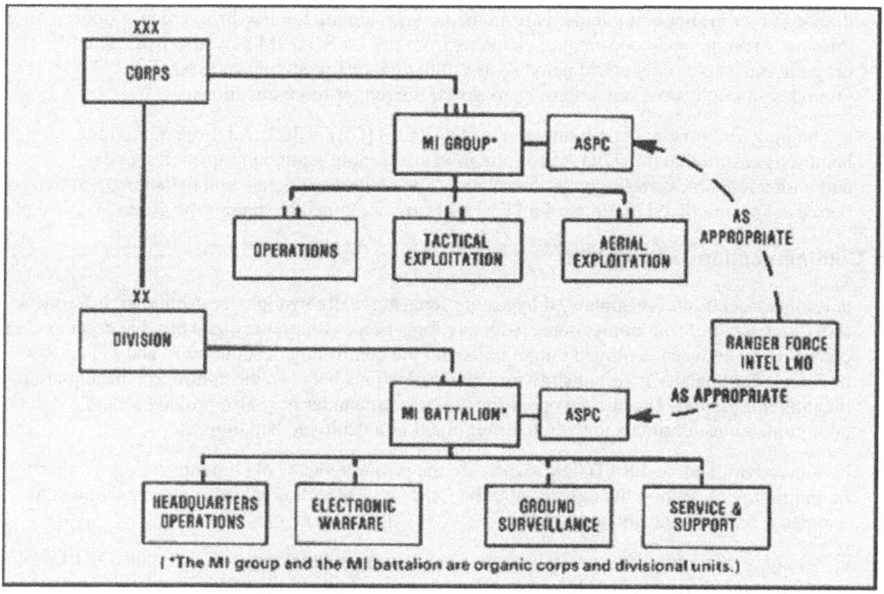

Figure 7-2. Intelligence source at corps and division.

7-2. Electronic warfare support.

a. Depending on the nature of the target and enemy abilities, a ranger mission may need support from EW units. This is mainly true during insertion. Electronic warfare operations disrupt or destroy the enemy's command and control of his forces and weapons systems, and retain friendly use of the electromagnetic spectrum. Electronic warfare is also used to support deception operations. They mislead the enemy by manipulation, distortion, or falsification of indicators, and persuade him to react in a manner prejudicial to his own interests.

b. Active jamming and chaff dispersal can prevent enemy early-warning radar from detecting the airborne force or determining its route. False transmissions can aid deception plans or feints used to help the ranger force's entry behind enemy lines. Selective jamming and imitative deception can disrupt the enemy command and control nets. This prevents the enemy from reacting to the presence of the ranger force in time to prevent mission accomplishment, or slows and interrupts his deployment of reaction forces.

c. Ranger elements, or other military forces, may conduct closely coordinated direct attacks against enemy EW sites during a ranger force's insertion or extraction. Not only does this reduce the enemy's ability to direct EW against the ranger force command, control, and communication system, but it misleads and confuses the enemy as to the true ranger objective.

d. The JTF or theater commander is responsible for planning the EW program to support a ranger mission. Planning and coordinating elements from the 1st SOCOM may also help. The EW program must be closely coordinated so as not to disrupt friendly air-to-ground or ground-to-ground communications, or to reveal the ranger force intentions.

e. The ranger regimental communications-electronics (CE) officer and those of the ranger battalions ensure that the required coordination concerning jammed frequencies, codes, authentication tables, and so on, is completed. The ranger regimental and battalion operations personnel ensure that the efforts of all EW units are combined to support the ground tactical plan.

7-3. Communications support.

a. Ranger operations are supported by secure long-range, lightweight, real-time, high-frequency (HF), and satellite communications. Effective long-range communications provide command and control links between deployed ranger units and the controlling headquarters. The JTF, ARFOR, or theater commander is responsible for communications between the controlling headquarters and the ranger regiment. The special operations force commander may also provide secure communications terminals to the ranger regiment or a deployed battalion.

b. Secure AM, FM, and SATCOM radios are the primary means of communication within the ranger regiment. Within the ranger battalion, AM and FM radios provide communications to company, platoon, and squad level.

c. During certain missions, specially trained and equipped quickreaction elements (QRE) deploy with the ranger force to provide secure communication links to the special operations force commander. These JCSE elements operate on either SATCOM or tactical satellite terminal (TACSAT) channels or through an airborne communications relay platform (see Figure 7-3). Depending on the mission, an airborne battlefield command and control center (ABCCC) aircraft or a joint airborne communications center/command post (JACC/CP) may be used. Their communications systems can operate at all levels of the national chain of command to permit a quick response to the tasking authority.

Figure 7-3. Airborne communications relay platform.

d. When only one ranger battalion is employed and the regimental headquarters is not the controlling headquarters, the ranger regiment provides a liaison cell to the controlling higher headquarters. This liaison cell includes a communications element from the regimental communications platoon. It can provide secure SATCOM, teletype, and facsimile transmission support to the ranger force in the objective area.

e. If two or more ranger battalions are employed, the ranger regimental headquarters deploys and acts as the command and control headquarters. The regimental communications platoon would then provide another communications link to the special operations force commander.

f. The ranger regimental CE officer ensures that the necessary communications links are set up and coordinated. The many communication means and channels available provide for effective control of a deployed ranger force. However, communication means must be closely coordinated at all levels of command to control the complex operations of a ranger mission. Planning and coordination with supporting aviation, transport, fire support, medical, and logistical elements before an operation are vital to efficient communications. The regimental CE officer must consider the communication systems linking the ranger force and other services. Air-to-ground and ship-to-shore communications are vital and must be set up early in ranger operations.

7-4. Fire support.

a. **Planning.**

(1) The ranger regiment has limited organic fire support assets. The six 60-mm mortars in each ranger battalion (two in each rifle company) provide the only organic indirect fire

available to the battalion commander. Each rifle company also has three 90-mm recoilless rifles and three Dragon missile systems that provide a small measure of organic direct fire support.

(2) The missions assigned to the ranger regiment and its subelements normally require fire support from organizations outside of the regiment. To plan, coordinate, and control these fires, the ranger regiment is organized with personnel dedicated to these tasks.

(3) The ranger regimental headquarters has a fire support element consisting of one fire support officer (FSO), two noncommissioned officers, and two fire support specialists. They prepare the fire support annex to the regiment's operation order. They ensure that all available fire support is planned to best support the regimental commander's ground tactical plan. They act as the primary coordinator for all external fire support, such as artillery, armed helicopter, close air support, naval gunfire, and short-range air defense. They also coordinate with each ranger battalion FSO to ensure unity of effort. The fire support element coordinates the efforts of air force, naval, and marine liaison officers when they are attached.

(4) The regimental commander may assign part of the fire support element to a committed ranger battalion as an attachment to help in the integration and planning of fires. The ranger battalions request through the fire support element that more or different fire support means be integrated into the ground tactical plan.

(5) The ranger battalion headquarters has a fire support team headquarters consisting of the battalion's FSO, two noncommissioned officers, and two fire support specialists. The FSO is the prime fire support coordinator for the ranger battalion commander. He maintains coordination with the regimental FSO and any attached fire support coordination elements, such as the USAF tactical air control party or naval shore fire control party. The FSO performs the following:

(a) Advises the ranger battalion commander on all fire support matters.

(b) Recommends allocation of fire support.

(c) Prepares fire support plans.

(d) Assigns target numbers.

(e) Processes target lists from the ranger company fire support teams to eliminate duplication.

(f) Monitors and functions as net control on the ranger battalion fire support coordination (FSCOORD) net.

(g) Reports changes in the status of fire support units to the ranger battalion commander, staff, and FISTS.

(6) When the ranger battalion is operating a single TOC, the FSO and his element are normally located there. When the ranger battalion is operating two TOCS, one TOC is normally manned by the FSO, the fire support sergeant, and a fire support specialist. The

senior fire support sergeant and the other fire support specialist would then monitor the FSCOORD net at the other TOC. The FSO ensures that any attached fire support elements also divide when needed to operate in a two-TOC configuration.

(7) The ranger battalion fire support team headquarters has three fire support teams assigned to it. These are normally allocated one to each of the ranger rifle companies. The company FSO and his team--

(a) Locate targets and request and adjust surface-to-surface fire support (mortar, field artillery, and naval gunfire).

(b) Plan fires to support the company ground tactical plan and prepare target lists.

(c) Coordinate fire support requests through the FSO for surface-to-surface and air-to-surface fires.

(d) Report battlefield information.

(e) Control air-to-surface fires of helicopters and prepare to assume control of fires from fixed-wing aircraft.

(f) Coordinate airspace use within the ranger rifle company operational area.

(g) Advise the ranger company commander on all fire support capabilities, limitations, and methods of employment.

(h) Inform all fire support units of target priority changes.

(i) Designate targets for laser-guided munitions.

(j) Assume operational control of, and administrative responsibility for, all augmentation fire support personnel (TACP, shore fire control party, and attack helicopter liaison teams).

(8) The company FSO is normally with the ranger rifle company commander. The FOs and their RATELOs are normally with the ranger rifle platoon leaders.

b. **Coordination.**

(1) Coordination of fires is a continuous process. It involves implementing the fire support plans of the ranger force commander and reacting to changing tactical situations. The success of a mission often hinges on how well fires are coordinated during the operation.

(2) The fire support planners within the many echelons of the ranger force become fire support coordinators during the actual execution of the operation. They use standard fire control coordination measures.

c. **Organic fire support.**

(1) **Indirect fires.**

(a) The 60-mortars constitute the only organic indirect fire support available within the ranger regiment. The mortar squads in each ranger rifle company normally operate as a section under the control of the mortar section sergeant. However, they can operate alone for short periods.

(b) The 60-mm mortar is used to destroy enemy equipment and light vehicles and to attack enemy personnel. Its main advantage is the ease with which it can be brought into operation against the enemy. This ease of employment allows the ranger force commander to concentrate combat power at the critical time and place. This helps to surprise and destroy the enemy.

(c) Mortar fires are employed by the ranger rifle company commander and coordinated by the company FSO. They are normally requested and adjusted by the FIST observers. The mortars are normally emplaced in a central position, and their fires are controlled by the fire direction center (FDC) using the indirect-fire method. The mortars may also use the direct-alignment or the direct-lay method.

(2) **Direct fires.**

(a) Each ranger battalion is authorized 90-mm recoilless rifles and Dragon missile launchers. These weapons constitute the heaviest organic direct-fire ability in the regiment. The antitank section in each ranger rifle company may operate as a complete section, but normally individual weapons are assigned throughout the company sector wherever their fires can best support the mission. These weapons can destroy most enemy armored vehicles and can penetrate the flank and rear armor of all enemy main battle tanks. The 90-mm recoilless rifle has a multiple projectile round that is effective against exposed enemy personnel. The HE round is highly effective against buildings and bunkers.

(b) The main disadvantage of direct-fire antiarmor weapons is their weight. These weapons, and their ammunition, are man-portable but place a strain on the unit when moving across rugged terrain.

(c) The ranger battalion is trained in the use of light antiarmor weapons, such as the US M72A2 LAW and the Warsaw Pact rocket-propelled grenade (RPG). The following provide protection for the ranger force against attack by enemy armored forces:

- The employment of light antiarmor weapons at all organizational levels.
- The placement of the 90-mm recoilless rifles or guided-missile systems on the most likely avenues of enemy-armor approach.
- An aggressive, offensively oriented employment.
- The ability to move into and across terrain impassable to armored vehicles.

d. **Field artillery fire support.**

(1) Many ranger unit operations are out of the range of supporting field artillery fires. Whenever such fires are available, they are planned for and integrated into the ranger force ground tactical plan.

(2) Coordination and exchange of call signs, frequencies, and target lists occur before insertion of the ranger force. Unless required by the tactical situation, ranger units do not normally receive augmentation forward observers from supporting field artillery units. The degree of mental and physical training necessary to execute certain ranger techniques requires more preparation than field artillery units are able to provide. The FIST from the ranger battalion requests and adjusts field artillery fires for the ranger force.

(3) Field artillery fires can be used to support the ranger force even if the objective area is out of range. Field artillery cannon fire and multiple rocket launch system (MRLS) fire can be used to suppress enemy air defenses. This helps the ranger force as it crosses the FEBA during airborne or air assault operations.

(4) Field artillery fires can be used to support the exflltration of ranger elements as they approach friendly lines. These fires can contribute to the deception plan and add combat power to feints used to support ranger operations.

(5) When in range of the objective area, field artillery units can be used to emplace field artillery delivered FASCAM to enhance the security of the ranger force.

(6) Copperhead rounds fired by 155-mm field artillery units can be terminally guided by the ranger FIST forward observers (FOs). They can attack hardened point targets or enemy armored vehicles by using a man-portable laser target designator (see Figure 7-4).

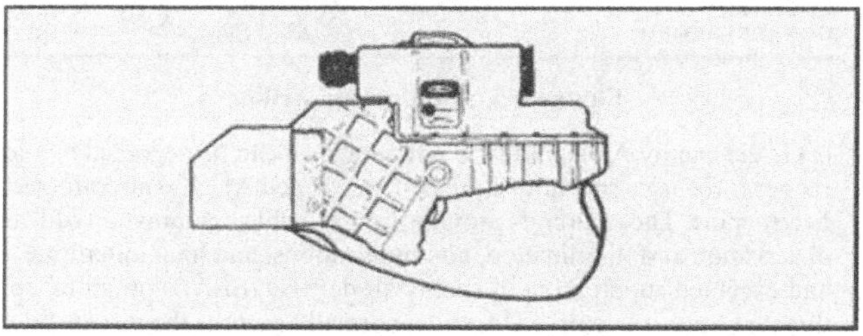

Figure 7-4. AN/PAQ-1 laser target designator.

e. **Aerial fires.**

(1) Aerial fire support is usually the prime means supporting the ranger force due to the distance behind enemy lines at which most ranger operations take place. Aerial fire support can be provided by either fixed-wing aircraft or helicopters.

(a) Fixed-wing aerial fire support may come from USAF, USN, or USMC units. The type of unit providing support, the aircraft, and the mix of ordnance carried all affect the fire support planning and coordination process. Some aircraft have a night and all-weather strike ability enabling them to support the ranger force during any level of visibility. Operations during poor weather that limits visibility to less than 3 nautical miles are still somewhat restricted. The ranger fire support coordinators must ensure that the correct aircraft are requested and employed effectively on the enemy. The

TACP directs and adjusts aerial fires in the objective area.

(b) The ranger force can use ground laser target designators to pinpoint targets for air strikes, as well as electronic navigation aids to permit nonvisual air strikes (beacon bombing). The ranger rifle company FIST or the TACP can control a laser-designated standoff air strike (see Figure 7-5).

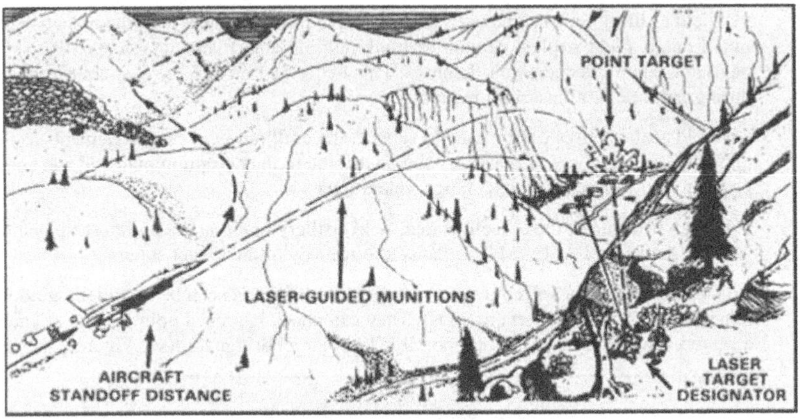

Figure 7-5. Standoff air strike.

(c) If the enemy ADA ability is not great or it can be degraded to a low level, the ranger force uses specially equipped and armed AC-130 aircraft (see Figure 7-6) for fire support. These aircraft provide an invaluable combination of firepower, night observation and illumination, communications, and long loiter time. A well-planned and executed suppression of enemy air defense (SEAD) program, coupled with ECM directed against enemy ADA units, normally permits the use of AC-130 aircraft.

Figure 7-6. AC-130 aircraft.

(2) The attack helicopter, armed with a mix of antitank guided missiles (ATGM), 2.75-inch rockets, 20-mm cannon, and 40mm grenade launchers, is accurate and responsive aerial fire support. The ranger force commander plans to use all sources of aerial fire to help accomplish his mission. However, the nature of ranger operations may preclude the extensive use of armed helicopters due to their limitations. Some attack helicopters are limited in range and lack an all-weather ability. They are restricted during operations at night by a lack of sophisticated night vision devices. They may lack the speed to go with the ranger force on some heliborne insertions. Attack helicopters may be used to escort and assist the ranger force as it crosses the FEBA. They may also be used to conduct feints and demonstrations to cover the insertion of the ranger force. As the MV-22 Osprey and the AH-64 Apache enter service, more use will be made of these aircraft to support ranger operations beyond the FERA, (See Figures 7-7 and 7-8).

Figure 7-7. MV-22 Osprey.

Figure 7-8. AH-64 Apache attack helicopter armed with Hellfire missiles.

(a) If attack helicopters are used to support a ranger operation, planned indirect fires are normally delivered along entry and exit corridors. Attack helicopters approach and depart the objective area using nap-of-the-earth flight profiles.

(b) Fires from armed helicopters are normally requested and controlled by the company FSO or one of his FOs, operating on a special ground-to-air net. The laser target designator may be used to precisely identify targets for the AH-64 Apache. Friendly unit locations may be marked by smoke, panels, lights, mirrors, or infrared sources.

f. **Naval gunfire.**

(1) During amphibious assaults or operations near a coastline, the ranger force may receive indirect fire support from naval gunfire. Normally, a ranger battalion would be supported by either a destroyer or a cruiser in a direct support role. However, the type and importance of the mission, the type of targets, the ships available, the hydrographic conditions, and the enemy capability determine how many and which type of ships are provided to support the ranger force.

(2) Naval gunfire is characterized by large volumes of highly destructive, flat-trajectory fire. Planned strikes in support of the ranger force may also include surface-to-surface missile fire. Some naval guns can fire a laser-guided projectile much like the Copperhead. This projectile, called a Deadeye round, is terminally guided the same as the Copperhead. Deadeyes can be guided either by an attached spotting party from the ANGLICO or by ranger LTD teams.

(3) When ranger units are being supported by naval gunfire, a shore fire control party is attached to the ranger force, This party normally consists of a liaison team and several spotter teams (forward observers). The liaison team is integrated into the operations of the fire support element (FSE) at the ranger battalion TOC. The spotter teams are attached to the ranger rifle companies. The shore fire control party LNO is the ship's representative to the ranger force commander, through the FSO. The spotter teams request and adjust fires from surface vessels. They can also request and control air strikes by carrier-based aircraft.

(4) The LNO and the spotter teams operate in the ground spot net, communicating with the ship by high frequency (HF) radio to request and adjust naval gunfire. The spotter teams communicate with the LNO using very high frequency (VHF) radios. The LNO also can communicate with aircraft using ultrahigh frequency (UHF) radios.

(5) Coordination and control measures that apply to naval gunfire are the same as for field artillery, with the addition of two terms, which are:

(a) Fire support area (FSA)--a sea area within which a ship may position or cruise while firing in support. It is labeled as FSA and numbered by a Roman numeral - for example, FSA VII.

(b) Fire support station (FSS)--a specified position at sea from which a ship must fire. This is very restrictive positioning guidance. It is labeled as FSS with a Roman numeral--for example, FSS VII.

g. **Air defense artillery.**

(1) Army air defense artillery units are rarely used to directly support the ranger force since

ranger operations normally take place deep behind enemy lines. However, it is possible that during the insertion or extraction phase of an operation, ADA units may support the ranger force as it crosses the FEBA.

(2) The ranger regiment's fire coordination element and the FSO in the ranger battalion are mainly concerned with adequate coordination of ADA fires. This prevents friendly units from being engaged. It allows the integration of ADA fires into the scheme of maneuver. The fire support planners ensure that enough coordination has been made so that ADA units are aware of the ranger force and know enough about its flight paths, routes, and altitudes to prevent engagement by friendly units. When conducting this coordination, care must be taken to pass only needed information to the ADA unit. To aid coordination, an ADA officer may be attached to the ranger force.

(3) The plan for a heliborne or airborne insertion and extraction of the ranger force must include the correct identification-friend-or-foe (IFF) codes, the flight profiles, the times of FEBA crossings, and the estimated number of aircraft making the crossing. There must also be a plan for crossing by straggler aircraft and battle-damaged aircraft that cannot reply to IFF interrogation. If possible, the routes selected should avoid concentrations of friendly ADA units.

(4) During an aerial crossing of the FEBA by a ranger force, the regimental headquarters may assign a liaison officer to the headquarters controlling the ADA fires in the crossing area. The LNO is then responsible for close and continual coordination. This ensures that the crossing is made without friendly ADA units engaging the ranger force while it is exiting or entering friendly lines.

(5) During ranger operations, the primary source of air defense fires is shoulder-fired antiaircraft missiles carried by specially trained individual rangers. These weapons provide the ranger force commander a limited defense against attack by enemy aircraft. The decision to take these weapons into the objective area depends on the factors of METT-T. Ranger operations should be completed before the enemy can react with ground forces. However, enemy aircraft may be able to react in time to threaten the ranger force. These lightweight missiles offer a highly effective, short-range counter to enemy air attack.

(6) The ranger force commander must define the weapons control status and rules of engagement for all short-range air defense (SHORAD) teams under his control. Priorities of target engagement must also be determined to prevent multiple engagements of a single enemy aircraft. The ranger battalion FSO and the attached ADA officer are responsible for the coordination of ADA fires within the objective area. Standard ADA weapons-control status and rules of engagement may be modified by the ranger force command.

7-5. Aviation support.

a. **Planning.**

(1) The ranger regiment may need extensive aviation support on some missions. The ranger battalion is organized and trained to move quickly to the objective area by air and to be resupplied by airdrop for short periods.

(2) Often the decision to execute a mission depends on the amount and type of available insertion and extraction aviation support. Normally, the theater commander or JTF commander allocates the type and amount of aviation assets to support a ranger operation. The theater or JTF commander's staff is responsible for the planning and coordination concerning aviation support. The ranger force commander must play an active part to ensure mission success.

(3) Due to the extensive training requirements and specialized operational techniques needed to support special operations, special air operations units in the Army and the USAF have been equipped and trained for the mission. These forces train with the ranger regiment and jointly develop tactics, techniques, and procedures of employment. These special units operate diverse aircraft that have unique capabilities. These help the ranger force commander to effectively employ his units. Command of these special units is maintained at a very high level in the chain of command. They are assigned to support a specific operation. Then they return to their parent unit's control.

b. **Coordination.** In addition to the staff of the theater commander or JTF commander, the coordination of aviation support occurs at all levels in the ranger force. These operations often are conducted in phases. A separate air mission commander is then designated. The AMC is responsible for the completion of each phase of the aviation mission, as well as the close and continuous coordination with the ranger force, commander. The ranger regimental air operations officer and the movements control officer are responsible for the planning, integrating, and controlling of aviation support.

c. **US Air Force.**

(1) The USAF provides special aviation support to the ranger regiment and other elements of the special operations force. The USAF has elements that are equipped with MC-130 Combat Talon aircraft (see Figure 7-9) and AC-130 Spectre gunships, as well as HH-53 Pave Low, UH-60, and UH-1 helicopters. These aircraft are specially equipped to conduct long-range infiltration, exfiltration, attack, and search-and rescue operations. The USAF also provides C-141B and C-130 aircraft for strategic and tactical airlift of the ranger force. There are specially trained USAF crews that can operate using AWADS and SKE. This gives them an excellent means of inserting the ranger force during reduced visibility.

Figure 7-9. MC-130 Combat Talon aircraft.

(2) The AFFOR commander provides the airlift control element (ALCE) support at the departure airfield. The ranger air operations officer ensures that coordination of aircraft load limits, air movement times, specific load plans, and manifests is conducted in an effective and timely manner. The air operations officer and the movements control officer ensure that a departure airfield control officer (DACO) from the deploying unit is designated to provide face-to-face coordination at the departure airfield.

(3) The USAF normally provides an airborne battle command and control craft. This is a specially equipped C-130-type aircraft that provides communications and data system down-links to ground stations.

(4) The terminal guidance needed for assault landings and air traffic control in the objective area is provided by a specially trained and equipped CCT. This element has habitual training and operations relationships with the ranger units. The USAF is also responsible for the specially trained and equipped weather teams that can be sent into the objective area with the CCT.

(5) The USAF Tactical Air Command (TAC) includes fighter, interceptor, ground attack, electronic countermeasure, early warning, reconnaissance, and airborne command and control squadrons, any of which may be used to support the ranger force. The Tactical Air Command provides the results of aerial reconnaissance to the AFFOR, who then processes it for use by the theater or JTF commander's staff. This reconnaissance may be by low-level photo, imaging infrared, side-looking airborne radar (SLAR), low-light-level television, or overhead photography.

(6) The Military Airlift Command (MAC) equips and trains special long-range SAR aircraft and crews. These aircraft, mainly helicopters, are equipped with sophisticated navigation, observation, and communications equipment, as well as extended-range fuel tanks and in-flight refueling capabilities (see Figure 7-10). The ranger regiment uses these aircraft for certain missions.

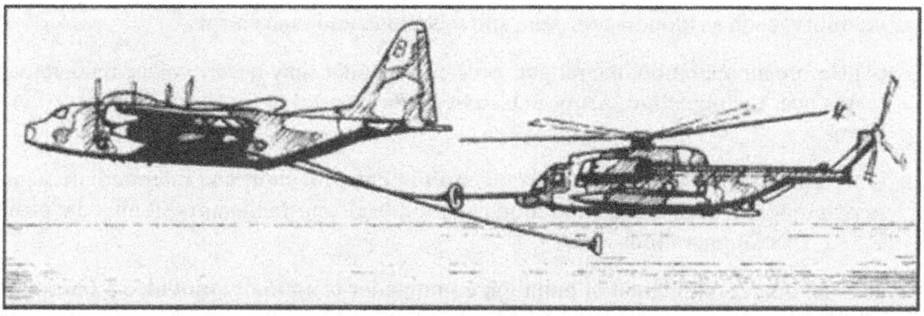

Figure 7-10. In-flight refueling.

d. **US Navy and US Marine Corps.** Both the USN and the USMC have units equipped and trained to support ground forces. The type of aviation support normally provided to the ranger regiment is close air support against targets in the objective area, or as suppression missions against enemy air defense installations. Long-range helicopters from the USMC may be used to support an amphibious operation or to insert special teams during ranger operations. The USN also has a search-and-rescue ability that may be used to support the ranger regiment.

e. **Special operations aviation units.**

(1) The US Army has specially equipped and trained SOA units tasked to support the ranger regiment, as well as other elements of the special operations force. These SOA units have a habitual training association with the ranger regiment and use dedicated aviation assets to -

(a) Insert, extract, and resupply the special operations force.

(b) Conduct armed escort, reconnaissance, surveillance, and electronic warfare in support of the special operations force missions.

(c) Provide airborne command, control, and communications for special operations force elements.

(d) Provide general support aviation during peacetime and contingency operations.

(2) Penetration of enemy airspace is normally done during limited visibility at very low altitudes. Penetrations are planned to maximize the use of existing terrain and exploit blind spots in enemy air defenses. On-board navigational aids are crucial to mission success. There normally is not any terminal guidance to the selected LZ or DZ. Deception operations of a tactical or operational nature help SOA units penetrate to the objective and reenter friendly lines. Advanced helicopters and short-takeoff-and-landing aircraft provide the help needed to do these missions.

(3) To survive enemy air defense systems and prevent mission compromise, SOA units avoid detection. Detected aircraft may reveal the location of supported special operations force ground forces or may be destroyed by enemy ADA fires. The SOA normally avoids detection by operating at night without illumination. This condition requires flight using sophisticated night vision devices. The SOA also operates in other conditions of reduced

visibility, such as clouds, fog, rain, snow, smoke, and dust storms.

(4) Before an operation, the ranger force commander may need to place reconnaissance teams near the objective. Army SOA assets can be used to transport ranger reconnaissance teams.

(5) Elements of SOA provide the ranger regiment with short and intermediate range penetration, resupply, and extraction capabilities. Long-range aircraft must be provided from USAF special operations assets.

(6) The ranger regimental or battalion commander is normally provided a liaison aircraft while the ranger force is in the REMAB. This ensures that he can travel quickly to the dispersed headquarters of the agencies providing support for the mission. This aircraft and crew are provided by the theater or JTF commander from EAC general support aviation assets. It may be a helicopter or fixed-wing aircraft, depending on the location of the REMAB to a suitable landing field.

(7) Army helicopters and fixed-wing aircraft may be used to provide SEMA aerial reconnaissance; radar, infrared and photographic imagery; and electronic intelligence support to the ranger force commander. This support is normally provided by the aerial exploitation battalion of the theater Army military intelligence group.

7-6. Engineer support.

a. The ranger regimental and battalion operations officers plan and request engineer support for ranger operations.

b. The most common engineer-type tasks needed during a ranger operation are the destruction of enemy facilities, equipment, and material by use of conventional explosives; and the interdiction of major lines of communication. To reduce the need for external support, the ranger regiment trains selected individuals in advanced demolition techniques. All members of the ranger regiment are well trained in the basic employment of conventional demolitions. Therefore, there is no need to attach combat engineers to conduct normal demolition missions. If specialized skill or knowledge is needed for a specific mission, qualified personnel can be attached.

c. During the occupation of the REMAB and the rehearsal for an operation, the ranger regiment needs construction engineer support to build simulated target areas. The simulated target areas should be camouflaged to prevent the enemy from recognizing the target. The construction engineers augment the ranger unit's self-help construction efforts. Limited-facilities construction and improvement may also be needed at the REMAB to provide security and isolation. The theater commander or JTF commander provides the construction engineer support.

d. Certain missions, such as those of interdiction, may need engineer special weapons teams. The teams are provided by the theater commander or JTF commander, or they may come from other special operations force assets. They are attached to the ranger force for a specific mission and released at the completion of the extraction and debriefing phase.

e. Some operations may need the ranger force to use captured enemy heavy construction equipment. Selected members of the ranger regiment are trained to operate such equipment.

However, construction engineer personnel may be attached to the ranger force for this mission.

CHAPTER 8

Combat Service Support

Combat service support consists of the logistical and administrative effort to maintain the battalion's ability to fight. The commander forecasts his needs for CSS. He employs CSS units to ensure accomplishment of the mission.

8-1. Ranger support element.

a. The ranger support element is a non-TOE organization formed from elements of CSS units collocated with CONUS-based ranger units augmented by selected personnel from the host installation. The mission of the RSE is to provide all the CSS needed to outload and deploy a ranger unit from CONUS. The personnel designated as part of the RSE have the same alert notification, reaction, and security requirements as members of the ranger unit they support. The RSE is normally built around a command and control group provided by a major CSS unit on the ranger unit's host installation. The remaining CSS elements and teams are then designated. An RSE consists of--

(1) An emergency operations center liaison group that provides timely coordination between the ranger unit and the functional elements of the RSE.

(2) A command and control group that coordinates and directs the actions of the functional elements.

(3) A supply section responsible for the timely issue of all Class II, III, IV, and VII items. This section maintains a stockage of contingency items marked for use by the resident ranger unit. It also issues items out of common stocks.

(4) A transportation section moves equipment and personnel from the unit area to the departure airfield and picks up equipment from local storage sites. It may also operate material-handling equipment and trailers to support loading of aircraft.

(5) An ammunition section responsible for the requisition, storage, issue, loading, and security of the ranger unit's basic load(s) of conventional munitions.

(6) A mess section provides all Class I support to the ranger unit. This includes operating a dining facility, providing food to remote sites, and drawing and issuing operational rations.

(7) A load team loads and unloads all equipment and supplies for the ranger unit. This team prepares pallets for loading, documents hazardous or dangerous cargo, and loads equipment pallets on USAF aircraft. It can also provide limited security for loads.

(8) A communications and electronics section provides secure communications among the RSE subelements, ranger unit, EOC, and DACG.

(9) An engineer platoon that is responsible for the construction of targets and rehearsal areas. This element of the RSE may or may not be activated, depending on the mission given the ranger unit. If the ranger unit deploys from CONUS to the objective area, this platoon has a key role to play in the premission preparation.

(10) A maintenance team whose primary mission is to ensure that the vehicles used by the RSE remain operational. It also provides predeployment maintenance and Class IX support for ranger unit vehicles and equipment.

(11) A medical team that provides unit-level medical care while the organic ranger medical section completes its outloading preparations. This team also provides any Class VIII supplies that the ranger unit may need.

(12) A security team responsible for security at the EOC, ranger unit location, and departure airfield. This team enforces stringent physical security precautions and maintains a constant counterintelligence (CI) surveillance of the outloading procedure. The security team provides armed guards for the protection of ammunition convoys, equipment movement, and vehicles.

(13) A rigger detachment that maintains, packs, rotates, and issues the personnel and equipment parachutes required by the ranger unit. It maintains enough stock of personnel parachutes to conduct a battalion-level parachute assault. The riggers lend technical skill to the rigging and aerial delivery of bulky, unusual, or delicate equipment.

b. The primary mission of the RSE is to provide the support necessary to quickly load and deploy the ranger units. The theater commander provides CSS to the ranger force once it has arrived at the OCONUS REMAB or ISB. The transition of support from the RSE to the CSS assets of the theater commander may be delayed for certain subelements in order to ensure a smooth deployment and enhance mission accomplishment. These subelements are released when the in-theater CSS assets can provide the support. Retained subelements of the RSE may be released from support requirements all at once, released in increments, directed to continue to provide support until mission completion, or augmented by in-theater CSS assets. The maintenance, medical, security, and rigger elements are those most likely to continue providing support OCONUS. This is likely if deployment is to an austere ISB pending immediate employment.

c. Once a ranger unit has been deployed OCONUS and has closed to a REMAB or ISB, the theater commander's staff coordinates CSS. If 1st SOCOM elements have deployed to the theater or operational area, the special operations support battalion (SOSB) of the 1st SOCOM coordinates CSS. Ranger logisticians at regiment and battalion level are mainly planners with limited logistics operations ability. Ranger logistical and personnel officers coordinate with CSS elements who provide such services.

d. The Theater Army Materiel Management Center (TAMMC) is the point of contact for information concerning in-theater logistical assets. In a developed theater of operations, a series of theater Army support groups will have been set up. One of these groups is assigned the mission of supporting the ranger force. The 1st SOCOM may also deploy certain assets available to provide CSS to the ranger force. The many CSS sources, the rapidly changing nature of SOF operations, and the unique CSS requirements of ranger units combine to make providing logistical and personnel support a challenge. The ranger unit's logistics officer must be included in the early phases of planning and deployment.

8-2. Logistics and personnel.

a. The echelons-above-corps CSS structure provides support to the ranger regiment as directed by the theater Army (TA) commander. In a low-intensity contingency involving a corps or less, there may be no echelons-above-corps logistics and personnel support elements. The corps support command (COSCOM) involved may be able to meet all needs from its own sources. However, in a combined operation in a high-intensity environment, the theater commander assembles a large CSS organization to provide support to the TA. This CSS base normally provides for all the ranger unit's requirements. During contingency operations in an area where there is no TA, the JTF commander provides the CSS elements. Until the buildup of these elements is complete, the ranger regiment must be prepared to operate from austere facilities with only the supplies it brings from CONUS.

b. The ranger regiment is normally assigned to the theater command for CSS. The TA commander, as directed by the theater commander, provides CSS to the ranger regiment and sets up support priorities. The EAC CSS organizations are under the command of the TA commander and consist of subordinate area and functional commands. The TA plans, coordinates, and executes its CSS functions through area-oriented support commands and specialized subordinate organizations. During the early stages of conflict, the TA may be organized as shown in Figure 8-1.

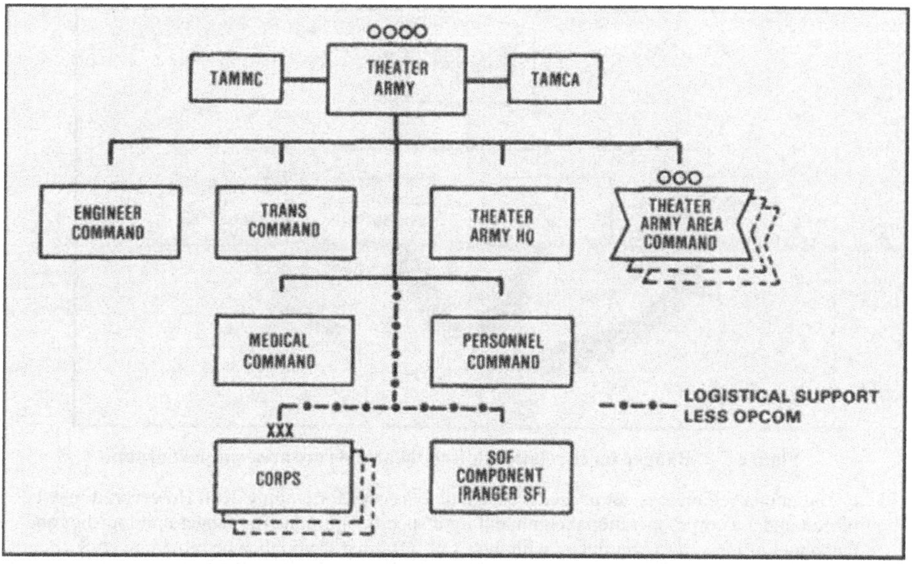

Figure 8-1. Ranger force relationship to theater Army.

c. The Theater Army Area Command (TAACOM) is the key CSS operator in the communications zone. It functions as a major subordinate command under the theater Army. The primary mission of the TAACOM is to provide direct support CSS to units such as the ranger regiment located in or passing through its assigned area. While located in an OCONUS REMAB, the ranger regiment receives most of its logistical support from the TAACOM.

d. The TAACOM area support group (ASG), which is subordinate to the TAACOM, is structured with operating units to provide direct support CSS to units located in or passing through its assigned geographical area of responsibility (see Figure 8-2). Area support groups are usually located along the lines of communication. The ASG is the prime source of CSS for the ranger regiment. It contains the units that actually provide services requested by the ranger unit logistics officer.

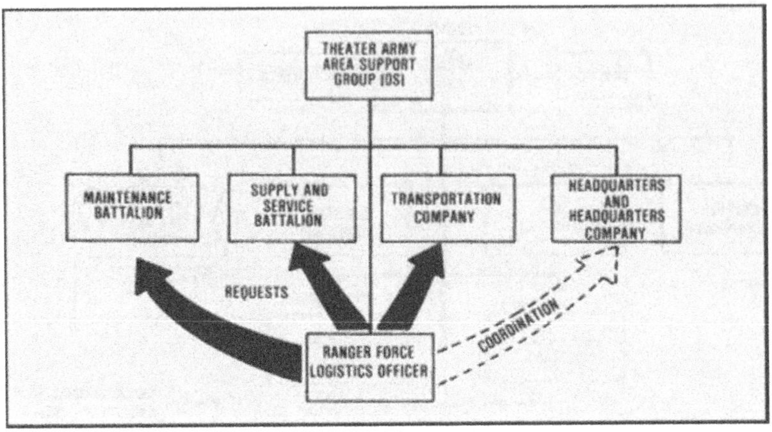

Figure 8-2. Ranger force relationship to theater Army area support group.

e. The ranger regiment is not normally assigned to a corps for employment. However, it may be placed under a corps' operational command for a specific mission. The ranger unit logistics and personnel officers must coordinate with the COSCOM that would then be providing CSS. (See Figure 8-3.) Corps support groups are major subordinate elements of the COSCOM. They provide command and control of COSCOM companies and battalions, which provide direct support (DS) and general support (GS) supply and field services to nondivisional units, such as the ranger battalion. (See Figure 8-4.)

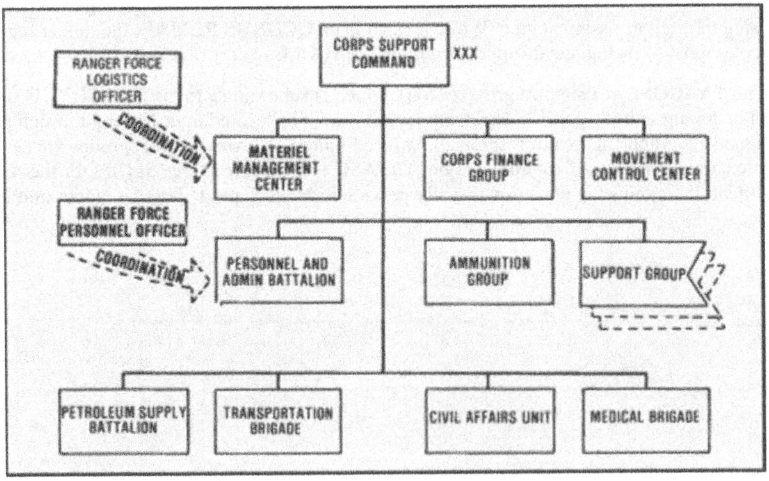

Figure 8-3. Ranger force relationship to corps support command.

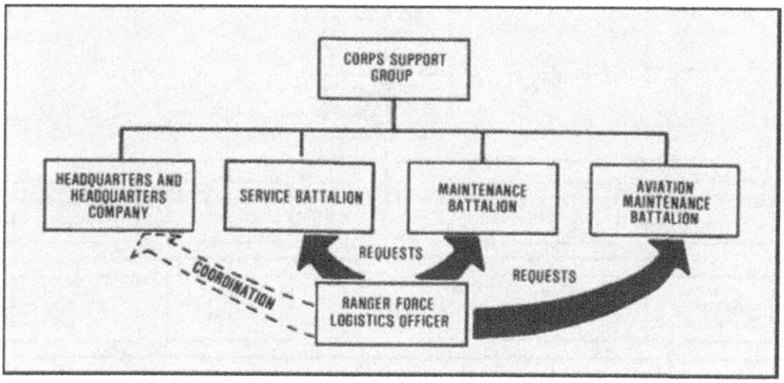

Figure 8-4. Ranger force relationship to corps support group.

f. Normally, the ranger regiment receives its logistical support from the corps support group responsible for the geographical area in which the REMAB or ISB is located. The corps support group provides all supplies except Classes V and VIII. The medical brigade provides Class VIII supply and medical maintenance support. The ammunition group provides Class V support. Financial and personnel services are provided by the corps finance group and the personnel and administration battalion, respectively.

g. Transportation needs are met by the transportation brigade. The ranger unit places ground transportation requests with either the Movement Control Center (MCC) or a designated movement control team. All airlift supply requirements are passed IAW FM 100-27. Standard and nonstandard supplies are normally requested from the supporting direct support unit (DSU). If authorized by the corps commander, the ranger regiment can place requirements for nonstandard items on the appropriate division within the corps MMC.

h. The COSCOM MMC (see Figure 8-5) performs supply and maintenance management for all classes of supply. The ranger unit logistics officer may coordinate with the MMC division chiefs or operations section.

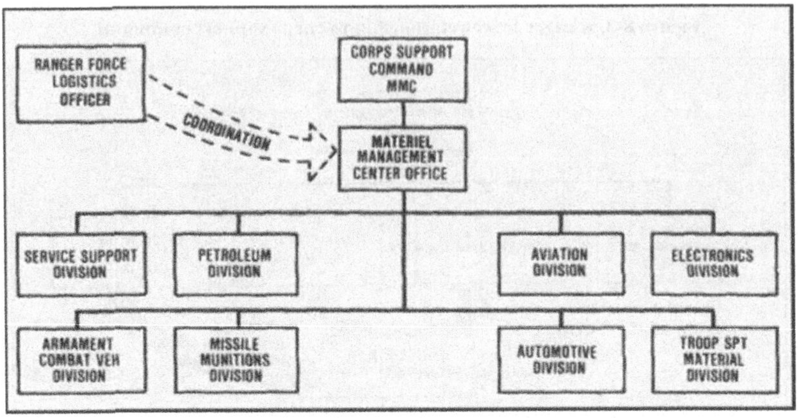

Figure 8-5. Ranger force relationship to COSCOM MMC.

i. Critical logistics operations, such as those supporting ranger and other SOF elements, need intensive management. They may be assigned to special staff sections that act as special operations force logistics managers. If designated, a logistics manager for special operations force elements would be a major point of contact for logistical support for the ranger regiment.

j. The ranger regiment may also receive logistical support from the SOSB. Normally, the SOSB provides administrative and logistical support to the Headquarters, Army Special Operations Forces (ARSOF). It complements the support provided by the Army component in the theater of operations by coordinating, augmenting, or assisting CSS forces. The SOSB provides unique quick-reaction support to the ARSOF or deploying SOF elements.

k. The ranger regiment does not make inordinate demands on the logistical system. However, a ranger unit may need certain nonstandard or low-density equipment. The concurrent need for nonstandard Class IX items requires management by logisticians to ensure support.

(1) Ranger battalion mess sections are responsible for Class I operations at the REMAB. The ranger regimental headquarters, which has no mess section, either collocates with one of the battalions or is supported by a mess facility operated by a nearby unit. It may share this facility on an area basis. The ranger unit personnel officer coordinates with the supporting DSU, which initiates the issue of Class I supplies.

(2) Class II, IV, and IX support to the ranger force is normally not of a great volume. It may consist of some nonstandard items. Depending on the need, the item, and the security, the SOSB may get some nonstandard items directly.

(3) Class III support needed by the ranger regiment is small and limited to standard items. The normal source of supply for Class III is the supply and service company providing area support.

(4) Class V support needed by the ranger regiment is small. It may include requests for nonstandard items of ammunition and explosives. The ranger logistics officer passes requests for standard items to the ordnance company (conventional ammunition) of the ammunition battalion assigned to the TAACOM ammunition group. Requests for nonstocked items are passed through the TAACOM Materiel Management Center or 1st SOCOM channels. The ranger unit must contact the ammunition company early to ensure a continuous supply of ammunition.

(5) Class VII supplies needed for the ranger regiment are often nonstandard or in excess of TOE. (See Appendix C for an equipment density listing.) All nonstandard or excess requests must be coordinated through the TAMMC or 1st SOCOM channels.

(6) Class VIII supplies for the ranger regiment are normally standard items supplied by the area support Class VIII point. The ranger regiment personnel officer sets up the Class VIII supply points of contact through the ranger battalion's medical officer.

(7) Aerial delivery equipment is provided by the airdrop equipment repair and supply company of the theater Army support group (TASG) (GS). There is normally one such company for each TAACOM. This unit supplies and repairs aerial delivery equipment. It does not pack and rig supplies for aerial delivery. The ranger regiment's requirements for aerial delivery supplies normally consist of standard items. Certain low-density items, such as HALO or HAHO parachute equipment, may also be provided through 1st SOCOM channels.

(8) The supporting motor transportation group of the transportation command is responsible for unit movement of the ranger regiment. The ranger regiment does not have any organic transportation assets. It depends on outside support whenever motor movement is needed.

8-3. Resupply.

a. Resupply operations for ranger units are normally covered during the planning and coordination phase of the mission. Ranger units try to complete their mission quickly and then be extracted or exfiltrate from the objective area. Ranger units carry into the objective area the equipment and munitions they need. They do not normally remain long enough to need extensive resupply. Certain operations may require the use of bulky or heavy items of equipment that cannot be carried by personnel. If the ranger force is airlanded on or near the objective, one or more of the aircraft may contain palletized supplies. These can be quickly off-loaded for future use.

b. If the ranger force is inserted into the objective area by parachute assault, aircraft can also drop resupply loads either just before or just after the personnel drop. These loads might include extra water, ammunition, demolitions, or barrier material. During peacetime, preloaded airdrop pallets are stored at CONUS depots for such contingency use. These mixed loads can be quickly moved to a departure airfield and inserted along with, or following, the ranger force. By dropping a resupply of water, food, equipment, and munitions with the ranger force insertion, the heavy-drop aircraft do not have to conduct a separate penetration of the enemy air defense zone. The ECM and SEAD programs used in the insertion of the personnel aircraft mask the penetration of the heavy-drop aircraft.

c. The aircraft dropping ranger personnel can also carry door bundles to deliver bulky or heavy items to the objective area. These door bundles, normally made up of A-7 and A-21 containers, may be color-coded to identify different loads or units. The most common door bundle loads are such items as 90-mm recoilless rifles and ammunition, 60-mm mortar ammunition, Dragon missiles, SHORAD missiles, or more demolition material. These bundles are pushed from the personnel aircraft just before the lead jumper's exit. They may be marked with lights or infrared sources for locating them at night.

d. If the tactical situation requires that a ranger force receive a resupply during the course of an operation, the most common method would be by airdrop. Airdrop could be needed if the operation takes longer than planned, initial supplies are lost, or changes occur in the operational plan. The ranger unit logistics officer and operations officers normally plan for resupply needs and schedule follow up loads of Classes I, II, IV, V, and VIII. There are two different airdrop requests channels for the ranger force to use:

(1) Nonemergency airdrop requests are sent through logistics channels. The ranger LOC sends requests for the airdrop of supplies and equipment to its supporting supply unit, and the supporting unit forwards the requests to the COSCOM or TAACOM MMC. The MMC coordinates with the movement control center (MCC). Jointly they direct the actions of the supply and transportation units. The MCC sends the request to the joint transportation board (JTB). The board validates it and assigns it a priority. The JTB then sends the request to the airlift control center for action. If the items requested are not on hand in the airdrop company, the MMC directs the proper supply activity to prepare the items for shipment to the airdrop unit. The MMC coordinates with the MCC, and the MCC ships the items. The airdrop unit prepares the supplies for the mode of delivery to be used. It also makes sure the supplies are delivered to the air terminal at the right time. (See Figure 8-6.)

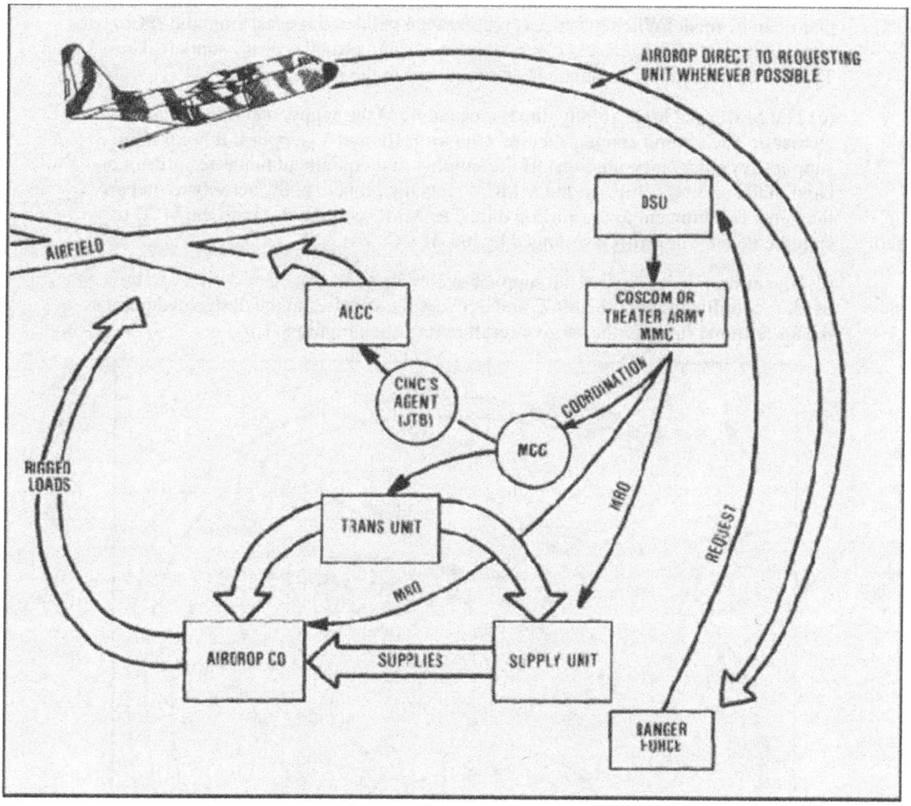

Figure 8-6. Ranger requests for nonemergency airdrop.

(2) Emergency requests for airdrop are passed through command channels in the quickest, most secure way.

(a) The request is usually sent to the command level that has authority to approve it, usually the tactical operations center at corps or above. The Air Force airlift officer at the direct air support center (DASC) coordinates with the Assistant Chief of Staff, Operations (G3) at the tactical operations center to determine whether airlift resources can be used for the mission.

(b) The TOC sends a request for an immediate airlift to the joint force commander's (JFC) designated agent for validation. At the same time, the airlift liaison officer advises the airlift control center (ALCC) that a request is coming through channels so plans can be made. When the ALCC receives the validated request from the JFC agent, it diverts or cancels less essential missions and provides airlift support. The TOC sends the request for supplies or equipment to the MMC.

(c) The MMC and MCC jointly direct the actions of the supply and transportation elements. The airdrop company stocks Classes I, III, and V supplies. It holds these supplies for emergency missions. If the supplies and equipment requested are not on hand in the airdrop company, the MMC directs the proper supply activity to prepare the items for shipment to the airdrop unit. The MMC coordinates with the MCC to ship the items. The airlift is arranged by the ALCC.

(d) The airdrop unit prepares the supplies according to the mode of delivery to be used. It coordinates with the MCC and delivers the supplies to the designated point at the air terminal (usually the cargo aircraft ramp). (See Figure 8-7.)

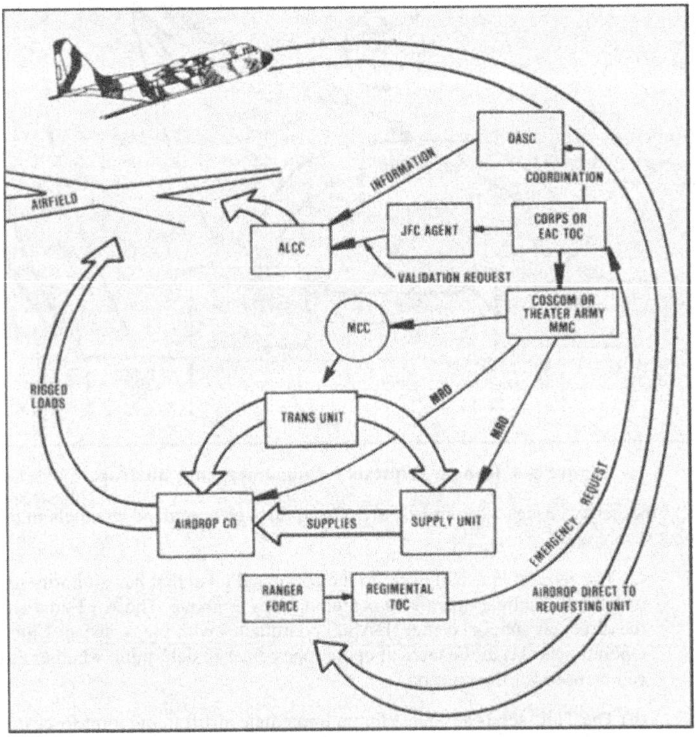

Figure 8-7. Request channels for emergency airdrop.

e. Aircraft delivering airdrop resupply deep behind enemy lines need to take precautions to avoid enemy detection and antiaircraft fire. The USAF is responsible for most of these precautions, such as ECM, suppression of enemy air defense (SEAD), and WILD WEASAL support. These do not concern the ranger commander. The choice of the type of parachute delivery system to use, however, may affect the ground commander's plan. The ranger force commander must specify any tactical restrictions that would require a specific type of delivery system. The ranger LOC and regimental logistics officer must advise the ranger force commander on the type of parachute system to be used. Conflicts must be quickly resolved so as not to interfere with the tactical operations.

(1) The safest way for the airdrop aircraft to penetrate enemy air defense and remain undetected is often by use of very low-level flight. There are four low-level parachute delivery systems that can be used without the aircraft being required to climb to a higher drop altitude and risk detection by enemy ADA.

(a) The low-altitude parachute extraction system (see Figure 8-8) needs a length of level ground with enough clearance for the delivery aircraft to make a very low pass over the extraction zone. The load is extracted by parachute from the aircraft and slides to a stop on a rugged, skid-type pallet. Heavy loads can be delivered using this system.

Figure 8-8. Low-altitude parachute extraction system.

(a) The containerized delivery system (CDS) (see Figure 8-9) consists of multiple, individually rigged A-22 containers, each with its own parachute, weighing up to 2,000 pounds each. The system provides single-pass delivery of up to 16 containers by C-130 aircraft and up to 28 containers by C-141 aircraft. The CDS loads may be delivered into drop zones using multiple points of impact to allow for tactical separation. The ranger force commander can direct the use of multiple points of impact, noting the advantages to be gained against the requirement for multiple passes.

Figure 8-9. Containerized delivery system.

(c) The high-speed, low-level, airdrop system (HSLLADS) (see Figure 8-10) is a single A-21 container specially rigged to withstand the parachute opening shock when airdropped from C-130 aircraft at high speed. This system can be used to deliver up to 600 pounds for each container, with a maximum of four containers for each pass, at speeds up to 250 knots.

Figure 8-10. High-speed, low-level, airdrop system.

(d) The CTU-2/A is a high-speed, aerial-delivery container (see Figure 8-11) that can be used to deliver supplies from high-performance aircraft flying at a minimum altitude of 300 feet above ground level and a maximum airspeed of 425 knots. The CTU-2/A is carried on the bomb racks of fighter or bomber-type aircraft. Upon release, a pilot parachute deploys the main chute and the container descends slowly.

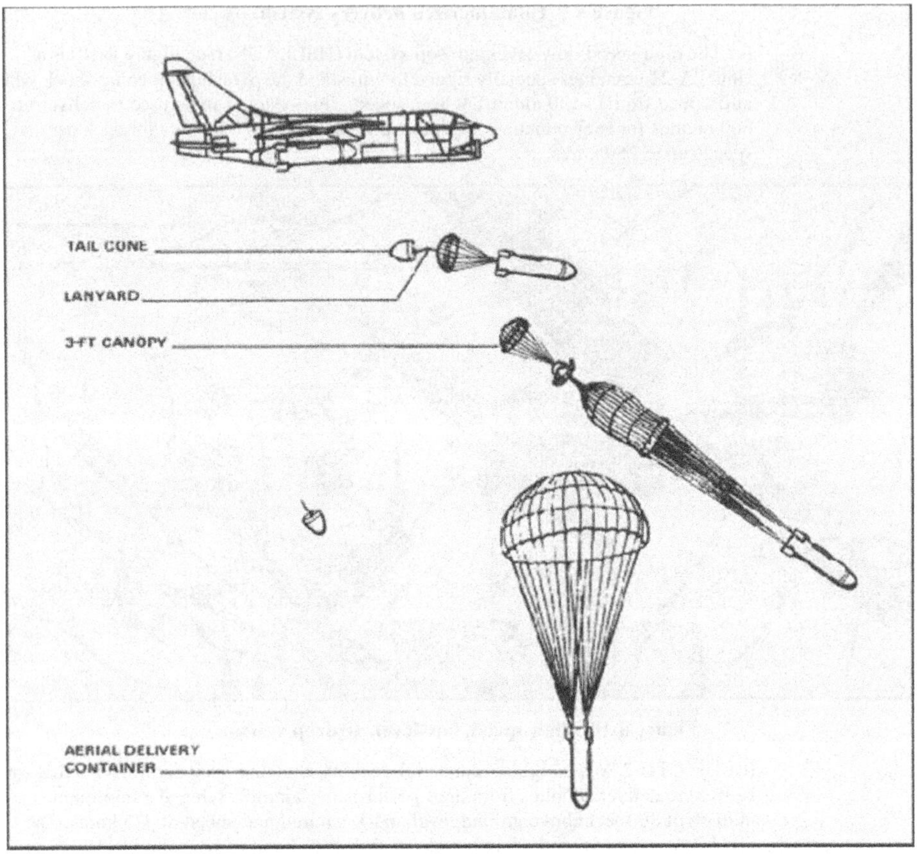

Figure 8-11. CTU-2/A high-speed, aerial-delivery container.

The CTU-2/A can be used to deliver up to 500 pounds of supplies such as weapons, water, food, or munitions. The container can be destroyed by burning. The main advantage of this system is that it can be delivered by high-performance aircraft deep behind enemy lines and in a dense air defense environment. The accuracy of this system is equal to that of a conventional bomb strike.

(2) Sometimes it may be better to drop resupply loads to the ranger force from a high altitude. The HAARS (see Figure 8-12) permits containerized unit loads weighing from 200 to 2,000 pounds to be delivered from aircraft at speeds up to 150 knots from up to 25,000 feet AGL. The HAARS consists of a cargo parachute, an airdrop container, an altitude

sensor, and a pilot chute. The pilot chute gives the descending bundle a speed slightly greater than an accompanying HALO parachutist. The system provides for the steady free-fall descent of loads from altitudes between 2,000 and 25,000 feet to an altitude at which a barometric sensor actuates deployment of the main parachute. This allows safe and accurate delivery of loads onto unprepared drop zones. The HAARS can deliver a payload to within 260 meters from a target impact point from a 10,000-foot altitude, with a proportional degree of accuracy from 25,000 feet. Ranger units can use the system for resupply of battalions and smaller units with rations, ammunitions, and medical supplies, breaking down the containerized material into man-packed loads.

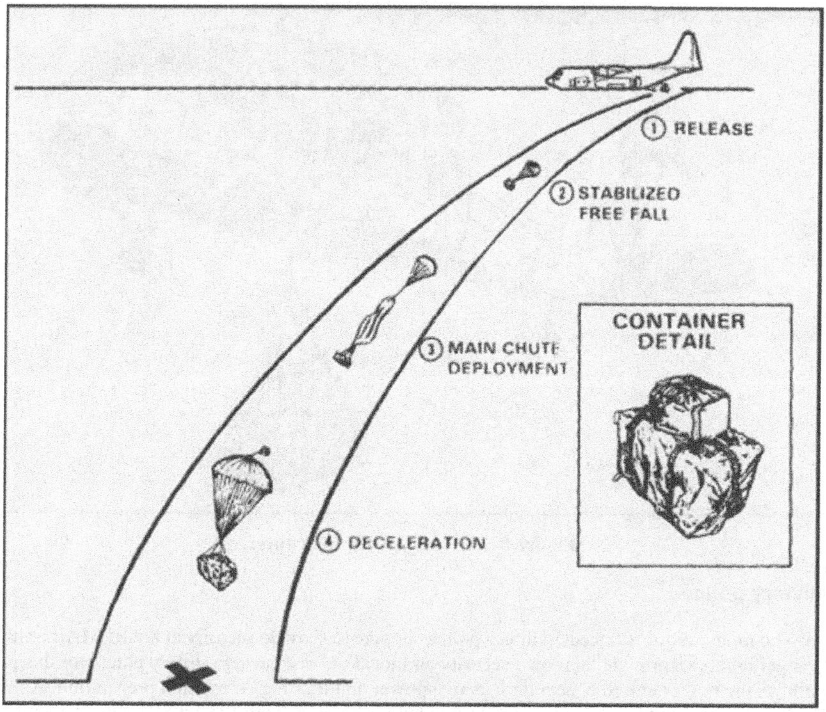

Figure 8-12. High-altitude, airdrop resupply system.

(3) The ranger force may need aircraft loads delivered during periods of poor visibility. The adverse weather aerial delivery system allows aircraft to drop resupply bundles accurately in adverse weather or darkness. These AWADS operations can be conducted safely and effectively with a minimum 300-foot ceiling AGL and a minimum visibility of less than half a mile. The minimum-drop altitude is 500 feet above the highest obstruction. These minimum ceiling or visibility restrictions do not apply to the Combat Talon aircraft.

f. Resupply to the ranger force can be provided by Army helicopters (see Figure 8-13). Normally this is done by SOA units. The danger to helicopters during long-range, low-level penetration of enemy rear areas requires that resupply be conducted when there is limited visibility.

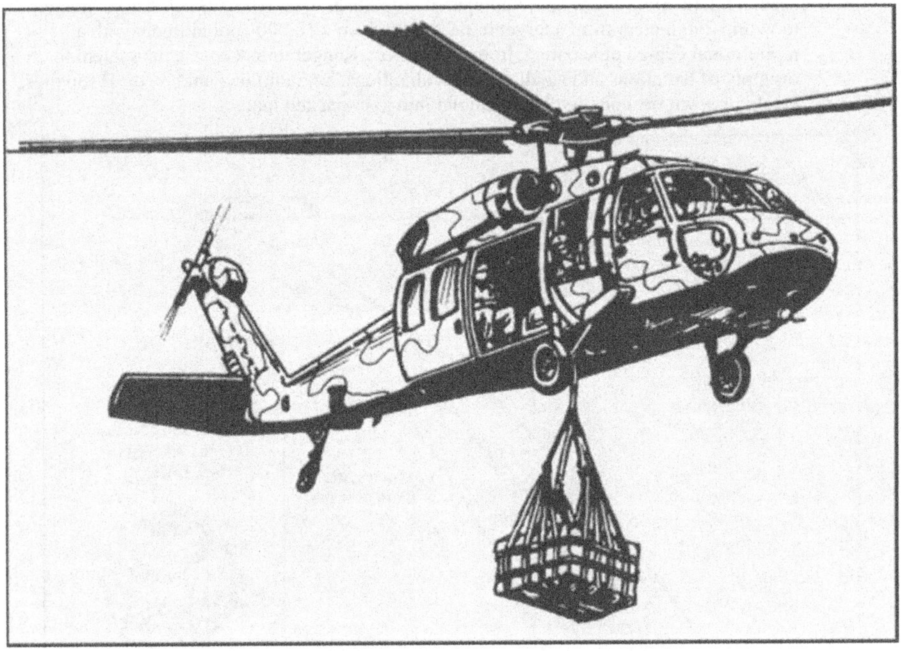

Figure 8-13. Resupply bt helicopter.

8-4. Military police.

a. The ranger regiment needs military police support to provide security at the REMAB. Although ranger units can provide their own security and local defense, using military police for this purpose allows the ranger unit to concentrate its manpower and training on mission preparation. A military police security company is normally designated to provide support to 1st SOCOM elements located in the communications zone. Military police provide security patrols of the REMAB, convoy security, and vehicular search. They provide entry and exit control of the sensitive compartmented information facility (SCIF).

b. A preemptive strike by enemy special operations forces is a threat to the ranger regiment while located in the REMAB; either by conventional attack or by terrorist-type bombing. The preparation and location of the headquarters, the type of buildings, the location of landing zones, the traffic flow, and the size of the local population all influence the size of the military police unit. A military police platoon, augmented with bomb and guard dog teams, should be enough security at the REMAB. These security forces would not normally deploy from the REMAB to an ISB with the ranger unit.

8-5. Counterintelligence.

Counterintelligence support to the ranger regiment is normally provided by a CI team from the military intelligence group at TA level. The CI team provides a liaison and informational flow from the CI agencies operating within the theater. It also acts as advisor to the ranger unit commander on OPSEC while performing CI surveillance operations in the vicinity of the REMAB. The CI support may also be provided through 1st SOCOM channels. It is through 1st SOCOM that liaison with national intelligence and counterintelligence agencies is achieved.

8-6. Personnel service support.

The personnel service support needs of a ranger unit are much like that of any other unit of the same size. To receive the support his unit needs, the ranger personnel officer must meet with the proper agencies within the theater of operations.

a. **Strength accounting.** Within the ranger regiment, strength accounting is conducted through normal personnel channels. Regimental strength figures are passed quickly to the TAACOM or COSCOM. These figures are crucial to the theater commander for planning future operations. When matched with casualty reports, they provide the information needed for replacements.

b. **Replacement operations.** Ranger units need exceptionally well-trained and physically conditioned personnel as replacements. Ranger operations require that replacements train with the unit before they can become an effective member of the team. There are five sources of replacements for ranger units:

(1) **In-theater volunteers.** Use of in-theater volunteers, preferably with experience in ranger units, is the quickest way to replace personnel. These volunteers have a knowledge of the basic skills required of rangers and have some operational experience. The ranger regimental personnel officer coordinates with the theater personnel officer to screen and select those volunteers with the required MOSs and grades. Volunteers from other special operations force components, or from other theaters of war, are identified by DA and screened for assignment to the ranger regiment. Even though these volunteers have individual training and experience, it takes several weeks to become fully integrated into the ranger unit's operations and to achieve the needed skill level.

(2) **Reorganization of units.** This method of reconstitution is used when two or more ranger units have high personnel losses. The remaining members of these units are reorganized into a single unit. This method provides the ranger regimental commander with enough ready-trained, physically fit personnel to form a full-strength unit. Extensive unit training must be conducted before such a reorganization regains the unity and esprit of its component parts. The high standards found among the subelements of the ranger regiment makes this a good method of reconstituting a ranger unit. The units that have lost all their personnel may be reconstituted later, using some other method, or left unreconstituted,

depending on the situation.

(3) **Hospital returnees.** These personnel, former members of the ranger regiment who were injured, evacuated, treated, and have now recovered, are valuable assets. They provide the unit with a major reconstitution ability. These personnel should be returned quickly to their original battalion. This enhances unity and increases morale and esprit. Returnees may not be physically able to return to full duty right away. While recuperating, they may be assigned to limited duty as instructors. The ranger unit uses the returnee's combat experience and individual skills to screen and train newly assigned volunteers. Lessons learned in combat can help prepare the unit for future operations.

(4) **Replacement volunteers from CONUS.** These personnel have training in the basic skills and may have completed the ranger course, but they do not have any ranger unit experience. They need several weeks of unit training to become prepared for combat operations.

(5) **Ranger indoctrination program.** During wartime, this program provides screening and training of newly assigned personnel. Volunteers who have neither served in a ranger unit before nor have graduated from the ranger course require extensive individual training before they are ready to begin unit training with a ranger battalion. This method of reconstitution of ranger units requires the longest training period. During this reconstitution phase, a ranger battalion needs extensive use of ranges, maneuver areas, and support by aviation and other special operations-type units. The ranger regimental commander may choose to move the program from CONUS to a location near the regiment's REMAB.

c. **Casualty reporting.** Casualty reports are handled through regimental personnel channels and must be timely and accurate. The reports are used to check strength accounting data, to assist graves registration personnel in the recovery and identification of remains, and to notify next of kin.

d. **Health service.** The health service support needed by the ranger units deployed OCONUS is usually small. Area medical support provided by host country or theater medical command assets usually meets their needs. Evacuation of wounded members of the ranger force begins upon extraction from the objective area. Personnel wounded in action are evacuated through USAF or Army channels to the nearest medical facility that can provide treatment. The ranger force medical officer and personnel officer coordinate with the area medical support facility providing preventive and routine medical care to the unit while at the REMAB. Vaccine needed for innoculation of the ranger force members and any special Class VIII supplies needed for the mission are drawn through normal medical channels.

e. **Financial service.** Financial support varies with the situation. In mid- and high-intensity conflicts, regular paydays are suspended in the theater of operations. As a nondivisional unit, the ranger regiment coordinates with the theater finance center or corps finance group for financial support. In the REMAB, ranger units are provided financial support from the closest finance support unit. The ranger unit personnel officer appoints Class A agents to help with the disbursement of funds. The ranger force may be given a specific amount of local currency or military scrip depending on the length of time in the REMAB and the operational situation. The ranger unit personnel officer also coordinates with the theater finance center to get any third-country currency that may be included as part of an escape-and-evasion kit provided to ranger elements.

f. **Postal service.** Postal service to the ranger unit is through normal channels. Mail for members of the deployed force is forwarded by the rear detachment commander to an established theater Army general support postal activity. This activity forwards the mail to the postal unit designated to support the REMAB site. The ranger unit commander appoints subordinate unit officers as censors. The exact location of the REMAB, and all aspects of the mission and its location, remains classified until released by an official Army spokesman. The ranger unit commander must ensure that the forwarding of mail, or the writing home of personal letters, does not endanger the ranger force mission. All outgoing mail is censored during wartime. For security reasons, outgoing mail may be held for a short time before it is forwarded. Mail from the ranger unit is never sent by host country or other nation's postal services.

g. **Chaplain activities.** Chaplain activities are provided by the chaplain assigned to the ranger battalion. He provides unit coverage through services, rites, and sacraments regardless of his or the unit member's religious affiliation.

h. **Legal service.** Ranger units do not need extensive legal service support. Legal service support is obtained through coordination with the theater Army staff judge advocate.

APPENDIX A

Organization and Equipment Introduction of the Ranger Regiment

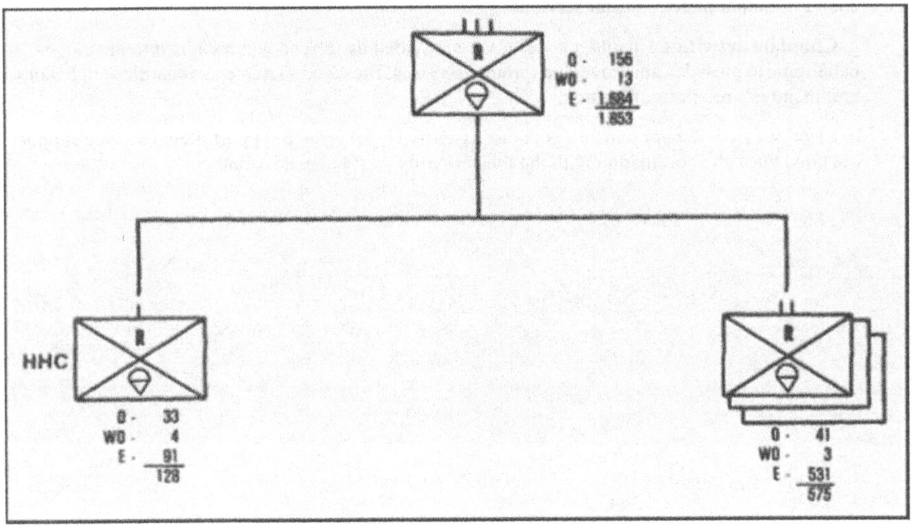

Figure A-1. Ranger regiment.

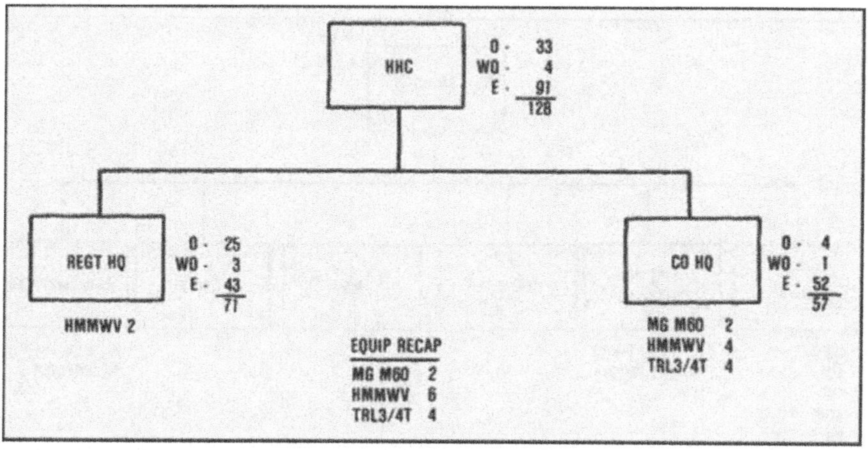

Figure A-2. Ranger regimental HHC.

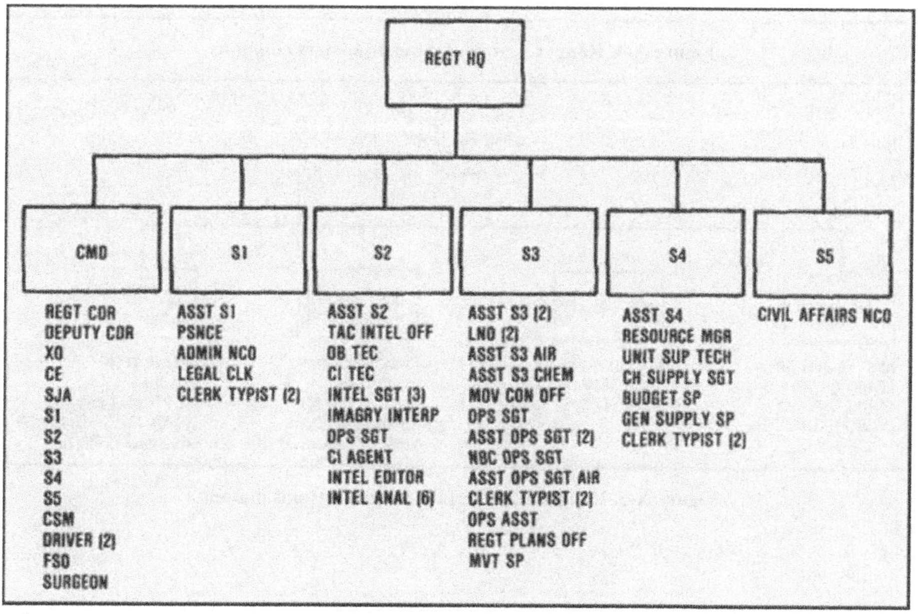

Figure A-3. Ranger regimental staff.

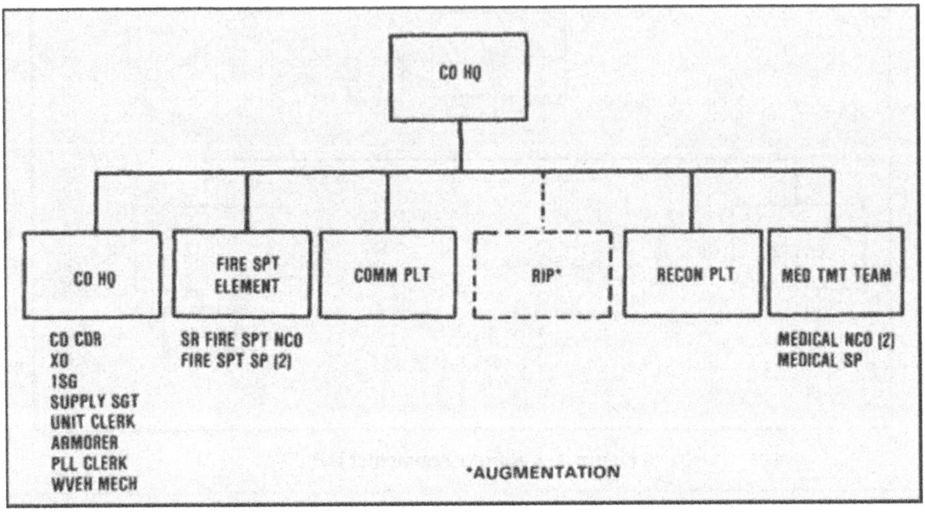

Figure A-4. Ranger regimental headquarters company.

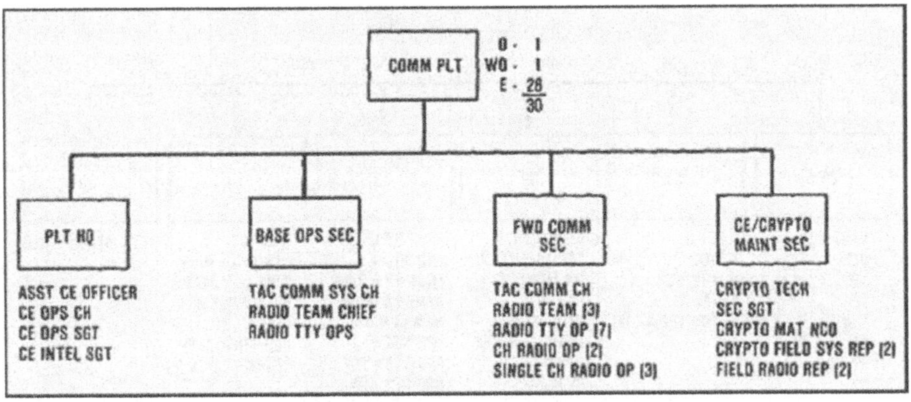

Figure A-5. Ranger regimental communications platoon.

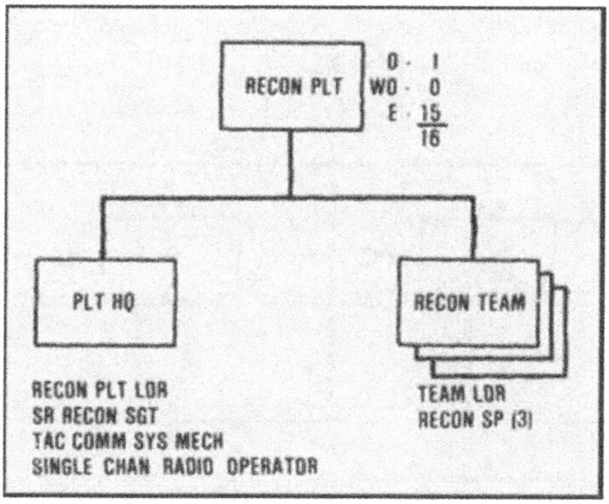

Figure A-6. Ranger regimental reconnaissance platoon.

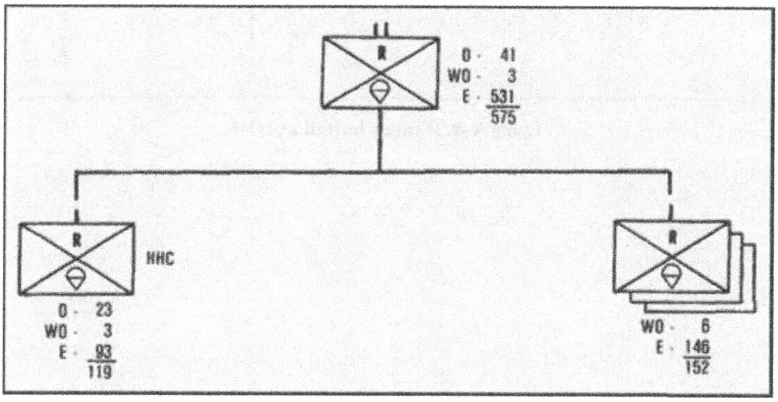

Figure A-7. The ranger battalion.

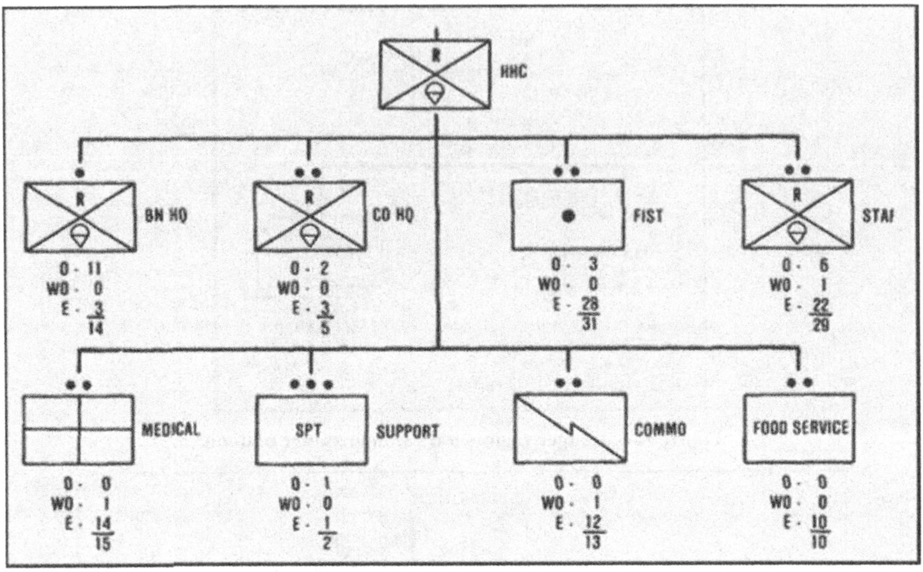

Figure A-8. Ranger battalion HHC.

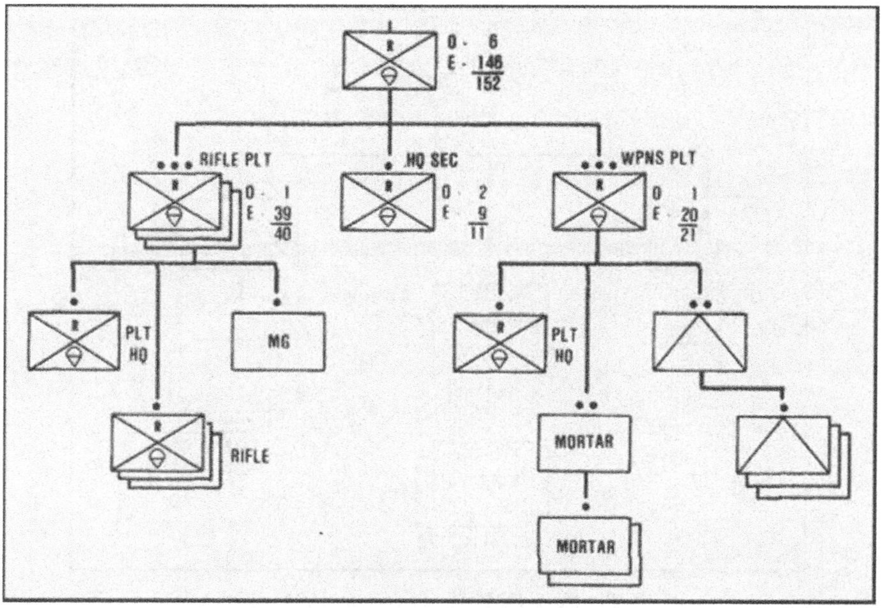

Figure A-9. Ranger rifle company.

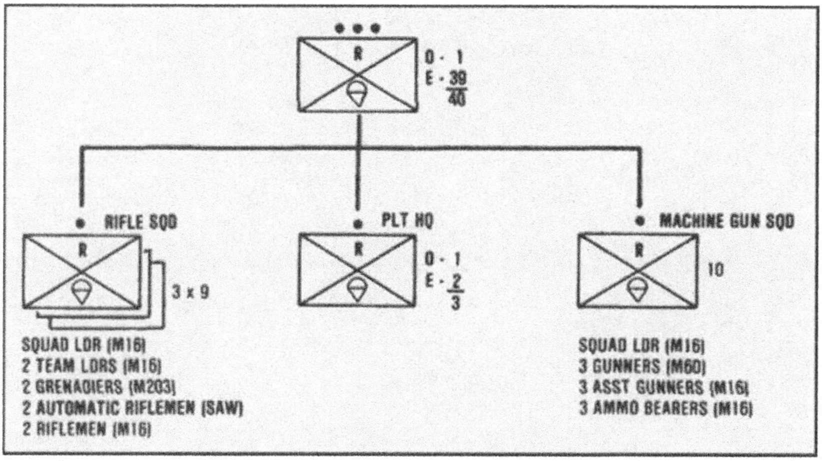

Figure A-10. Ranger rifle platoon.

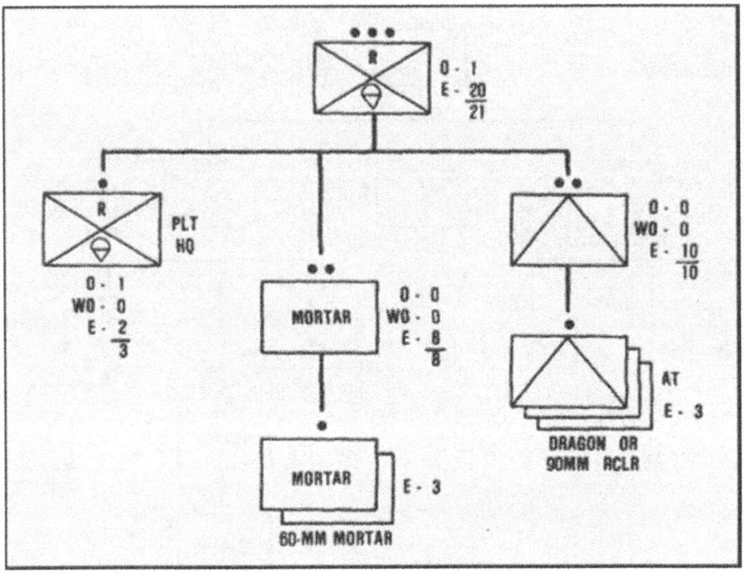

Figure A-11. Ranger weapons platoon.

APPENDIX B

Composition and Equipment of Ranger Regiment Liaison Teams

Team Red	**RANK**
Liaison Officer	MAJ
Senior Intelligence Analyst	SSG
Clerk-Typist	SP4
Tactical Communications System Operator/Mechanic	SGT
Combat Telecommunications Center Operator	SP4
Tactical Communications System Operator/Mechanic	SP4
Team White	
Liaison Officer	CPT
Senior Intelligence Analyst	SSG
Clerk-Typist	SP4
Tactical Communications System Operator/Mechanic	SGT
Combat Telecommunications Center Operator	SP4
Tactical Communications System Operator/Mechanic	SP4

Equipment (for Each Team)

HMMWV with 3/4-ton trailer	5-kw generator
Communications shelter	Tent, GP, small
AN/PCS-3 radio set (3)	AN/VSC-7 net control station (1)
MX 300 (4)	KY 57 with battery case (5)
HYX 57 with battery case (4)	HYP 57 (5)

KYK 13 (2)

KY 65 with battery case (1)

KG 84 (1)

H189 (2)

GRC 193 (1)

VGC-74 (1)

PP 1648 (1)

SB-22 switchboard (1)

KYI 18 (1)

KY 65 WLA (1)

Digital message device (1)

AN/PRC-77 radio (1)

Facsimile transmission device (1)

TA 312 with WD1 (5)

APPENDIX C

Equipment Density

C-1. Communication equipment.

	Regimental Headquarters	Battalions (x3)
AN/VSC-7 net control station	7	2
AN/PSC-3 radio set	23	13
AN/PRC-41 radio set	2	5
AN/PRC-104 radio set	9	26
AN/GRC-193A radio set	4	0
AN/PRC-70 radio set	6	0
AN/PRC-77 radio set	22	50
AN/VRC-46 radio set	2	0
AN/VRC-49 radio set	0	2
AN/PRC-68 radio set	0	66
TSEC/KY-57 speech security device	60	68
TSEC/KY-58 speech security device	6	0
TSEC/KY-65 speech security device	11	0
AO-8990/P digital message device group	15	23
AN/GXC-7A tactical FAX device	7	2
AN/UGC-74A(V)3 communications terminal	7	0
SB-22 switchboard	2	0
SB-3614(V)/TT switchboard	2	1

TSEC/KG-84	5	0
SB-993/GT switchboard	0	12

C-2. Weapons and equipment each battalion.

Infrared aiming light	96	Interrogator set, Stinger missile	3
Rifle, M16	451	Grenade launcher, M203	67
Rocket launcher, M202	13	Machine gun, M60	27
Mortar, 60-mm M224	6	Rifle, sniper, M21	24
Recoilless rifle, 90-mm	18	Viewer, infrared	9
Tracker, Dragon	9	Night vision devices (all types)	322
Truck, 1/4-ton	2	Laser target designators	11
Radio (all types)	164	Tactical satellite radios (all types)	15
Squad automatic weapon	54		
Laser range finder	14		

APPENDIX D

Standing Orders, Rogers' Rangers

1. Don't forget nothing.
2. Have your musket clean as a whistle, hatchet scoured, sixty rounds powder and ball, and be ready to march at a minute's warning.
3. When you're on the march, act the way you would if you was sneaking up on a deer. See the enemy first.
4. Tell the truth about what you see and what you do. There is an army depending on us for correct information. You can lie all you please when you tell other folks about the Rangers, but don't never lie to a Ranger or officer.
5. Don't never take a chance you don't have to.
6. When we're on the march we march single file, far enough apart so one shot can't go through two men.
7. If we strike swamps, or soft ground, we spread out abreast, so it's hard to track us.
8. When we march, we keep moving till dark, so as to give the enemy the least possible chance at us.
9. When we camp, half the party stays awake while the other half sleeps.
10. If we take prisoners, we keep 'em separate till we have had time to examine them, so they can't cook up a story between 'em.
11. Don't ever march home the same way. Take a different route so you won't be ambushed.
12. No matter whether we travel in big parties or little ones, each party has to keep a scout 20 yards ahead, 20 yards on each flank, and 20 yards in the rear so the main body can't be surprised and wiped out.
13. Every night you'll be told where to meet if surrounded by a superior force.
14. Don't sit down to eat without posting sentries.
15. Don't sleep beyond dawn. Dawn's when the French and Indians attack.
16. Don't cross a river by a regular ford.

17. If somebody's trailing you, make a circle, come back onto your own tracks, and ambush the folks that aim to ambush you.

18. Don't stand up when the enemy's coming against you. Kneel down, lie down, hide behind a tree.

19. Let the enemy come till he's almost close enough to touch. Then let him have it and jump out and finish him up with your hatchet.

MAJ Robert Rogers

APPENDIX E

The Ranger Creed

R Recognizing that I volunteered as a Ranger, fully knowing the hazards of my chosen profession, I will always endeavor to uphold the prestige, honor, and high "esprit de corps" of the Ranger Regiment.

A Acknowledging the fact that a Ranger is a more elite soldier who arrives at the cutting edge of a battle by land, sea, or air, I accept the fact that as a Ranger, my country expects me to move farther, faster, and fight harder than any other soldier.

N Never shall I fail my comrades. I will always keep myself mentally alert, physically strong, and morally straight and I will shoulder more than my share of the task, whatever it may be. One hundred percent and then some.

G Gallantly will I show the world that I am a specially selected and well-trained soldier. My courtesy to superiors, my neatness of dress, and my care of equipment shall set the example for others to follow.

E Energetically will I meet the enemies of my country. I shall defeat them on the field of battle, for I am better trained and will fight with all my might. Surrender is not a Ranger word. I will never leave a fallen comrade to fall into the hands of the enemy and under no circumstances will I ever embarrass my country.

R Readily will I display the intestinal fortitude required to fight on to the Ranger objective and complete the mission, though I be the lone survivor.

RANGERS LEAD THE WAY

APPENDIX G

Training

G-1. Application and guidance.

a. To prepare for combat, ranger units train on missions derived from their analysis of operational and contingency plans and the ranger mission training plans (MTPs). Training guidance is in doctrinal manuals, field circulars, and soldier's training publications. Training must ensure that the ranger unit is skilled in all basic light infantry skills and in ranger-unique skills. In the final analysis, good training depends upon the leadership and teamwork of ranger unit officers and NCOs.

b. Everything rangers do should prepare them to perform well in combat. Ranger units train IAW the battalion training management system (BTMS) concentrating on basic skills and on live-fire, hands-on training.

c. An effective training program breeds high morale and few discipline-related problems. Some qualities of ranger units that have good training programs are as follows:

(1) Leaders who can solve problems.

(2) A thorough and continuous analysis of the unit's mission.

(3) Validated training objectives.

(4) Attention to the basics (move, shoot, and communicate).

(5) Multiechelon sustainment training.

(6) Standardized procedures.

(7) Physical and mental strength.

(8) Live fire as an integral part of training.

(9) Safety as a basis to all operations.

(10) The use of night and adverse weather and difficult terrain as combat force multipliers,

d. When all training is planned with realism, rangers will be in training situations that require the same attention as in battle. Realism and stress in training result in faster learning and better retention of skills. Training objectives must be clear with measurable standards to determine proficiency. Rangers and ranger units must train to the standards that will be expected in combat.

e. Examples of unit training goals include:

(1) Developing each member of the chain of command to perform his duties with competence, confidence, and pride.

(2) Maintaining small unit integrity during all training and support activities. Accomplishing tasks as units.

(3) Ensuring that all tasks for training are keyed to mission needs and weaknesses. Assets--ranger's time, money, and materiel--are expended only after thorough planning.

(4) Training to standard.

(5) Developing and maintaining discipline and esprit to ensure the abilities of each ranger and all units are achieved.

(6) Emphasizing use of the ranger mission training plan (MTP) for each echelon of the organization.

(7) Emphasizing the basics.

(8) Maintaining a high level of physical fitness.

(9) Working together within the ranger regiment and other elements of the special operations forces community.

(10) Developing a unit atmosphere of quick reaction under stress. Encouraging and rewarding individual initiative. Accomplishing unit missions IAW the commander's intent.

G-2. Philosophy.

a. To be effective, training should be physically and mentally challenging. Each ranger will then have a sense of purpose, accomplishment, and satisfaction. Training is evaluated by the results rather than hours of training. Units that are well trained today need reinforcement training to sustain readiness. Training management must be continuous. Training managers must use the time available to improve training efficiency. Cyclic training will not produce the desired results. The "train-to-correct deficiencies" method is needed. The training managers must analyze weaknesses, conduct prescriptive and reinforcement training, and reevaluate their unit's training status. The process that is used to develop and maintain a continuous training and evaluation program is shown in Figure G-1.

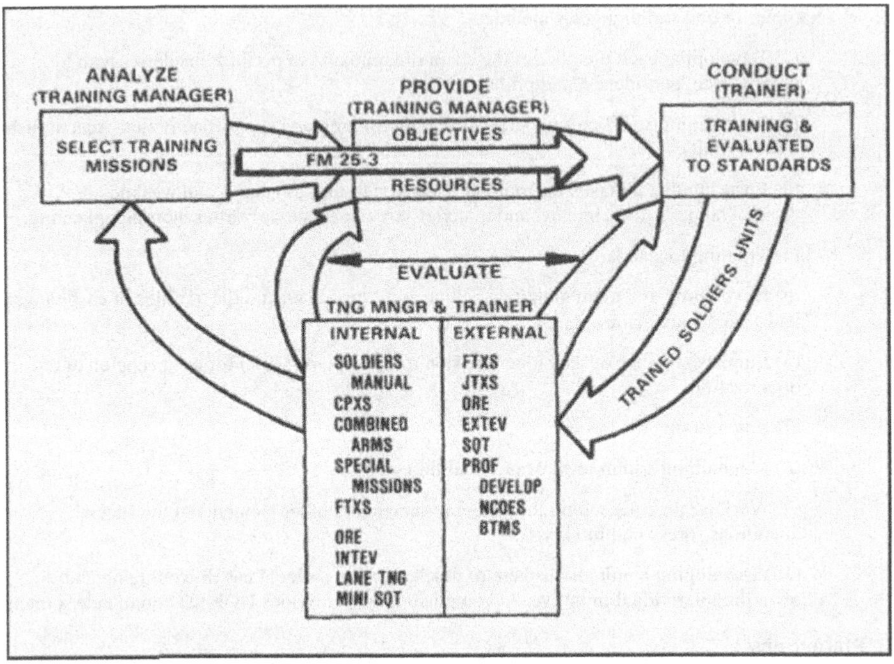

Figure G-1. The Army training and evaluation program.

b. Individual and collective training must be integrated within a unit training program. Training of the individual ranger, leaders, and units at each level must occur at the same time. The ranger commander or leader plans and conducts training for every member of his unit.

c. Ranger commanders analyze operational and contingency plans to determine unit missions and translate the associated tasks into training objectives. These objectives form the basis of the unit training program. The regimental headquarters provides training guidance in a formal regimental training plan. Commanders at lower levels program, supervise, and conduct training within their units.

d. All members of the chain of command are responsible for training. The NCO leader or first-line supervisor conducts individual and small unit collective task training. The officer plans individual training in his unit and directs and conducts collective training.

e. Centralized development of training objectives and decentralized training builds well-trained units and outstanding leaders. Authority and responsibility for the detailed planning, scheduling, conducting, evaluating, and supervising of training are delegated to the lowest element capable of managing this training. Higher headquarters is responsible for establishing objectives, monitoring and evaluating training, and issuing general guidelines.

f. Multiechelon training and evaluation means using all assets and time available to improve combat readiness. The multiechelon structure of the ranger MTP provides a basis for units to be continually trained and evaluated by internal evaluation. Each leader must know his mission tasks and train to accomplish those tasks to the set standards.

G-3. Mission analysis.

a. Mission analysis is continuous for a good training program. The selection of training missions or tasks results from analysis of the unit's overall mission. (See Figure G-1 above.) Tasks and subtasks are then placed in order of priority.

b. The trainer must identify countertasks for the opposing force (OPFOR). The countertasks should be such that an OPFOR that meets the standards for its countertask will be successful. Both sides have a chance to win.

c. The trainer should then develop combat drills and situational training exercises (STXs), if not furnished by the proponent school.

G-4. Unit training.

a. **Responsibility.** Ranger commanders at all levels are responsible for all training and management described herein and current Army training doctrine. This includes individual and collective training at all levels. Units must train to the standards specified in training objectives.

b. **Management.** The training management model for commanders is:

- Set objectives.
- Provide resources.
- Coach subordinates.
- Measure results.
- Provide feedback.

G-5. Individual training.

There are three elements basic to a well-trained individual ranger. He must be physically fit, be proficient in the use of his weapon, and have the psychological commitment to fight the enemy.

a. Individual training is the responsibility of the first-line supervisor under the supervision of the chain of command. It is conducted using performance-oriented techniques under battlefield conditions. New skills must be developed and practiced, and acquired skills must be reinforced throughout the training year. A high level of proficiency must be maintained through continuous training and practice.

b. Trainers should use peer instruction. Commanders ensure needed resources are available to conduct challenging individual training. Ranger units schedule and conduct this training and evaluate individual proficiency.

c. Flexibility in planning and scheduling is necessary. The goal of all individual training must provide for reinforcement training to offset the effects of the forgetting curve. (See Figure G-2).

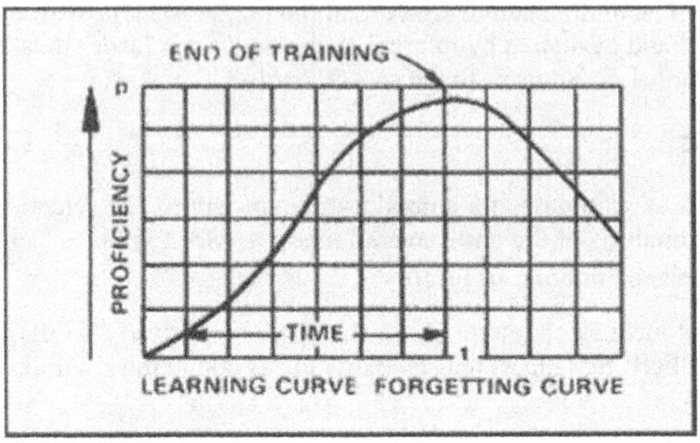

Figure G-2. The learning curve and forgetting curve.

d. The command sergeants major, unit first sergeants, and staff section sergeants are the keys to an effective training program. These senior NCOs must counsel, guide, and critique junior NCOs in training preparation and execution. Fire team leaders, section chiefs, and squad leaders coordinate and present individual training.

e. The NCO supervisors must ensure that the ranger is trained in the skills required for the specific military occupational specialty (MOS) and in those skills used daily. Not all training can be acquired in an office. For example, all rangers must maintain a high skill level with their weapons, be trained in NBC subjects, and be capable of performing contingency missions.

G-6. Collective training.

a. **Guidance.** Collective training trains crews, teams, squads, platoons, companies, and battalions to do their missions. The ranger MTP is used to plan training and evaluate the unit's weaknesses and strengths. If a ranger commander receives new training requirements or missions, they should become part of the collective training program. The ranger MTP describes the combat tasks to be performed by every echelon of the ranger battalion. It is a plan for organizing, conducting, and evaluating the year-round unit training program.

b. **Responsibility.** The ranger unit commander is responsible for the conduct of collective training and for all other training in his unit. Practical exercises emphasize the techniques of terrain use, movement, and effective employment of weapon systems. Training under stress will be conducted to move the unit to a higher plane of combat readiness.

c. **Drills.** Drills are written for small units--for example, fire team and squad, crew and section--and describe an action that is shorter than an entire STX. Drill training objectives cover both individual and collective skills (critical collective tasks) and list them together for effective group action. Drills must be mastered in order to react to rapidly changing situations in combat. The trainer can interrupt the execution of a drill to provide necessary training on individual or collective skills.

d. **Situational training exercise.**

(1) The STXs are short, mission-oriented tactical exercises in which a group of closely related collective tasks are trained. They are like drills but are more complex and flexible. They usually involve a larger unit than does a drill. They integrate leader training, leader tasks, drills, and separate individual tasks into a realistic setting. Trainers may execute STXs as written or tailor them to meet specific unit requirements. An STX is not standardized, but it must have a firm doctrinal base.

(2) Due to their design and purpose, STXs apply mainly to platoon- and company-size units. This does not preclude the development of STXs or similar exercises for unit staffs and headquarters. They combine "how-to-train" specifics for selected collective tasks with the "what-to-train" specifics found in doctrinal manuals and other documents.

G-7. Performance-oriented training.

a. Performance-oriented training emphasizes hands-on performance. Training is conducted to ensure a level of proficiency, rather than established number of hours of required training.

b. Performance-oriented training has precise training objectives and makes good use of resources. Training objectives serve as the basis for preparing, conducting, and evaluating all training.

G-8. Evaluation.

a. A training program must include an evaluation of training (planned and conducted). An evaluation plan must:

(1) Provide immediate feedback to the trainer.

(2) Recognize effective trainers.

(3) Evaluate all units in the conduct of training to provide for a comparison of training programs.

(4) Be flexible to unit needs; provide general and specific evaluations based on the level of training within the unit.

(5) Be in effect at all times.

b. Major considerations for training evaluations include:

(1) Ranger unit commanders establish and maintain an effective internal and external evaluation program.

(2) The director of each staff element is responsible for training evaluations within his area of staff responsibility with emphasis on low-density MOS groups.

(3) The operations officer of the ranger organization develops and validates training and evaluation plans.

c. Evaluation of training must be done vertically from the major task (mission) down through subtasks (collective tasks) to the leader and individual tasks. In selection of individual tasks for evaluation, priority should be given to critical individual tasks that impact on the success of the collective task or mission.

d. The critique and after-action review, after each major training event, are the most important aspects of training. The after-action review takes place immediately after completion of a training task (individual or collective). The individual(s) self-analyzes and critiques the training event. Then after a formal critique of the training, the individuals) trains to standard (not to time).

G-9. Field training.

a. **Considerations.** Ranger commanders schedule training under field conditions to permit maximum training opportunities. Field training emphasis is on mastery of a series of individual (soldier and leader) tasks and collective training tasks. These tasks are repeated until established standards are met. The goal of ranger field training is to ensure realistic, demanding, and meaningful training.

(1) When training is conducted under conditions of moderate stress, most combat critical skills are learned quicker and retained longer. (See Figure G-3 on the following page.) Stress can be incorporated in a training situation by physical deprivation (sleep, food), by live-fire exercises, by operation against an OPFOR under simulated battlefield conditions for an extended period, by the intensity of activity including demands for decision making, or by physically demanding activities. Ranger units train under stress so that they will react quickly in combat.

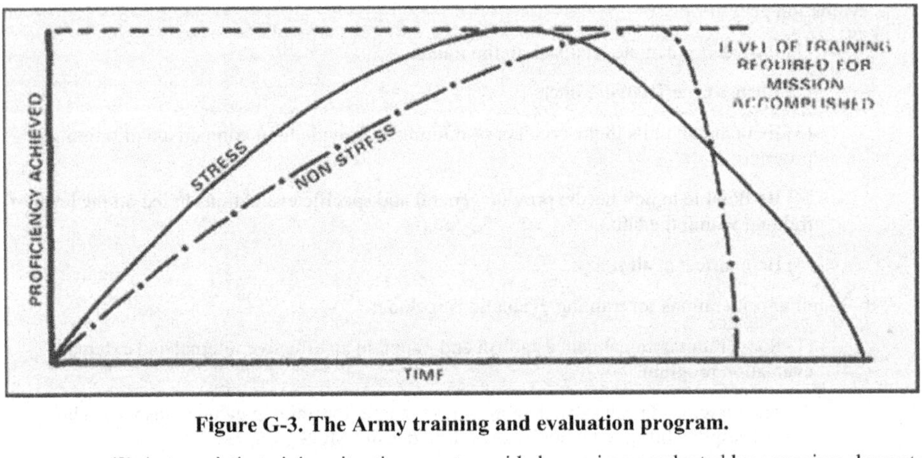

Figure G-3. The Army training and evaluation program.

(2) Antagonistic training situations are two-sided exercises conducted by opposing elements. Each element has a training objective (task and standards). The opposing force has a corresponding countertask. This training may be progressive--for example, starting with a

crawl-through, then a walk-through, and progressing to both elements operating at a fast pace. The use of multiple integrated laser equipment system (MILES) greatly enhances task and countertask training. The units should also be considered for a live-fire exercise against a realistic OPFOR target array when the basics have been mastered.

b. **Methods of training.**

(1) **Considerations.** The chance for success in combat can be increased by becoming proficient in training situations that depict realistic combat conditions. (See Figure G-4.) Meeting the standards of a realistic live-fire training exercise assures the commander that his unit could perform well in combat. Figure G-4 shows the relationship between realistic training and the challenge to the trainer.

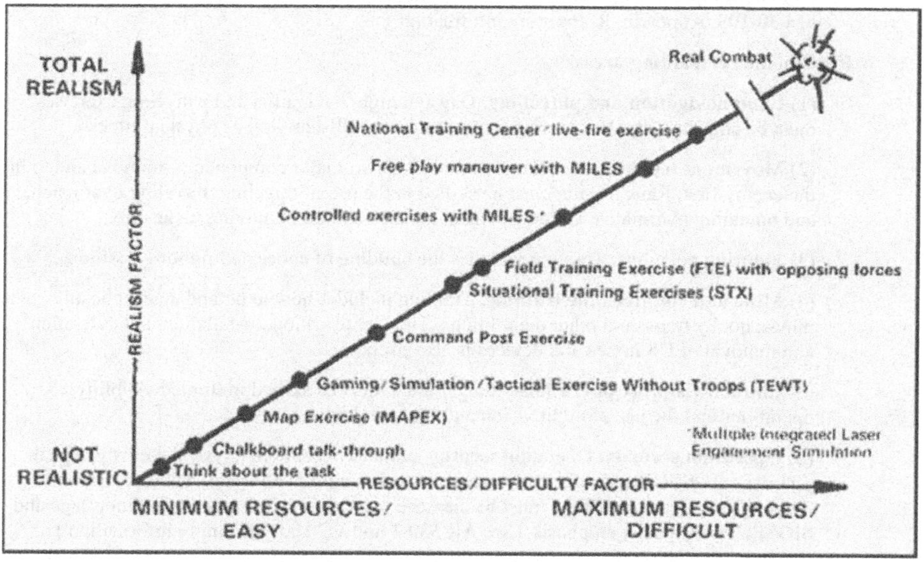

Figure G-4. Training proficiency scale.

(2) **Ranger cohesiveness training.** Ranger units promote teamwork by undergoing tough, stressful, often hazardous training. This challenges the unit, which gains in cohesiveness when it meets these challenges. This includes serving together in live-fire exercises, athletic events, precision parades, high-standard inspections, tough physical training, and combat.

(3) **Field training exercise.** An FTX is training normally conducted under simulated, realistic combat conditions. This training normally exercises command and control systems and the administrative and logistical needs of sustained combat. Stress training is part of this exercise, focusing on live-fire and MILES drills and STXS.

(4) **Combined arms training.** Combined arms training should include all means of support such as tactical air, armed helicopters, and indirect fire with emphasis on fire support coordination and airspace control.

(5) **Live-fire exercises.** Including live-fire exercises in drills and STXs offers a chance to evaluate units in their response to simulated combat missions. Ranger units train under pressure and always anticipate being selected to conduct a live-fire exercise. Live-fire exercises develop confidence and esprit, and reinforce unit discipline.

(6) **Opposing force.** Committing one unit against another adds a competitive spirit to a field training exercise (FTX). The use of MILES and squad combat operations exercises (simulation) (SCOPES) provides realism to two-sided field exercises. Trainers should use OPFOR equipment and training devices available, as well as making their own devices. (See FM 30-103, Appendix R, for more information.)

c. **Key functional training areas.**

(1) **Land navigation and patrolling.** Day and night navigation and patrolling exercises must be emphasized. Orienteering improves these skills, as well as physical fitness.

(2) **Movement techniques.** Infantry combat has two major components--to move and to hit the enemy first. Ranger units must be skilled in the use of traveling, traveling overmatch, and bounding overmatch. Units in contact must be skilled in moving under fire.

(3) **Fighting** positions. Training includes the building of concealed fighting positions.

(4) **Mine and countermine training.** Training includes how to defend against hostile mines, booby traps, and other demolitions. The identification, installation, neutralization, and removal of US mines and devices is also stressed.

(5) **Limited visibility operations.** Ranger units must be skilled in limited visibility operations and the use of night vision and thermal devices.

(6) **Operation security.** Operation security includes deception, physical security, signal security (SIGSEC), and information security. It is taught in all training and at all levels. Although all areas of OPSEC must be measured in training, techniques of camouflage and SIGSEC need special emphasis. (See AR 530-1 and TC 100-1 for more information.)

G-10. Professional development.

Ranger units at battalion and regimental levels conduct many officer and NCO professional development programs. This ensures that all officers and NCOs are skilled in their field, confident, and wellrounded. Professional readings and book reports, tactics seminars, and tactical exercises without troops are good development methods.

G-11. Physical training.

a. This program ensures that each ranger maintains the physical ability and stamina needed to do his mission. Other objectives are to promote unit cohesiveness and discipline. These are met by combining individual testing, formation, running, athletic competition, weight control, proper diet, teamwork drills, and freedom from drug and alcohol abuse. Mental fitness and toughness come from experience and success. Good physical conditioning increases a unit's chances of success.

b. Physical training (PT) programs are designed to include the elements of overload, progression, balance, variety, and regularity.

c. Ranger unit physical fitness standards are:

(1) Score 80 points for each Army physical readiness test (APRT) event and do six chinups.

(2) Pass the ranger swimming test.

(3) Complete an 8-kilometer run in 40 minutes.

(4) Complete a 12-kilometer road march in three hours (with rucksack, helmet, and weapon).

(5) Meet the Army height and weight standards.

G-12. Environmental and specialized training.

a. Rangers conduct training in environments as close to real combat missions as possible. This conditions rangers to survive and fight in a specific environment. Such training includes combat-critical tasks and technique training for that environment.

b. Each ranger battalion's environmental and specialized training includes at least the following:

(1) Jungle (once every 12 months).

(2) Desert (once every 12 months).

(3) Urban (once every 6 months).

(4) Extreme cold (once every 18 months).

(5) Mountain (once every 12 months).

(6) Amphibious (once every 18 months).

G-13. Ranger indoctrination program.

a. The RIP is a training and selection program. It is run by specially selected NCOs from the regimental headquarters. It has as its mission the training and indoctrination of newly assigned soldiers. It physically toughens and mentally sharpens the newly assigned soldier. It teaches basic skills and techniques. The RIP identifies and eliminates in three weeks any applicant who does not show dedication, motivation, physical fitness, and emotional stability.

b. An indoctrination program for newly assigned ranger-qualified officers and NCOs prepares them to assume duty as leaders within the ranger regiment. These personnel monitor and participate in RIP training, as well as undergoing individual ranger leadership training. The RIP provides them an overview of the regiment's history, its mission, and its goals. (See Appendixes C,

D, and E for further information about the history and goals of the ranger regiment.) The RIP also provides a review of the skills required in a ranger unit. Refresher training in airborne and pathfinder operations, use of indirect fires, and light infantry doctrine is conducted. During RIP all newly assigned officers and NCOs must meet the ranger physical fitness standard.

c. The RIP cadre also conducts a preranger program to prepare soldiers to attend the US Army Ranger School. It is a physically demanding course that teaches troop-leading procedures, operation orders, and small unit patrol actions.

Glossary

The following acronyms, abbreviations, and definitions are provided to aid the reader in understanding this publication.

Section I

ACRONYMS AND ABBREVIATIONS

AAT	Army assault team
ABCCC	airborne battlefield command and control center
ACL	allowable cargo load
ADA	air defense artillery
AFFOR	the USAF element of a joint task force
AGL	above ground level
AKZ	armor kill zone
ALCC	airlift control center
ALCE	airlift control element
AM	amplitude modulation
AMC	air mission commander
APERS	antipersonnel
APRT	Army physical readiness test
ARFOR	the Army element of a joint task force
ARSOC	Army special operations command
ARSOF	Army Special Operations Forces
ASG	area support group
ASPC	all-source production center

ATGM	antitank guided missile
AWADS	adverse weather aerial delivery system
BCE	battle coordination element
BTMS	battalion training management system
CA	civil affairs
CAS	close air support
CCT	combat control team
CDS	container delivery system
CE	communications-electronics
CI	counterintelligence
CINC	commander in chief
COMINT	communications intelligence
COMSEC	communications security
CONUS	continental United States
COSCOM	corps support command
CP	command post
CS	combat support; a chemical agent
CSS	combat service support
DACG	departure airfield control group
DACO	departure airfield control officer
DASC	direct air support center
DMDG	digital message device group
DOD	Department of Defense
DS	direct support
DSU	direct support unit
DZ	drop zone
EAC	echelons above corps

EACIC	Echelons Above Corps Intelligence Center
ECCM	electronic counter-countermeasure
ECM	electronic countermeasures
E&E	escape and evasion
EEFI	essential elements of friendly information
EOC	emergency operations center
EW	electronic warfare
FARP	forward area rearm and refuel point
FASCAM	family of scatterable mines
FAX	facsimile transmission
FDC	fire direction center
FEBA	forward edge of the battle area
FFL	friendly front lines
FIST	fire support team
FLOT	forward line of own troops
FM	field manual; frequency modulation
FO	forward observer
FORSCOM	US Army Forces Command
FPF	final protective fire
FPL	final protective line
FRAGO	fragmentary order
FSA	fire support area
FSCOORD	fire support coordination or coordinator
FSE	fire support element
FSO	fire support officer
FSS	fire support station
FTX	field training exercise

G3	Assistant Chief of Staff, Operations
GS	general support
GSR	ground surveillance radar
HAARS	high-altitude, airdrop resupply system
HAHO	high-altitude, high-opening
HALO	high-altitude, low-opening
HE	high explosive
HF	high frequency
HHC	headquarters and headquarters company
HSLLADS	high-speed, low-level airdrop system
HUMINT	human intelligence
IAW	in accordance with
IFF	identification-friend-or-foe
IP	initial point
IRP	initial rally point
ISB	intermediate staging base
JACC/CP	joint airborne communications center/command post
JCS	Joint Chiefs of Staff
JCSE	joint communications support element
JFC	joint force commander
JSOC	joint special operations command
JTB	joint transportation board
JTF	joint task force
kw	kilowatt
LAPES	low altitude parachute extraction system
LAW	light antitank weapon
LNO	liaison officer

LOC	logistical operations center
LRRP	long-range reconnaissance patrol
LRSU	long-range surveillance unit
LTD	laser target designator
LZ	landing zone
MAC	Military Airlift Command
MANPADS	man-portable air defense system
MCC	movement control center
METT-T	mission, enemy, terrain, troops and time available
MI	military intelligence
MILES	multiple integrated laser engagement system
mm	millimeter
MMC	Materiel Management Center
MMEE	minimum mission-essential equipment
MOS	military occupational specialty
MOUT	military operations on urbanized terrain
MRLS	multiple rocket launch system
MTP	mission training plan
NBC	nuclear, biological, chemical
NCA	National Command Authority
NCO	noncommissioned officer
NOD	night observation device
OCOKA	observation, concealment, obstacles, key terrain, and avenues of approach
OCONUS	outside continental United States
OPCOM	operational command
OPCON	operational control
OPFOR	opposing force

OPORD	operation order
OPSEC	operations security
ORP	objective rally point
PAC	personnel administration center
PDF	primary direction of fire
PIR	priority intelligence requirements
PLL	prescribed load list
POM	preparation for oversea movement
PSYOP	psychological operations
PT	physical training
PZ	pickup zone
QRE	quick-reaction element
RATELO	radiotelephone operator
REMAB	remote marshalling base
RFL	restrictive fire line
RIP	ranger indoctrination program
RP	rally point; release point
RPG	rocket-propelled grenade
RRF	ranger ready force
RRP	reentry rally point
RSE	ranger support element
RSTA	reconnaissance, surveillance, and target acquisition
S1	adjutant
S2	intelligence officer
S3	operations and training officer
S4	supply officer
S5	civil affairs officer

SAR	search and rescue
SATCOM	satellite communications)
SCIF	sensitive compartmented information facility
SCOPES	squad combat operations' exercises (simulation)
SCUBA	self-contained underwater breathing apparatus
SEAD	suppression of enemy air defense
SEMA	special electronic mission aircraft
SERE	survival, evasion, resistance, and escape
SF	special forces
SHORAD	short-range air defense
SIGSEC	signal security
SKE	station-keeping equipment
SLAR	side-looking airborne radar
SOA	special operations aviation
SOCB	special operations communications battalion
SOCOM	special operations command (usually, 1st SOCOM)
SOP	standing operating procedures
SOSB	special operations support battalion
SOTF	special operations task force
SPL	support platoon leader
STX	situational training exercise
TA	theater Army
TAACOM	Theater Army Area Command
TAC	Tactical Air Command
TACP	tactical air control party
TACSAT	tactical satellite terminal
TAMMC	Theater Army Materiel Management Center

TASG	theater Army support group
TDA	table of distribution and allowances
TOA	transfer of authority
TOC	tactical operations center
TOE	table(s) of organization and equipment
TRP	target reference point
TTY	teletype
UHF	ultrahigh frequency
US	United States
USAF	United States Air Force
USMC	United States Marine Corps
USN	United States Navy
VHF	very high frequency
WP	white phosphorus
XO	executive officer

Section II

DEFINITIONS

ADVANCE FORCE. An operation requiring the immediate deployment of US forces with the intent to negate a threat before the deployment of follow-on forces is required.

AIRLAND BATTLE DOCTRINE. An approach to military operations that realizes the full potential of US forces. Two notions (extending the battlefield and the ability to integrate conventional, nuclear, chemical, and electronics means) are combined to describe a battlefield where the enemy is attacked to the full depth of his formation.

AREA OF INFLUENCE. A geographical area wherein a commander is directly capable of influencing operations by maneuver or fire support systems normally under his command or control.

AREA OF INTEREST. That area of concern to the commander, including the area of influence, areas adjacent thereto, and extending into enemy territory to the objectives of current or planned operations. This area also includes areas occupied by enemy forces who would jeopardize the accomplishment of the mission.

AREA OF OPERATIONS. That portion of an area of conflict necessary for military operations. Areas of operations are geographical areas assigned to commanders for which they have responsibility and in which they have authority to conduct military operations.

CHALK. One specific aircraft load.

CLANDESTINE OPERATIONS. Activities to accomplish intelligence, counterintelligence, and other similar activities sponsored or conducted by governmental departments or agencies, in such a way as to ensure secrecy or concealment. (They differ from covert operations in that emphasis is placed on concealment of the operation rather than on concealment of the identity of the sponsor.)

COMMUNICATIONS INTELLIGENCE (COMINT). The interception and processing of foreign communications passed by radio, wire, or other electromagnetic means, and by the processing of foreign encrypted communications, however transmitted. Interception comprises search, intercept, operator identification, signal analysis, traffic analysis, crypto-analysis, study of plain text, the fusion of these processes, and reporting the results.

COMMUNICATIONS SECURITY. The protection resulting from all measures designed to deny to unauthorized persons information of value that might be derived from the possession and study of telecommunications, or to mislead unauthorized persons in their interpretations of the results of such study. COMSEC includes crypto-security, physical security, and transmission security.

COUNTERINTELLIGENCE. That aspect of intelligence activities, both offensive and defensive, designed to detect and neutralize or destroy the effectiveness of hostile foreign intelligence activities and to protect information against espionage, personnel against subversion, and installations and materiel against sabotage.

COVERT OPERATIONS. Operations that are so planned and executed as to conceal the identity of or permit plausible denial by the sponsor. They differ from clandestine operations in that emphasis is placed on concealment of identity of the sponsor rather than on concealment of the operation.

DECEPTION. Those measures designed to mislead the enemy by manipulation, distortion, or falsification of evidence to induce him to react in a manner prejudicial to his interests.

DEMONSTRATION. A show of force in an area where a decision is not sought. It is similar to a feint but does not make contact with the enemy.

DRILL. A drill focuses on a "chunk" or "slice" of battle and is targeted at small units. Drills are critical collective tasks that require a high degree of proficiency from a small unit. Generally, small units "drill" collective tasks that require rapid responses by unit members in the absence of detailed orders from unit leaders.

ELECTRONIC COUNTERMEASURES. That division of electronic warfare involving actions taken to prevent or reduce an enemy's effective use of the electromagnetic spectrum.

ELECTRONIC INTELLIGENCE (ELINT). The technical and intelligence information derived from foreign noncommunications electromagnetic radiations emanating from other than atomic detonations or radioactive sources.

ELECTRONIC WARFARE. Military action involving the use of electromagnetic energy to determine, exploit, reduce, or prevent hostile use of the electromagnetic spectrum and action that retains friendly use of the electromagnetic spectrum.

ELECTRONIC WARFARE SUPPORT MEASURES (ESM). That division of EW involving actions taken to search for, intercept, identify, and locate sources of radiated electromagnetic energy for the purpose of immediate threat recognition. ESM provides a source of information needed for immediate action involving electronic countermeasures, electronic counter-countermeasures, avoidance, targeting, and other tactical employment of forces.

ESPIONAGE. Actions directed toward the acquisition of information through clandestine operations.

EVACUATION/SAFEGUARD. The use of US forces to evacuate US citizens or allies from a hostile or potentially hostile area, or to safeguard US personnel or property.

EXFILTRATION. The removal of personnel or units from areas under enemy control by stealth, deception, surprise, or clandestine means. Exfiltration is a method of extracting ranger units.

EXTRACTION. The removal of personnel or units from areas under enemy control by any means, to include exfiltration, airland, airmobile, or amphibious.

FEINT. A diversionary supporting attack conducted to draw the enemy's attention from the main effort. It is normally executed by brigades and small units. Feints are usually shallow, limited objective attacks that go in before or during the main attack.

GUERRILLA WARFARE. Military and paramilitary operations conducted in enemy-held or hostile territory by irregular and predominantly indigenous forces.

HIGH-INTENSITY CONFLICT. A war between the US and any other nation or nations in which both belligerents use the most modern technology and all resources in intelligence, mobility, firepower (including NBC), command, communications and control, and service support.

HUMAN INTELLIGENCE. Intelligence derived from information provided by human resources.

INFILTRATION. The movement through or into an area of territory occupied by either friendly or enemy troops or organizations. The movement is made either by small groups or by individuals, at extended or irregular intervals. When used in conjunction with the enemy, it implies that contact is avoided. Infiltration is a subset of insertion.

INSERTION. Placement of troops and equipment into an operational area by any means.

INTERDICT. To isolate or seal off an area by military means; to prevent, hinder, or delay the use of an area or route by enemy forces.

INTERDICT LINE OF COMMUNICATIONS. An attack to seal off an area or to deny use of a route or approach.

JOINT TASK FORCE. A JTF may be constituted and designated by the Secretary of Defense or by the commander of a unified command, specified command, or an existing JTF. Normally, it performs missions having specific, limited objectives or missions of short duration. It dissolves when it has achieved its purpose. The joint task force commander is responsible to the commander, known as the establishing authority, who created the JTF. The JTF is composed of elements of two or more services operating under a single JTF commander. The JTF commander has operational control over the entire force and may have direct command of his own staff with representatives from the other services. He exercises logistics coordination or control only as necessary to meet his subordinate commanders' logistics needs.

LIAISON. That personal contact or communication maintained between elements of military forces to ensure mutual understanding and unity of purpose and effort.

LINE OF COMMUNICATIONS. All routes (land, water, and air) that connect an operating military force with one or more bases of operations and along which supplies and military forces move.

LOW-INTENSITY CONFLICT. Limited politico-military operations conducted to achieve political, social, economic, or psychological objectives. It is generally confined to a specific geographic area and is often characterized by operational constraints imposed on ranger units by political and economic considerations. It may involve intermittent combat against regular threat military forces, but it more often involves conflict with irregular, security, or paramilitary forces of a country or group involved in an operation that is counter to the best interests and national security of the United States. It is characterized by relatively low levels of combat and constraints placed on the weaponry, tactics, and movements of the ranger force. It may involve counterterrorist or antiterrorist operations and generally occurs before a pro forma declaration of war.

MID-INTENSITY CONFLICT. War, declared or undeclared, between the US and the regular forces of one or more nations and their respective allies, if any, in which the belligerents use the most modern technology and all resources in intelligence, mobility, firepower (excluding nuclear, chemical, and biological weapons), command, communications and control, and support. They are used for limited objectives under definitive policy limitations as to the extent of destructive power that can be used or the extent of geographical area that might be involved.

MILITARY STRATEGY. Art and science of using the armed forces of a nation to secure the objectives of national policy by application of force or the threat of force. Military strategy sets the fundamental conditions for operations.

NATIONAL COMMAND AUTHORITY. This includes the President and Secretary of Defense or the duly deputized alternates or successors. Commonly referred to as NCA.

OPERATIONAL COMMAND/CONTROL. Operational command (OPCOM) and operational control in joint force terminology both refer to the authority exercised by joint commanders over subordinate service components. Those terms are not interchangeable in joint operations. OPCOM applies to the authority exercised by commanders of unified commands. OPCON is the authority that subordinate joint task force commanders exercise in the conduct of specific operations. The authority that military departments exercise over their respective components is commonly referred to as command less operational command.

OPERATIONAL LEVEL OF WAR. The operational level of war uses available military resources to attain strategic goals within a theater of war. It is the theory of larger unit operations. It also involves planning and conducting campaigns. Campaigns are sustained operations designed to defeat an enemy force in a specified space and time with simultaneous and sequential battles. The disposition of forces, selection of objectives, and actions taken to weaken or to outmaneuver the enemy all set the terms of the next battle and exploit tactical gains. They are all part of the operational level of war. In AirLand Battle doctrine, this level includes the marshalling of forces and logistical support, providing direction to ground and air maneuver, applying conventional and nuclear fires in depth, and employing unconventional and psychological warfare.

OPERATIONS SECURITY. Those measures designed to protect information concerning planned, ongoing, and completed operations from unauthorized disclosure. It includes all actions a command takes to deny the enemy information about friendly units and their operations.

ORDER OF BATTLE (OB). The identification, strength, command structure, and disposition of the personnel, units, and equipment of any military force.

RAID. Usually a small-scale operation, involving a swift penetration of hostile territory to secure information, confuse the enemy, or destroy his installations. It ends with a planned withdrawal upon completion of the assigned mission.

RANGER MISSION. To conduct strike, raid, or special light infantry operations. Execution of ranger missions is normally directed by the National Command Authority. Rangers also support operational-level commanders during operations of limited duration when allocated or assigned by the theater commander.

RECONNAISSANCE. A mission undertaken to obtain, by visual observation or other detection methods, information about the activities and resources of an enemy or potential enemy; or to secure data concerning the meteorological, hydrographical, or geographical characteristics of a particular area.

SEARCH AND RESCUE. The use of aircraft, surface craft, submarines, specialized rescue teams, and equipment to search for and rescue personnel in distress on land or at sea.

SEIZURE. As an operational purpose, the capture of a voluntarily restricted portion of an enemy-controlled territory. Capture of an isolated land mass such as an island is usually categorized as a seizure.

SEIZE KEY FACILITY. An attack on an installation with the intent of holding it for a short period. Normally, rangers would make the assault to seize and secure an installation for follow-on forces.

SERVICE COMPONENT. Each service component commander is responsible for recommending the proper use of his forces and for accomplishing operational tasks assigned by the joint commander. He is also responsible for his service in matters of internal administration and discipline; training in own service doctrine, techniques, and tactics; designation of specific units to meet joint requirements; logistics functions normal to the component; tactical employment of service component forces; and service intelligence matters.

SHOW OF FORCE. A mission carried out to demonstrate US resolve, whereby US forces are deployed to defuse a situation that may be harmful to US interests or national objectives.

SIGNAL INTELLIGENCE (SIGINT). A generic term that includes COMINT and ELINT. Activities concerned with the collection and processing of foreign electromagnetic emissions for the purpose of producing intelligence information.

SITUATIONAL TRAINING EXERCISE. An STX is a short, mission-oriented exercise in which a group of closely-related collective tasks are trained, using a doctrinally preferred method of executing these tasks to established standards. An STX is drill-like in nature, but it is more complex and flexible than a drill. It usually involves a larger unit than a drill.

SPECIAL OPERATIONS. Military operations conducted by specially trained, equipped, and organized DOD forces against strategic or tactical targets in pursuit of national, military, political, economic, or psychological objectives. Special operations may support conventional military operations, or they may occur independently when the use of conventional forces is either inappropriate or infeasible. Sensitive peacetime operations, except for training, are normally authorized by the NCA and conducted under the direction of the NCA or designated commander. Special operations may include unconventional warfare, counter-terrorist operations, collective security, psychological operations, and civil affairs measures.

SPECIAL OPERATIONS FORCES. All civil affairs, psychological operations, ranger (active component), and special forces units within the US Army, US Army Reserve, and National Guard.

SPECIAL OPERATIONS COMMAND. The major subordinate command to FORSCOM exercising command and control of all active component Army special operations forces not placed under the operational command of unified or specified commands.

STRATEGIC LEVEL OF WAR. Military strategy uses the armed forces of a nation to secure the objectives of national policy by applying force or the threat of force. Military strategy sets the fundamental conditions for operations.

STRIKE. An attack that is intended to inflict damage on, seize, or destroy an objective. Sometimes referred to as direct action missions.

SURVEILLANCE. The systematic observation of aerospace, surface or subsurface areas, places, persons, or things by visual, aural, electronic, photographic, or other means.

TACTICAL. Pertaining to the employment of units in combat.

TACTICAL LEVEL OF WAR. Tactics are the specific techniques smaller units use to win battles and engagements that support operational objectives. Tactics employ all available combat, combat support, and combat service support. Tactics involve the movement and positioning of forces on the battlefield in relation to the enemy, the provisioning of fire support, and the logistical support of force before, during, and following engagements with the enemy. At corps and division, operational and tactical levels are not clearly separable, but they are guided by the same principles. An operation designed to defeat any enemy force in an extended area does so through operational maneuver and a series of tactical actions.

TACTICAL COMBAT INTELLIGENCE. That knowledge of the enemy, weather, and geographical features needed by a commander in the planning and conduct of combat operations. It is derived from the interpretation of information on the enemy and the environment.

TARGET ANALYSIS. A detailed and systematic examination of processed intelligence to identify and locate targets.

UNIFIED COMMAND. A unified command is established and designated by direction of the President. The command is organized under a single unified CINC and has a broad, continuing mission. The CINC is responsible to the Secretary of Defense for accomplishing his broad and continuing mission and operates under the strategic and operational direction of the JCS. A unified command is composed of assigned forces of two or more services. These forces, as service components, are under the operational command of the unified commander. The CINC exercises operational command authority through service component commanders. When tasked by higher authority, he may establish a subordinate JTF to conduct specific missions. Under emergency conditions, the CINC has the authority to use all facilities and supplies of assigned forces to accomplish his mission.

References

REQUIRED PUBLICATIONS

Required publications are sources that users must read in order to understand or to comply with this publication.

Field Manuals (FM)

FM 7-8	The Infantry Platoon and Squad (Infantry, Airborne, Air Assault, Ranger)
FM 7-10	The Infantry Rifle Company (Infantry, Airborne, Air Assault, Ranger)
FM 7-20	The Infantry Battalion (Infantry, Airborne, and Air Assault)

RELATED PUBLICATIONS

Related publications are sources of additional information. They are not required in order to understand this publication.

Army Regulations (AR)

AR 350-1	Army Training

Field Manuals (FM)

FM 10-52-1	Commander's Handbook for Water Usage in Desert Operations
FM 10-501	Airdrop of Supplies and Equipment: Rigging Containers
FM 10-512	Airdrop of Supplies and Equipment: Rigging Typical Supply Loads
FM 10-542	Airdrop of Supplies and Equipment: Rigging Loads for Special Operations
FM 10-547	Airdrop of Supplies and Equipment: Rigging the High Speed Aerial Delivery Container CTU-2/A
FM 10-553	Airdrop of Supplies and Equipment: Rigging Ammunition
FM 12-3-4	Echelons Above Corps (EAC) Personnel and Administrative Doctrine

FM 14-6	Comptroller/Finance Services in Theaters of Operations
FM 19-15	Civil Disturbances
FM 20-11-1	Military Diving (Volume 1)
FM 20-11-2	Military Diving (Volume 2)
FM 21-20	Physical Training Program
FM 21-26	Map Reading
FM 21-75	Combat Skills of the Soldier
FM 21-76	Survival, Evasion, and Escape
FM 21-77A(S)	Joint Worldwide Evasion and Escape Manual
FM 22-9	Soldier Performance in Continuous Operations
FM 23-67	Machinegun, 7.62-mm, M60
FM 25-1	Training
FM 25-2	Unit Training Management
FM 25-3	Training in Units
FM 25-4	How to Conduct Training Exercises
FM 25-5	Training for Mobilization and War
FM 25-7	Training Ranges
FM 26-2	Management of Stress in Army Operations
FM 29-47	Main Supply and Service Company, Forward Supply and Service Detachments, Supply and Service Battalion, Airborne Division
FM 29-51	Division Supply and Field Service Operations
FM 30-5	Combat Intelligence
FM 30-17	Counterintelligence Operations
FM 31-18	Long-Range Reconnaissance Ranger Company
FM 31-19	Special Forces Military Free-Fall Parachuting
FM 1-24	Special Forces Air Operations

FM 31-25	Special Forces Waterborne Operations
FM 31-70	Basic Cold Weather Manual
FM 34-1	Intelligence and Electronic Warfare Operations
FM 34-60A	Counterintelligence Operations (U)
FM 55-9	Unit Air Movement Planning
FM 57-38	Pathfinder Operations
FM 57-220	Basic Parachuting Techniques and Training
FM 63-3J	Combat Service Support Operations--Corps
FM 63-4	Combat Service Support Operations--Theater Army Area Command
FM 63-5	Combat Service Support Operations--Theater Army
FM 90-3	Desert Operations
FM 90-4	Airmobile Operations
FM 90-5	Jungle Operations
FM 90-6	Mountain Operations
FM 90-8	Counterguerilla Operations
FM 90-10	Military Operations on Urbanized Terrain (MOUT)
FM 90-10-1	An Infantryman's Guide to Urban Combat
FM 100-5	Operations
FM 100-20	Low Intensity Conflict
FM 100-26	The Air-Ground Operations System
FM 101-5	Staff Organization and Operations
FM 101-10-1	Staff Officers' Field Manual: Organization, Technical and Logistical Data (U)

Firing Table (FT)

FT 60-P-1	Firing Tables for Mortar, 60-mm: M224, Firing Cartridge, HE, M720, Cartridge, HE, M49A4, Cartridge, TP, M5OA3, Cartridge, WP, M302AI, Cartridge, Illuminating, M83A3

Soldier's Training Publication (STP)

STP 7-11B24-SM	Soldier's Manual MOS 11B, Infantryman, Skill Level 2/3/4

Technical Manuals (TM)

TM 10-1670-267-12&P	Operator's and Organizational Maintenance Manual Including Repair Parts and Special Tools List for High Altitude Airdrop Resupply System: 500 Pound Capacity

Training Circulars (TC)

TC 23-14	Sniper Training and Employment
TC 23-90	Mortar Training
TC 24-1	Communications-Electronics Operation Instructions, the CEOI
TC 31-20-5(C)	Special Forces Communications (U)
TC 90-6-1	Military Mountaineering

Other Related Publications

King, Michael J. Dr., *Leavenworth Papers, No. 11*, "Rangers Selected Combat Operations in World War II," Combat Studies Institute, USACGSC, Fort Leavenworth, KS

PROJECTED PUBLICATIONS

Projected publications are sources of additional information that are scheduled for printing but are not yet available. Upon print, they will be distributed automatically via pinpoint distribution. They may not be obtained from the USA AG Publications Center until indexed in DA Pamphlet 310-1.

Field Manuals (FM)

FM 7-70	Light Infantry Squad/Platoon
FM 7-71	Light Infantry Company
FM 7-72	Light Infantry Battalion
FM 7-73	Light Infantry Brigade

FM 23-36 60-mm Mortar, M224

TECHNICAL MANUAL
ARMY NO. 9-1005-319-23&P
AIR FORCE TO 11W3-5-5-42

***ARMY TM 9-1005-319-23&P**
AIR FORCE TO 11W3-5-5-42

DEPARTMENTS OF THE ARMY
AND AIR FORCE
Washington, DC, 1 May 1991

Technical Manual for 5.56MM,

M16A2, M4 and M4A1 Carbine

DISTRIBUTION STATEMENT. APPROVED FOR PUBLIC RELEASE; DISTRIBUTION IS UNLIMITED.

REPORTING ERRORS AND RECOMMENDING IMPROVEMENTS

You can help improve this manual. If you find any mistakes or if you know of a way to improve the procedures please let us know.

Army users mail your letter, DA Form 2028 (recommended Changes to Publications and Blank Forms), or DA Form 2028-2 located in the back of this manual direct to: Director, Armament and Chemical Acquisition and Logistics Activity, ATTN: AMSTA-AC-NML, Rock Island, IL 61299-7630.

Air Force users submit AFTO Form 22, Technical Order System Publication Improvement Report and Reply, to: WR-ALC/MMDET, Robins AFB, GA 31098-5609.

A reply will be furnished to you.

			Page
		HOW TO USE THIS MANUAL	iii
CHAPTER	**1**	**INTRODUCTION**	1-1
		External View of 5.56mm Rifle, M16A2	1-0
		External View of 5.56mm Carbine, M4	1-0.1
		External View of 5.56mm Carbine, M4A1	1-0.1
		Chapter Overview	1-1
Section	I	General Information	1-1
Section	II	Equipment Description and Data	1-3
Section	III	Principles of Operation	1-6

*This manual supersedes ARMY TM 9-1005-319-23&P dated 28 August 1987, including all changes.

			Page	Illus Figure
CHAPTER	**2**	**UNIT MAINTENANCE INSTRUCTIONS**	2-1	
		Chapter Overview	2-1	
Section	I	Repair Parts, Special Tools, TMDE, and Support Equipment	2-1	
Section	II	Service Upon Receipt	2-1	
Section	III	Preventive Maintenance Checks and Services (PMCS)	2-3	
Section	IV	Troubleshooting	2-20	
Section	V	Maintenance Procedures	2-33	
Section	VI	Preparation for Storage or Shipment	2-70	
CHAPTER	**3**	**DIRECT SUPPORT MAINTENANCE INSTRUCTIONS**	3-1	
		Chapter Overview	3-1	
Section	I	Repair Parts, Special Tools, TMDE, and Support Equipment	3-1	
Section	II	Troubleshooting	3-2	
Section	III	Maintenance Procedures for the M16A2 Rifle, M4 and M4A1 Carbine	3-15	
Section	IV	Preembarkation Inspection of Materiel in Units Alerted for Overseas Movement (A.F. Only)	3-95	
CHAPTER	**4**	**MAINTENANCE OF AUXILIARY EQUIPMENT**	4-1	
		Chapter Overview	4-1	
Section	I	Auxiliary Equipment Repair	4-1	
APPENDIX	**A.**	**REFERENCES**	A-1	
APPENDIX	**B.**	**MAINTENANCE ALLOCATION CHART**	B-1	
APPENDIX	**C.**	**REPAIR PARTS AND SPECIAL TOOLS LIST**	C-1	
Section	I	Introduction	C-1	
Section	II	Repair Parts List	C-1-1	
Group	00	5.56MM Rifle, M16A2, Carbine, M4 and M4A1	C-1-1	C-1
	00.1	M4 and M4A1 Carrying Handle Assembly	C-1A-1	C-1A
	00.101	M4 and M4A1 Rear Sight Assembly	C-1B-1	C-1B
Group	01	Bolt carrier assembly	C-2-1	C-2
	0101	Bolt assembly	C-3-1	C-3
	0102	Key and bolt carrier assembly	C-4-1	C-4
Group	02	Handle assembly	C-5-1	C-5
Group	03	Upper receiver and barrel assembly	C-6-1	C-6
	0301	Barrel assembly, M16A2	C-7-1	C-7
	0301	Replacement Barrel & Front Sight Assembly, M4/M4A1	C-7-1	C-7
	0302	Upper receiver assembly	C-8-1	C-8
	030201	Forward assist assembly	C-9-1	C-9
	030202	Rear sight assembly, M16A2	C-10-1	C-10
Group	04	Lower receiver and buttstock assembly	C-11-1	C-11
	0401	Buttstock assembly, M16A2	C-12-1	C-12
	0401A	Buttstock assembly, M4/M4A1	C12A-1	C12A
	0402	Hammer assembly	C-13-1	C-13
	0403	Trigger assembly	C-14-1	C-14
	040301	Trigger subassembly,M16A2 and M4	C-15-1	C-15
	0404	Lower receiver and receiver extension assembly, M16A2	C-16-1	C-16
	0404A	Lower receiver and receiver extension assembly, M4 and M4A1	C16A-1	C16A

			Page	Illus Figure
APPENDIX	**C.**	**REPAIR PARTS AND SPECIAL TOOLS LIST--Cont.**		
Section	III	Special Tools List	C-17-1	C-17
Section	IV	National Stock Number and Part Number Index	I-1	
APPENDIX	**D.**	**EXPENDABLE/DURABLE SUPPLIES AND MATERIALS LIST**	D-1	
APPENDIX	**E.**	**ILLUSTRATED LIST OF MANUFACTURED ITEMS**	E-1	
		ALPHABETICAL INDEX	Index-1	

WARNING

ALL WARNINGS in this technical manual pertain to both the rifle and the carbines unless otherwise specified.

Before starting an inspection, be sure to clear the rifle. Do not pull the trigger until the rifle has been cleared. Inspect the chamber to ensure that it is empty and no ammunition is in position to be chambered.

Do not keep live ammunition near work area.

To avoid injury to your eyes, use care when removing and installing spring-loaded parts.

All M1 6A2 rifles and M4 carbines must be inspected and gaged at least once annually for safety and serviceability. Initial gaging is required 1 year from receipt of the weapons. Air Force users refer to inspection requirements in Air Force Manual (AFM) 36-2227, Volume 1.

All Army Reserve and Army National Guard M16A2 rifles and M4 carbines must be inspected and gaged at least once every 2 years, after the initial inspection/gaging procedures have been accomplished. This initial gaging procedure is required 1 year from receipt of the weapons. This 2 year interval may be maintained unless preventive maintenance checks and services (PMCS) or other physical evidence indicates that an individual unit's M16A2 rifles and M4 carbines require inspection/ gaging at a more frequent interval. If it is determined that a yearly inspection is necessary for an individual unit, only that unit will be affected. This will not affect other units in regard to the interval of inspection.

It is recommended that training units inspect/gage all rifles and carbines at the end of each training cycle. Training units will inspect/gage all rifles and carbines at least once annually.

Below direct support maintenance, **DO NOT** interchange bolt assemblies from one rifle/carbine to another. Doing so may result in injury to, or death of, personnel.

Bolt cam pin must be installed or rifle/carbine will blow up while firing the first round. If the bolt cam pin is not installed, injury to or death of, personnel may result.

Dry cleaning solvent is flammable and toxic and should be used in a well ventilated area. The use of rubber gloves is necessary to protect the skin when washing rifle parts.

When using solid film lubricant or dichloromethane, be sure the area is well ventilated.

When using carbon removing compound (item 8, app D), avoid skin contact. If carbon removing compound comes In contact with the skin, wash thoroughly with running water. using a good lanolin base cream after exposure to the compound is helpful. Using gloves and protective equipment is required.

The lock plate prevents the selector lever from being placed in BURST and will be installed at the discretion of the unit commander. It is mandatory for use in civil disturbance (riot control).

Only blank cartridge M200 is to be used when the blank firing attachment is attached to the rifle/carbine.

WARNING (CONT)

Do not fire blank ammunition at a representative enemy at distances of less than 20 feet (6.10 Om). The unburned propellant grains can cause injury within this distance.

For further information on safety, care, and handling of ammunition: Army and Air Force users refer to M16A2 Rifle Operator's Manual.

For additional first aid data, see Field Manual (FM) 21-11.

HOW TO USE THIS MANUAL

Read this manual carefully before performing required maintenance. This manual will be referred to for Inspection/Maintenance and Repair procedures.

GENERAL

There are several things you need to know to use this manual efficiently.

1. All references in the manual are to pages only. Reference to maintenance procedures is to the page where the respective initial setup appears.

2. Illustrations for the maintenance procedures show only those parts affected by the operation being performed.

3. Whenever the male gender is mentioned in the manual (i.e., crewman, repairman), it also pertains to females.

4. When the term "evacuate to support maintenance" is used, the entire rifle must be evacuated.

5. When a procedure is common to M16A2 rifle, and M4/M4A1 carbine, **ONLY** the M16A2 configuration will be depicted. If a procedure is not common to both weapons, the procedure will be incorporated.

6. When the word rifle is referenced in text, it will reference the rifle and the carbines.

INDEXES

This manual is organized to help you find the information you need quickly. There are several useful indexes.

1. **Table of Contents**. Lists in order all chapters, sections, and appendixes. Gives page references.

2. **Nomenclature Cross-References List.**

3. **Chapter Overviews**. Summarize material covered in the chapter. Are located at the beginning of each chapter.

4. **Symptom Index**. Located just before the troubleshooting table in each maintenance chapter. Lists, in alphabetical order, parts of the rifle with possible malfunctions. References pages of the troubleshooting table.

5. **Alphabetical Index**. Located at the end of the manual. An extensive subject index for everything in the manual. Gives page references.

MAINTENANCE PROCEDURES

There are two maintenance chapters:

Army personnel use chapter two for unit maintenance procedures and chapter three for direct support maintenance procedures.

Air Force personnel: Only Air Force Specialty Code 753XX Combat Arms Training and Maintenance (CATM) specialists, technicians, and gunsmiths are authorized to perform maintenance procedures contained in this manual.

Each maintenance task has an initial setup containing a list of the following things you will need in order to do your maintenance task.

1. **Tools and Special Tools**. For standard and special tools, see appendixes B and C. Army users are to use the Tool Set, Gage Set, and/or Shop Set listed in the initial setup.

2. **Materials/Parts**. Lists expendable materials and 100 percent replaceable parts. Each material or part is followed by a part number or appendix reference.

3. **References**. Lists other publications containing necessary information.

4. **Equipment Condition**. Lists conditions to be met before starting the procedure. The reference on the left of the condition is a page reference to instructions for setting up the condition.

5. **General Safety Instructions**. Lists safety instructions to follow before performing maintenance procedures.

EXTERNAL VIEW OF 5.56 MM RIFLE M16A2

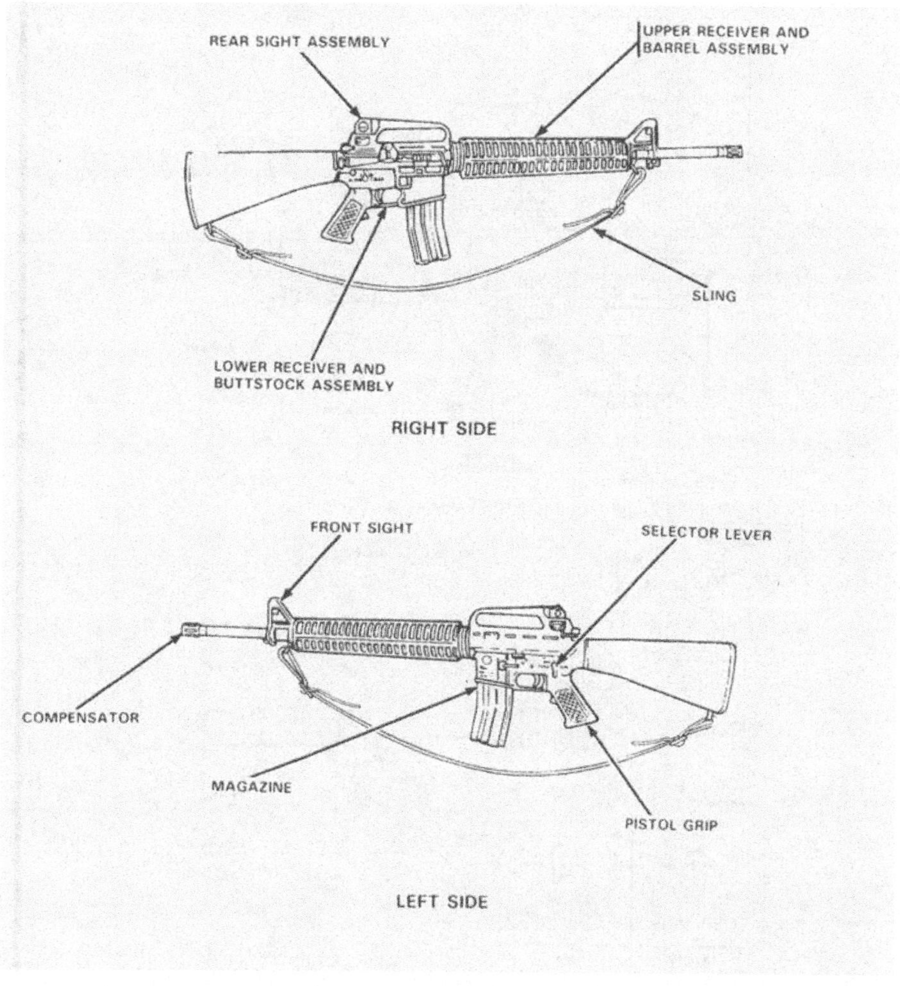

EXTERNAL VIEW OF 5.56MM CARBINE, M4/M4A1

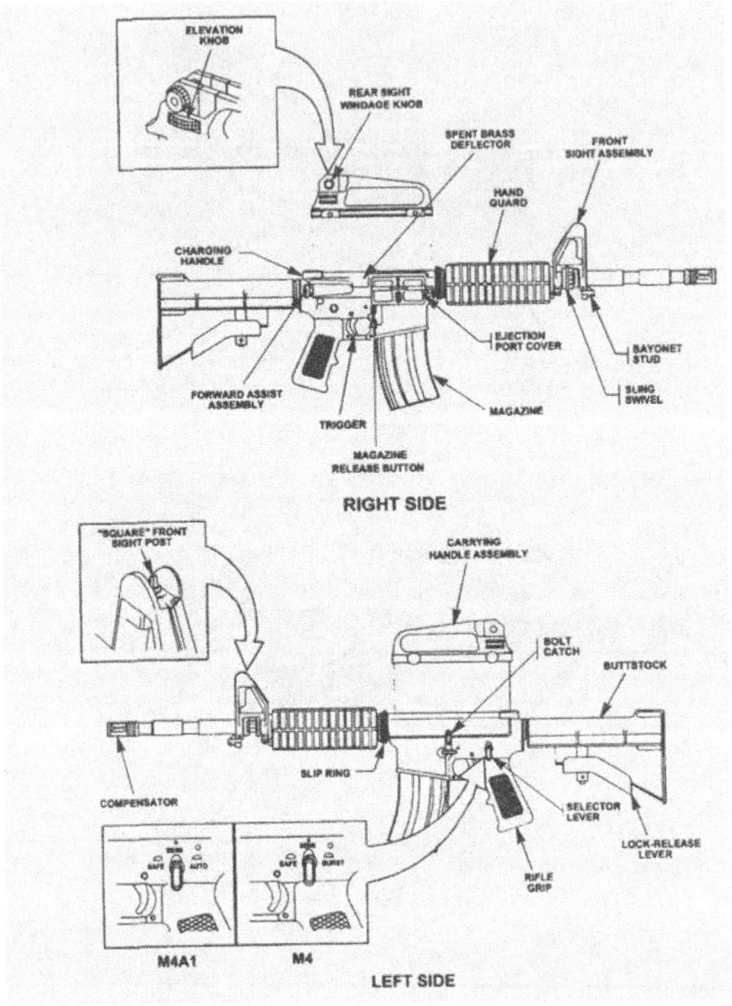

CHAPTER 1
INTRODUCTION

CHAPTER OVERVIEW

This chapter contains general information, equipment description and data, and principles of operation for the M16A2 rifle and M4/M4A1 carbines.

Section I. GENERAL INFORMATION

1-1. SCOPE.

a. Type of Manual: Unit and Direct Support Maintenance.

b. Model Number and Equipment Name: 5.56mm Rifle M16A2, M4 and M4A1 Carbines.

c. Purpose of Equipment. Provides personnel an offensive/defensive capability to engage targets with small arms fire.

1-2. MAINTENANCE FORMS, RECORDS, AND REPORTS. Department of the Army forms and procedures used for equipment maintenance will be those prescribed by DA PAM 738-750, The Army Maintenance Management System.

Air Force users refer to TO 11W-1-10 for applicable forms and records.

1-3. DESTRUCTION OF ARMY MATERIEL TO PREVENT ENEMY USE. See TM 750-244-7.

1-4. PREPARATION FOR STORAGE OR SHIPMENT. Refer to page 2-70.

Air Force users refer to Special Package Instruction (SPI) 00-856-6885.

1-5. OFFICIAL NOMENCLATURE, NAMES AND DESIGNATIONS.

NOMENCLATURE CROSS-REFERENCE LIST

Common Name	Official Nomenclature
Action Spring	Compression Helical Spring
Ball Bearing	Bearing Ball
Bolt Catch Spring	Compression Helical Spring
Bolt Carrier Key Tool	Machine Key
Burst Disconnector	Lock-Release Lever
Cam Clutch Spring	Helical Spring
Carbine	M4/M4A1 Carbine
Charging Handle Assembly	Handle Assembly

1-5. OFFICIAL NOMENCLATURE, NAMES AND DESIGNATIONS (CONT).

NOMENCLATURE CROSS-REFERENCE LIST

Common Name	Official Nomenclature
Disconnector Springs	Compression Helical Spring
Ejector Spring	Helical Spring
Extractor Spring Assembly	Spring Assembly
Hammer Spring	Torsion Helical Spring
Lower Receiver Extension	Spring Receiver Holder
Magazine	Cartridge Magazine
Magazine Catch Spring	Compression Helical Spring
Peel Washer	Shim
Pistol Grip	Rifle Grip
Pivot Pin Detent	Takedown Pin Detent
Rifle	Rifle, 5.56mm, M16A2
Rifle Barrel Assembly	Barrel Assembly
Selector Lever	Fire Control Selector
Semiautomatic Disconnector	Lock-Release Lever
Sling	Small Arms Sling
Trigger Spring	Torsion Helical Spring
Upper Receiver	Upper Cartridge Receiver

1-6. REPORTING EQUIPMENT IMPROVEMENT RECOMMENDATIONS (EIR). If your M16A2 rifle needs improvement, let us know. Send us an EIR. You, the user, are the only one who can tell us what you don't like about your equipment. Let us know why you don't like the design.

Army users submit SF 368 (Product Quality Deficiency Report) to: Commander, U.S. Army Armament Research, Development and Engineering Center, ATTN: AMSTA-AR-QAW (R)/Customer Feedback Center, Rock Island, IL 61299-7300.

Air Force users submit Materiel Deficiency Report (MDR) to: DIR MAT MGT ROBINS AFB GA//MMIBTC// and Product Quality Deficiency Report to: DIR MAT MGT ROBINS AFB GA//MMQA// IAW Technical Order 00-35D-54.

A reply will be sent to you.

1-7. CORROSION PREVENTION AND CONTROL (CPC). CPC of Army materiel is a continuing concern. It is important that any corrosion problems with this item be reported so that the problem can be corrected and improvements can be made to prevent the problem in future items,

While corrosion is typically associated with rusting of metals, it can also include deterioration of other materials such as rubber and plastic. Unusual cracking, softening, swelling, or breaking of these materials may be a corrosion problem.

If a corrosion problem is identified, it can be reported using Standard Form 368, Product Quality Deficiency Report. Use of key words such as "corrosion", "rust", " deterioration", or "cracking" will assure that the information is identified as a CPC problem.

Army users submit Product Quality Deficiency Report (SF 368) to:

Commander
U.S. Army Armament Research, Development and Engineering Center
ATTN: AMSTA-AR-QAW (R)
Rock Island, IL 61299-7300

Air Force users submit Materiel Deficiency Report (MDR) to:

DIR MAT MGT
ATTN: MMIBTC
Robins AFB, GA

and Product Quality Deficiency Report to:

DIR MAT MGT
ATTN: MMQA
Robins AFB, GA

Section II. EQUIPMENT DESCRIPTION AND DATA

1-8. EQUIPMENT CHARACTERISTICS, CAPABILITIES, AND FEATURES.

a. Characteristics.

(1) Light weight
(2) Air-cooled
(3) Gas-operated
(4) Magazine-fed
(5) Semiautomatic or burst fire

b. Capabilities. Provides personnel an offensive/defensive capability to engage targets with direct small arms fire.

c. Features.

(1) Receivers are made of light-weight aluminum alloys; however, the safety , durability, and function of the rifles are in no way reduced. The portability and logistical values are greatly increased, particularly when air transport is used.

(2) The bolt locking action is one of the mechanical features of the rifle. The bolt assembly and barrel extension contain locking lugs which engage and lock the bolt assembly firmly in the barrel extension. The initial force of the explosion of the cartridge is absorbed by the barrel, barrel extension, and bolt assembly.

1-8. EQUIPMENT CHARACTERISTICS, CAPABILITIES, AND FEATURES (CONT).

(3) The trigger guard is easily adaptable to winter operations. A spring-loaded retaining pin is depressed to allow ready access to the trigger when wearing arctic mittens.

(4) The ejection port cover prevents dirt or sand from getting into the ejection port. The ejection port cover must be closed during periods when firing is not anticipated. It opens automatically by the forward or rearward movement of the bolt carrier assembly.

1-9. LOCATION AND DESCRIPTION OF MAJOR COMPONENTS.

(A) **MAGAZINE**. 30 cartridge capacity.

(B) **SLING**. The sling is adjustable and provides a means to carry the weapon.

(C) **BOLT CARRIER ASSEMBLY**. Carries bolt assembly to chamber and fires the weapon. Contains the firing pin, cartridge extractor, bolt assembly, cartridge ejector, and bolt cam pin.

(D) **CHARGING HANDLE ASSEMBLY**. Provides a means of charging the weapon.

(E) **M16A2 UPPER RECEIVER AND BARREL ASSEMBLY**. Upper receiver contains rear sight assembly, ejection port, ejection port cover, and a housing for the key and bolt carrier assembly and bolt assembly. Rifle barrel assembly is air-cooled, contains compensator and front sight assembly, and holds the two handguard assemblies and the sling swivel.

(F) **LOWER RECEIVER AND BUTTSTOCK ASSEMBLY**. Lower receiver contains the trigger assembly, sear, hammer assembly, selector lever, rifle grip, bolt catch, and buttstock assembly. The buttstock assembly houses the action spring, buffer assembly, and extension assembly.

(G) **M4/M4A1 CARRYING HANDLE**. Provides a means of carrying carbine.

(H) **M4/M4A1 UPPER RECEIVER AND BARREL ASSEMBLY**. Upper receiver contains, ejection port, ejection port cover, a housing for key and bolt carrier assembly and bolt assembly, and mounting surface for the carrying handle assembly. Carbine barrel assembly is air-cooled, contains compensator and front sight assembly, and holds the two handguard assemblies and the sling swivel.

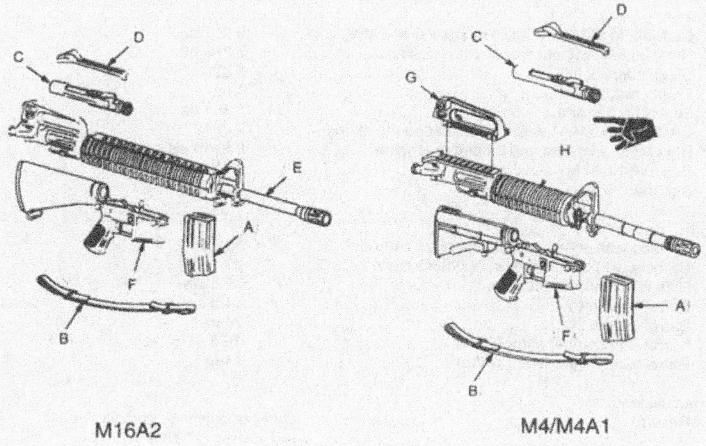
D
C
E
A
F
B
M16A2
D
C
G
H
A
F
B
M4/M4A1

1-10. EQUIPMENT DATA.

	US CUSTOMARY	METRIC
Weight:		
Carbine, M4/M4A1 without magazine and sling	6 lb 7 oz	2.91 kg
Rifle, MI 6A2 without magazine and sling	7 lb 8 oz	3.40 kg
Sling, adjustable	4 oz	0.11 kg
Empty magazine	4 oz	0.11 kg
Loaded magazine	1 lb 1 oz	0.48 kg
Carbine, M4/M4A1 w/sling and loaded magazine	7 lb 12 oz	3.51 kg
Rifle M16A2 w/sling and loaded magazine	8 lb 13 oz	4.00 kg
Bayonet-Knife M7	10.5 oz	0.30 kg
Scabbard M10 o	5 oz	0.14 kg
Length:		
Carbine with compensator, buttstock extended	33.0 in	83.82 cm
Carbine with compensator, buttstock closed	29.75 in	75.57 cm
Rifle with compensator	39.63 in	100.66 cm
Barrel (Carbine)	14.5 in	36.83 cm
Barrel (Rifle)	20 in	50.8 cm
Barrel with compensator (Carbine)	15.5 in	39.37 cm
Barrel with compensator (Rifle)	21 in	53.34 cm
Mechanical features:		
Rifling	right-hand twist 6 grooves, 1 turn in 7 inches (17.78 cm)	
Method of operation	direct gas	
Type of breech mechanism	rotating bolt	
Method of feeding	magazine	
Cooling	air	
Trigger pull (M16A2 & M4)	5.5 to 9.5 lb	2.49 to 4.31kg
Trigger pull (M4A1)	5.5 to 8.5 lb	2.49 to 3.86 kg
Ammunition:		
Caliber	223	5.56mm
Type	ball, blank, dummy, and tracer	
Firing characteristics:		
Muzzle velocity (Carbine) (approximate)	2,970 fps	905.85 mps
Muzzle velocity (Rifle) (approximate)	3,100 fps	945.5 mps
Chamber pressure	52,000 psi	358,540 kPa
Cyclic rate of fire (Carbine) (approximate)	700-970 rds/m	
Cyclic rate of fire (Rifle) (approximate)	700-900 rds/m	

1-10. EQUIPMENT DATA (CONT).

	US CUSTOMARY	METRIC
Maximum rate of fire:		
Semiautomatic	45 rds/m	
Burst	90 rds/m	
Sustained rate of fire	12/15 rds/m	
Maximum range	3,938 yards	Approximately 3,600 meters
Maximum effective range:		
Individual/point targets (Carbine)	547 yards	500 meters
Individual/point targets (Rifle)	602 yards	550 meters
Area targets (Carbine)	650 yards	600 meters
Area targets (Rifle)	875 yards	800 meters

Section III. PRINCIPLES OF OPERATION

1-11. GENERAL. The 5.56mm M16A2 and M4/M4A1 carbine:

a. Is gas-operated. It fires in either the semiautomatic or burst mode.

b. Has positive locking of the bolt. Firing pin is part of the bolt carrier assembly and cannot strike the primer until the bolt assembly is fully locked.

Section III. PRINCIPLES OF OPERATION (CONT).

1-12. PRINCIPLES OF OPERATION.

(A) **MAGAZINE**. Holds cartridges ready for feeding and provides a guide for positioning cartridges for stripping. Provides quick reload capabilities for sustained firing.

(B) **SLING**. Provides the means for carrying the weapon.

(C) **BOLT CARRIER ASSEMBLY**. Provides stripping, chambering, locking, firing, extraction, and ejection of cartridges using the drive springs and projectile propelling gases for power.

(D) **CHARGING HANDLE ASSEMBLY**. Provides initial charging of the weapon. The handle latch locks the charging handle assembly in the forward position during sustained fire to prevent injury to the operator.

(E) **M16A2 UPPER RECEIVER AND BARREL ASSEMBLY**. Provides support for the bolt carrier assembly. The barrel chambers the cartridge for firing and directs the projectile.

(F) **LOWER RECEIVER AND BUTTSTOCK ASSEMBLY**. Provides firing control for the rifle and carbine. M16A2 ONLY provides storage for basic cleaning materials.

(G) **M4/M4A1 CARRYING HANDLE ASSEMBLY**. Provides a means of carrying the carbine, contains rear sight assembly, and can be removed for mounting various optics.

(H) **M4/M4A1 UPPER RECEIVER AND BARREL ASSEMBLY**. Provides support for the bolt carrier assembly. The barrel chambers the cartridge for firing and directs the projectile.

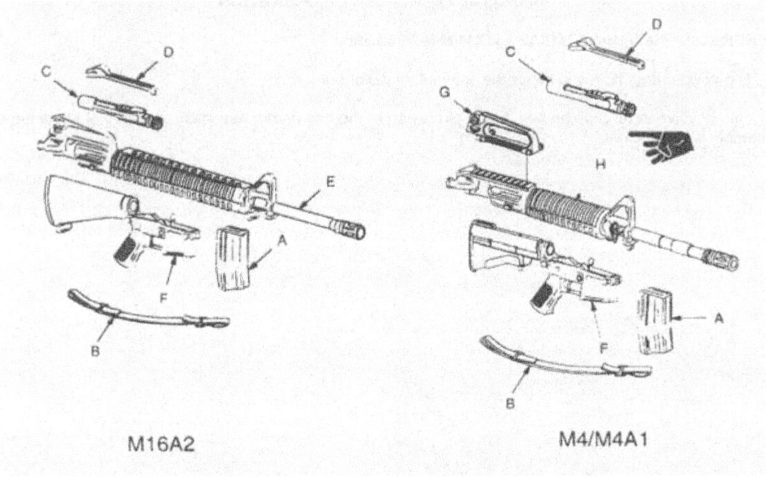

CHAPTER 2
UNIT MAINTENANCE INSTRUCTIONS

CHAPTER OVERVIEW

This chapter provides information and Instructions to help keep the rifle in good repair and contains the following sections:

a. Repair Parts, Special Tools, TMDE, and Support Equipment
b. Service Upon Receipt
c. Preventive Maintenance Checks and Services (PMCS)
d. Troubleshooting
e. Maintenance Procedures

Section I. REPAIR PARTS, SPECIAL TOOLS, TMDE, AND SUPPORT EQUIPMENT

2-1. COMMON TOOLS AND EQUIPMENT. For authorized common tools and equipment, refer to the Modified Table of Organization and Equipment (MTOE) applicable to your unit.

Air Force users must maintain the following common tools:

3-ounce soft-brass hammer
Vise
Flat tip screwdriver
Punch
Tweezers/round nose pliers
Hammer
Needle nose pliers

2-2. SPECIAL TOOLS, TMDE, AND SUPPORT EQUIPMENT. Special tools required for unit support are listed in appendixes B and C. Fabricated tools are listed and illustrated in appendix E.

2-3. REPAIR PARTS. Repair parts are listed and illustrated in appendix C of this manual.

Section II. SERVICE UPON RECEIPT

2-4. GENERAL.

a. Inspect the rifle for damage Incurred during shipment. If rifle has been damaged, report the damage on SF 364, Report of Discrepancy (ROD).

b. Check the rifle against the packing slip to see If shipment is complete. Army users report all discrepancies in accordance with DA PAM 738-750.

Air Force users submit Materiel Deficiency Report (MDR) to: DIR MAT MGT ROBINS AFB GA//MMIBTC// and Product Quality Deficiency Report to: DIR MAT MGT ROBINS AFB GA//MMQA//. IAW Technical Order 00-35D-54.

c. Check to see whether the equipment has been modified.

2-5. SERVICE UPON RECEIPT OF MATERIEL.

WARNING

Before starting an inspection, be sure to clear the rifle. Do not actuate the trigger before clearing the rifle. Inspect the chamber to make sure it is empty and free of obstructions. Check to see there are no obstructions in the barrel and no ammunition is in position to be chambered.

SERVICE UPON RECEIPT

LOCATION	ITEM	ACTION	REMARKS
1. Container	a. M16A2 rifle or M4/M4A1 carbine	a. Remove rifle from containers.	
		b. Inspect the equipment for damage incurred during shipment	If the equipment has been damaged, report the damage on SF Form 364, Report of Discrepancy (ROD).
		c. Check the equipment against the packing list to see if the shipment is complete	Report all discrepancies in accordance with the instructions of DA PAM 738-750.
	b. Basic issue items	Check for missing items	Refer to TM 9-1005-319-10 (operator's manual).
2. M16A2 rifle or M4/M4A1 carbine	a. Barrel assembly	If volatile corrosion inhibitor (VCI) is in barrel, remove and discard.	
	b. All parts	a. Field-strip rifle and inspect for missing, damaged, and rusted or corroded parts.	Refer to operator's manual.
		b. Clean and lubricate	Refer to operator's manual.
		c. Reassemble	Refer to operator's manual.
		d. Function check	Refer to page 2-69.

SERVICE UPON RECEIPT (CONT)

LOCATION	ITEM	ACTION	REMARKS
		e. Check to see whether the equipment has been modified	Refer to DA PAM 25-30.
	c. Magazine	Check for positive retention and functioning of bolt catch	Refer to operator's manual.

Section III. PREVENTIVE MAINTENANCE CHECKS AND SERVICES (PMCS)

2-6. GENERAL This section contains the procedures and instructions necessary to perform unit preventive maintenance checks and services. These services are to be performed by unit maintenance personnel with the assistance of the operator where practical.

2-7. PREVENTIVE MAINTENANCE CHECKS AND SERVICES

WARNING

Before starting an inspection, be sure to clear the rifle. Do not keep live ammunition near the work area.

a. General. The PMCS procedures are contained in the table following. They are arranged in logical sequence requiring a minimum amount of time and motion on the part of the persons performing them and are arranged so that there will be minimum interference between persons performing checks simultaneously on the same end item.

b. Item No. Column. Checks and services are numbered in disassembly sequence. This column shall be used as a source of item numbers for the "TM Number" column on DA Form 2404, Equipment Inspection and Maintenance Worksheet, in recording results of PMCS.

c. Interval Column. This column gives the designated interval when each check is to be performed.

d. Item To Be Checked Or Serviced Column. This column lists the items to be checked or serviced.

e. Procedure Column. This column contains a brief description of the procedure by which the check is to be performed. It contains all the information required to accomplish the checks and services. Information marked SH Indicates a specific equipment shortcoming and the procedure needed to correct the shortcoming.

NOTE

For the purpose of this technical manual, the following definition is supplied. This definition is not intended to apply to any other document

Shortcoming (SH): A fault that requires maintenance or supply action on a piece of equipment, but does not render equipment Not Mission Capable

f. Not Fully Mission Capable If: Column. This column contains a brief statement of the condition (e.g., malfunction, shortage) that would cause the covered equipment to be less than fully ready to perform its assigned mission.

2-7. PREVENTION MAINTENANCE CHECKS AND SERVICES (CONT).

PREVENTIVE MAINTENANCE CHECKS AND SERVICES FOR M16A2 RIFLE (CONT)

Item No.	Interval	Item to be Checked or Serviced	Procedure	Not Fully Mission Capable if:

WARNING

Before starting an inspection, be sure to clear the weapon. Do not pull the trigger until the weapon has been cleared. Inspect the chamber to ensure that it is empty and no ammunition is in position to be chambered. Do not keep live ammunition near work area.

NOTE

An inactive weapon is a weapon which has been stored in an arms room for a period of 90 days without use. The weapon may or may not have been assigned to an individual.

Inactive weapons shall receive quarterly PMCS unless inspection reveals more frequent servicing is necessary.

Normal cleaning (PMCS) of an inactive weapon will be performed every 90 days. Should the unit armorer detect corrosion on a weapon prior to the end of the 90-day period, the PMCS should be performed immediately.

Solid Film Lubricant (SFL) is the authorized touch up for the M16A2 Rifle and M4/M4A1 Carbine and may be used on up to one third of the exterior finish of the weapon.

FOR ARMY CONUS USE ONLY AND AIR FORCE TRAINING WEAPONS ONLY: Solid Film Lubricant may be used as a touch up without limitation on the upper receiver and barrel assembly. This is to say that units which **DO NOT** fall under the category of Divisional Combat Units or rapid deployment type units may have up to 100 percent of the exterior surface of the upper receiver and barrel assembly protected with SFL.. Prior to application of SFL, the surface must be thoroughly cleaned and inspected for corrosion and/or damage. If corroded or damaged, the part must be repaired or replaced prior to application of SFL. Continued use under combat conditions would result in an unprotected surface when the SFL wears off. This would result in a large light reflecting surface and accelerated deterioration of the unprotected surface. Therefore, Divisional Combat Units and units which fall under the definition of Rapid Deployment type must adhere to the limitation of **NOT** over one third of their exterior surface covered by SFL.

When determining mission capability, deadline if it is a deficiency.

PREVENTIVE MAINTENANCE CHECKS AND SERVICES FOR M16A2 RIFLE (CONT)

Item No.	Interval	Item to be Checked or Serviced	Procedure	Not Fully Mission Capable if:
1	Quarterly	Magazine (serviceability check)	Disassemble as In TM 9-1005-319-10 (operator's manual). Inspect tube (1) for bulges, dents, or damaged feeder lips (2). Inspect spring (3) and follower (4) for kinks or damage. SH-Replace the magazine if any of these conditions exist. Reassemble magazine and check for binding during operation of follower (4). SH-Replace the magazine if the follower binds. 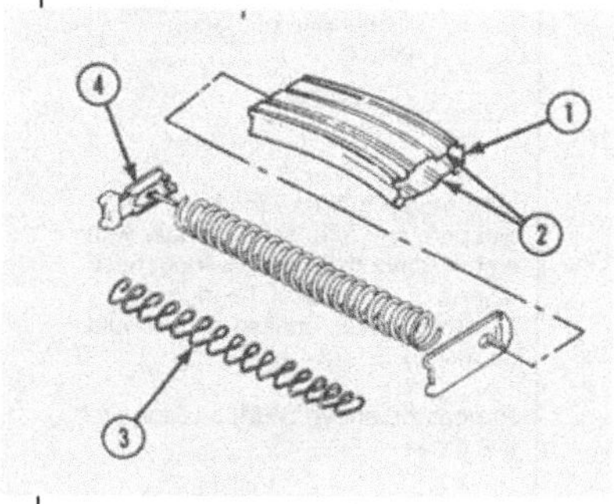	A magazine is not available fu; use with the rifle.
2	Quarterly	Charging handle assembly and selector lever	**WARNING** If the rifle falls any of the following selector lever tests, evacuate It to support maintenance. Continued use of the rifle could result In injury to, or death of, personnel.	

2-7. PREVENTIVE MAINTENANCE CHECKS AND SERVICES (CONT).

PREVENTIVE MAINTENANCE CHECKS AND SERVICES FOR M16A2 RIFLE (CONT)

Item No.	Interval	Item to be Checked or Serviced	Procedure	Not Fully Mission Capable if:
2	Quarterly (cont)	Charging handle assembly and selector lever (cont)	Pull charging handle (1) to rear. Check that chamber is clear. Let bolt carrier assembly (2) close. Leave hammer in cocked position. Do not pull trigger	Charging handle does not lock in place when in the forward position.
			Place selector lever (3) in SAFE position Pull trigger. Place selector lever (3) in SEMI position.	Hammer falls.
			NOTE For the purpose of the following test, "SLOW" is defined as 1/4 to 1/2 the normal rate of trigger release.	
			Pull trigger	Hammer does not fall.
			Hold trigger to the rear, charge weapon, and release the trigger with a slow, smooth motion, without hesitations or stops, until the trigger is fully forward (an audible click should be heard).	Hammer falls.
			Repeat the above SEMI position test five times	The weapon malfunctions during any of these five tests.

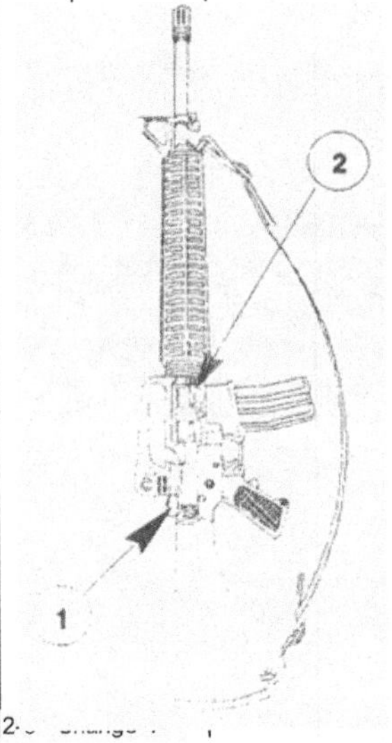

PREVENTIVE MAINTENANCE CHECKS AND SERVICES FOR M16A2 RIFLE (CONT)

Item No.	Interval	Item to be Checked or Serviced	Procedure	Not Fully Mission Capable if:
2	Quarterly (cont)	Charging handle assembly and selector lever (cont)	**M16A2 and M4 ONLY** Place selector lever (3) in BURST position. Charge weapon and squeeze trigger. Hold trigger to the rear, pull the charging handle to the rear and release it three times. Release trigger. Squeeze trigger. **NOTE** The burst disconnector should have held the hammer to the rear when it engaged the deep notch of the burst cam. **M4A1 ONLY** Place selector lever (3) in AUTO position Charge carbine and squeeze trigger Hammer should fall. Hold trigger to the rear, charge carbine, and release trigger. Squeeze trigger. Hammer should not fall.	Hammer does not fall. Hammer does not fall. Hammer does not fall. Hammer falls.

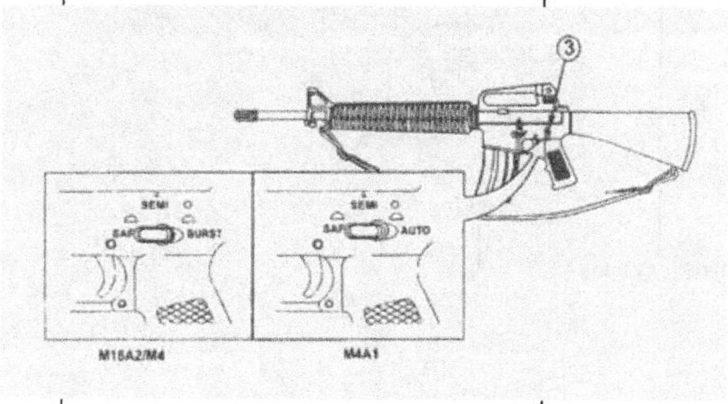

2-7. PREVENTIVE MAINTENANCE CHECKS AND SERVICES (CONT).

PREVENTIVE MAINTENANCE CHECKS AND SERVICES FOR M16A2 RIFLE (CONT)

Item No.	Interval	Item to be Checked or Serviced	Procedure	Not Fully Mission Capable if:
2	Quarterly (cont)	Charging handle assembly and selector lever (cont)	**NOTE** Automatic sear should have released hammer while holding trigger in the squeezed position before releasing and resqueezing the trigger. **All weapons** With hammer in forward position, using moderate finger/thumb pressure attempt to place the selector lever (3) in SAFE position	Moderate finger/thumb pressure moves selector lever to SAFE position.

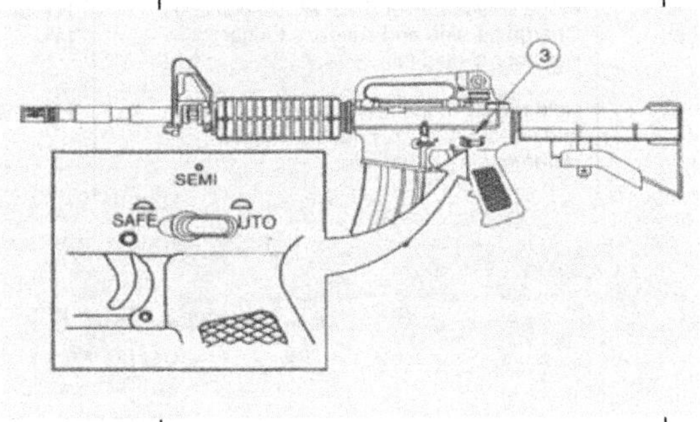

2-7. PREVENTIVE MAINTENANCE CHECKS AND SERVICES (CONT).

PREVENTIVE MAINTENANCE CHECKS AND SERVICES FOR M16A2 RIFLE (CONT)

Item No.	Interval	Item to be Checked or Serviced	Procedure	Not Fully Mission Capable if:
3	Quarterly	Upper receiver and barrel assembly (handguard assemblies) 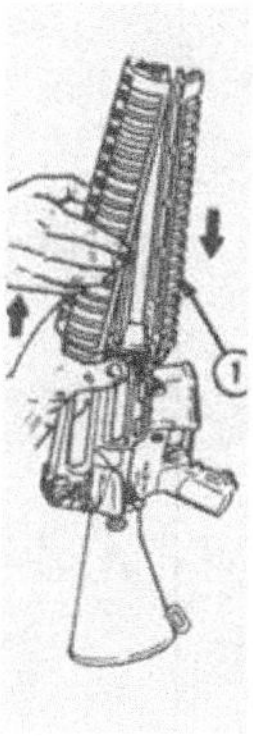	**CAUTION** Do not use screwdriver or any other tool when removing the handguard assemblies, doing so may damage the handguard assemblies and/or slip. **NOTE** Refer to operator's manual for "buddy system" procedure on removing handguard assemblies. Remove and Inspect handguard assemblies (1) internally and externally for cracks and/or damage. Cracks are acceptable providing they do not extend into the handguard retaining flange, or adversely affect rifle operation or operator safety or proper retention of handguard assembly. Discard and replace the handguard assembly (1) If the heatshield is loose enough to rattle when Installed on rifle.	Handguard missing or unserviceable.

PREVENTIVE MAINTENANCE CHECKS AND SERVICES FOR M16A2 RIFLE (CONT)

Item No.	Interval	Item to be Checked or Serviced	Procedure	Not Fully Mission Capable if:
4	Quarterly	Upper receiver and barrel assembly (serviceability check) 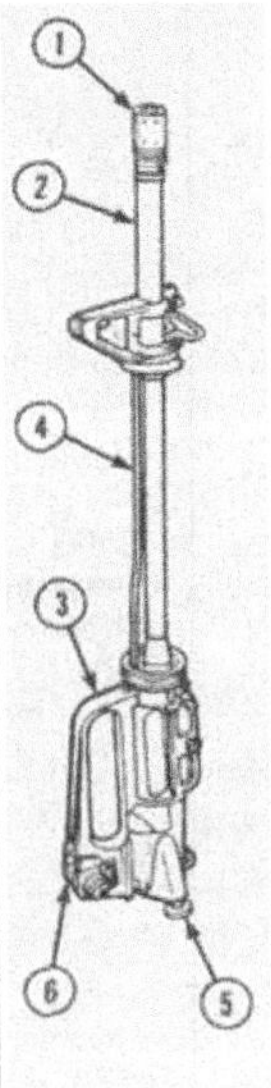	**WARNING** Dry cleaning solvent is flammable and toxic and should be used in a well ventilated area The use of rubber gloves is necessary to protect the skin when washing rifle parts **CAUTION** Damage may occur if excessive force is used to release takedown pin or pivot pin. Use hand pressure ONLY. Release takedown pins and open and separate receivers Hand check compensator (1) for looseness on barrel (2), then hand check barrel (2) for looseness on upper receiver (3). Check center slot of compensator for alignment (p 2-50). If compensator or barrel is loose, evacuate to support maintenance Check gas tube (4), forward assist assembly (5), and rear sight assembly (6) for damage, The rear sight spring should retain the rear sight assembly (6) In either position with firmness SH - If damaged, evacuate to support maintenance.	Compensator or barrel is loose

2-7. PREVENTIVE MAINTENANCE CHECKS AND SERVICES (CONT).

PREVENTIVE MAINTENANCE CHECKS AND SERVICES FOR M16A2 RIFLE (CONT)

Item No.	Interval	Item to be Checked or Serviced	Procedure	Not Fully Mission Capable if:
4	Quarterly (cont)	Upper receiver and barrel as sembly (service ability check) (cont)	**NOTE** If front or rear sight is moved, return to original position Check front sight post, de tent, and spring (7) for dam age and corrosion Clean and lubricate Check charg ing handle (8) and election port cover (9) for defects and proper function Check sling swivel (10) and rivet (11) for damage and proper function SH Other corn ponents are defective, re place as necessary.	Charging handle (8) is defective.

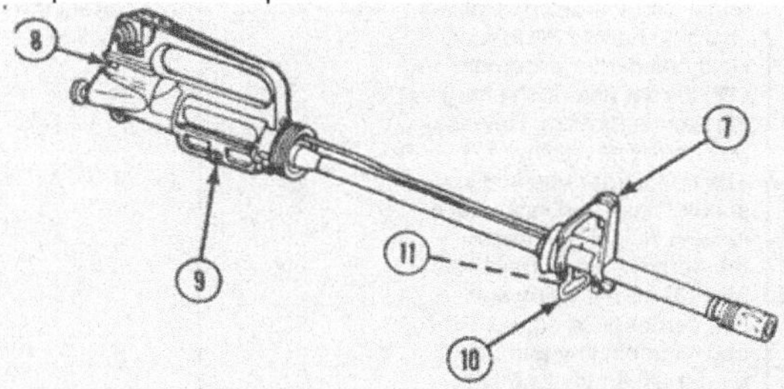

CAUTION

Do not use a wire brush to roughen surfaces Use a well ventilated area during cleaning and application of solid film lubricant. If solid film lubricant comes in contact with moving parts or functioning surfaces of the rifle, remove lubricant immediately by washing with dry cleaning solvent.

NOTE

Shiny metal exterior surfaces of the rifle should be recoated with solid film lubricant (Item 21, app D). Clean surface with dry cleaning solvent (item 16, app D); dry, roughen with abrasive cloth (Item 13, app D) and apply solid film lubricant.

PREVENTIVE MAINTENANCE CHECKS AND SERVICES FOR M16A2 RIFLE (CONT)

Item No.	Interval	Item to be Checked or Serviced	Procedure	Not Fully Mission Capable if:
4	Quarterly (cont)	Upper receiver and barrel assembly (serviceability check) (cont) 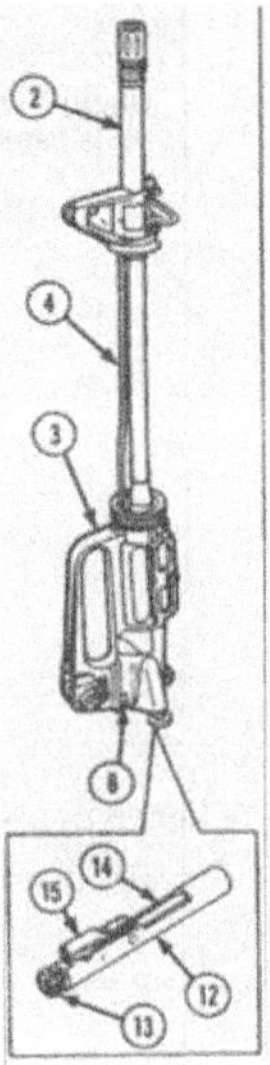	Inspect upper receiver (3) finish for scratches or worn shiny spots. If scratched or worn shiny in spots, disassemble and remove all lubricant from surface with dry cleaning solvent (item 16, app D). Wear rubber gloves (item 18, app D) and use a wash pan (item 24, app D) to apply solvent. Let parts dry thoroughly. Roughen the surface using abrasive cloth (item 1 3, app D) and apply solid film lubricant (item 16, app D). Allow 16 to 24 hours to dry before handling. Hold barrel (2) at 40-degree angle (muzzle down). Pull charging handle (8) to rear. Hold bolt carrier assembly (12) to rear and push charging handle forward. Release bolt carrier assembly (12). The bolt carrier assembly should close and lock under its own weight. If It does not, remove the bolt assembly (13) from the key and bolt carrier assembly (14) and slide the key and bolt carrier assembly (without bolt) back and forth In the upper receiver and barrel assembly. If the gas tube (4) hits the carrier key (1 5), or If the gas tube binds in the carrier key, try to correct the malfunction by adjusting (slightly bending) the gas tube In the area of the handguard assemblies. If unable to adjust, evacuate to support maintenance.	Adjustment does not correct the malfunction.

2-7. PREVENTIVE MAINTENANCE CHECKS AND SERVICES (CONT).

PREVENTIVE MAINTENANCE CHECKS AND SERVICES FOR M16A2 RIFLE (CONT)

Item No.	Interval	Item to be Checked or Serviced	Procedure	Not Fully Mission Capable if:
4	Quarterly (cont)	Upper receiver and barrel assembly (serviceability) check) (cont)	**M4/M4A1** Inspect carrying handle assembly (16) and mounting surface (17) of upper receiver for damage. If the carrying handle is missing or can not be correctly mounted, repair as authorized or evacuate to support maintenance. Inspect carrying handle assembly (16) to insure unit applied identification (ID) code matches unit applied ID code on carbine. If it doesn't match locate correct carrying handle assembly and match up to correct M4/ M4A1 carbine. If a match can not be found, the weapon should be re-zeroed by the operator.	

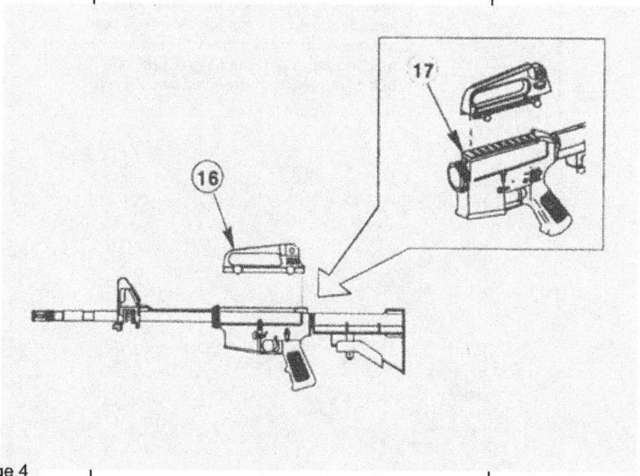

2-7. PREVENTIVE MAINTENANCE CHECKS AND SERVICES (CONT).

PREVENTIVE MAINTENANCE CHECKS AND SERVICES FOR M16A2 RIFLE (CONT)

Item No.	Interval	Item to be Checked or Serviced	Procedure	Not Fully Mission Capable if:
			WARNING Below direct support maintenance, do not interchange bolt assemblies from one rifle to another. Doing so may result in injury to, or death of, personnel.	
5	Quarterly	Key and bolt carrier assembly and bolt assembly (serviceability check)	Remove and disassemble. Visually inspect bolt assembly (1) for cracks, especially in the area of the bolt cam pin hole (2). Check for cracks on locking lugs (3), for a cluster of pits or chipped bolt face (4), and for an elongated firing pin hole (5). If cracked or broken, evacuate to support maintenance for repair.	Defects are found.
			Check for worn or missing bolt rings (6) Check for proper staggering of bolt rings Insert the bolt assembly (1) into the key and bolt carrier assembly (7). Turn key and bolt carrier assembly (7) so the bolt assembly (1) points down. The bolt assembly must not drop out. Remove bolt assembly (p 2-35). Check for broken or missing firing pin retaining pin (8) and bolt cam pin (9); replace as necessary.	The bolt assembly drops out of the key and bolt carrier assembly due to its own weight. Missing or broken firing pin retaining pin or bolt cam pin.

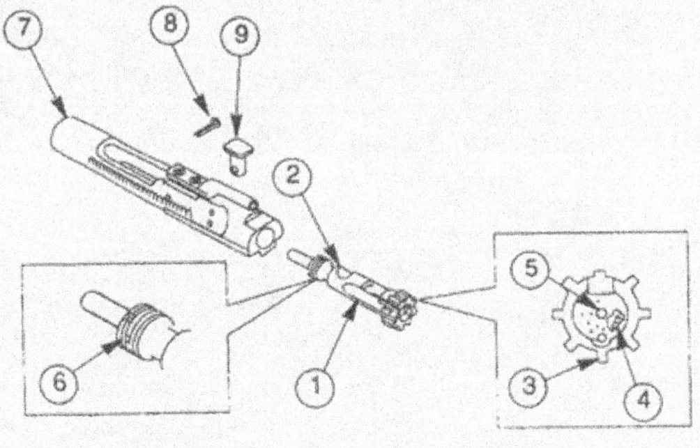

PREVENTIVE MAINTENANCE CHECKS AND SERVICES FOR M16A2 RIFLE (CONT)

Item No.	Interval	Item to be Checked or Serviced	Procedure	Not Fully Mission Capable if:
5	Quarterly (cont)	Key and bolt carrier assembly and bolt assembly (serviceability check) (cont)	Check cartridge extractor (10), extractor spring assembly (11), cartridge ejector (12), and ejector spring (13) for dirt and serviceability. If dirty, clean, lubricate and assemble. If unserviceable, replace as necessary.	Parts are missing or unserviceable
			Check key and bolt carrier assembly (7) and carrier key (14) for damage and looseness. If damaged or loose, evacuate to support maintenance	Key and bolt carrier assembly or carrier key is damaged, or carrier key is loose.
			NOTE If carrier key Is dented, evacuate to support maintenance.	
			Check firing pin 15) for chips or breaks If damaged, evacuate to support maintenance.	Firing pin is damaged.
			Pits or wear in area Illustrated (16) is permissable	

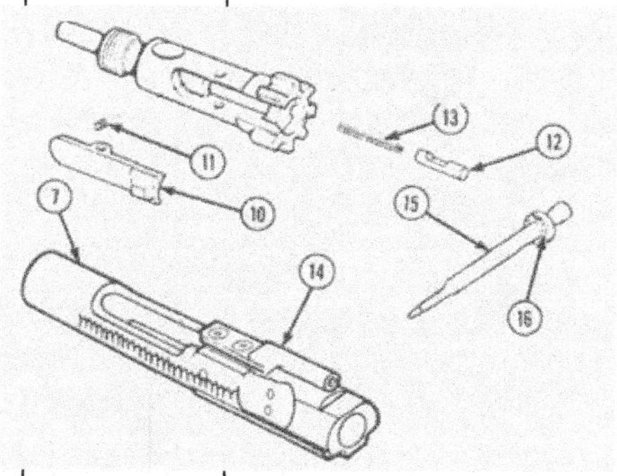

2-7. PREVENTIVE MAINTENANCE CHECKS AND SERVICES (CONT).

PREVENTIVE MAINTENANCE CHECKS AND SERVICES FOR M16A2 RIFLE (CONT)

Item No.	Interval	Item to be Checked or Serviced	Procedure	Not Fully Mission Capable if:
6	Quarterly	Lower receiver and buttstock assembly (serviceability check)	Remove buffer assembly (1) and action spring (2) Check buffer assembly for cracks SH - Buffer assembly is cracked Check action spring (2) for kinks and free length Free length should be **RIFLE**: 11 3/4 Inches (29 85 cm) minimum to 13 1./2 inches (34 29 cm) maximum SH - If action spring is kinked or does not meet free length requirements **CARBINE**: 10 1/16 inches (25.56 cm) minimum to 11 1/4 inches (28.58 cm) maximum. Do not attempt to adjust spring length SH - If action spring is kinked or does not meet free length requirements	

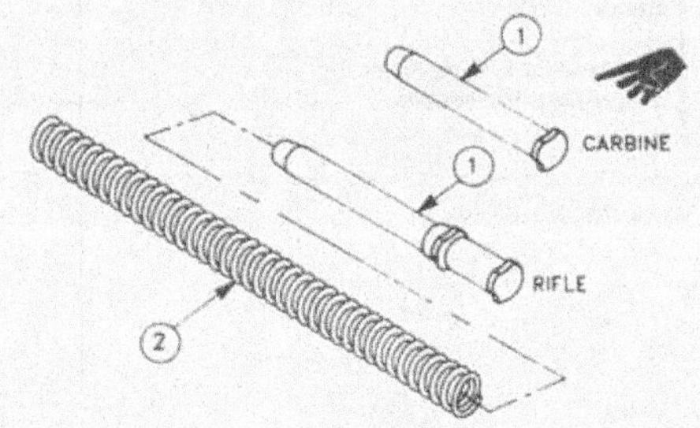

PREVENTIVE MAINTENANCE CHECKS AND SERVICES FOR M16A2 RIFLE (CONT)

Item No.	Interval	Item to be Checked or Serviced	Procedure	Not Fully Mission Capable if:
6	Quarterly (cont)	Lower receiver and buttstock assembly (serviceability check) (cont)	Remove pistol grip screw (3), lockwasher (4), pistol grip (5), helical spring (6), safety detent (7), pivot pin (8), pivot pin detent (9), and helical spring (10). Clean and lubricate metal components. Also clean and generously lubricate pivot pin holes and spring/detent holes. Replace defective/damaged components as necessary.	Components are defective/damaged.
			Disengage takedown pin (11) and pull out, push back in to re-engage takedown pin (an audible click should be heard). If an audible click is not heard, see page 2-57 for repair.	Components are defective/damaged.

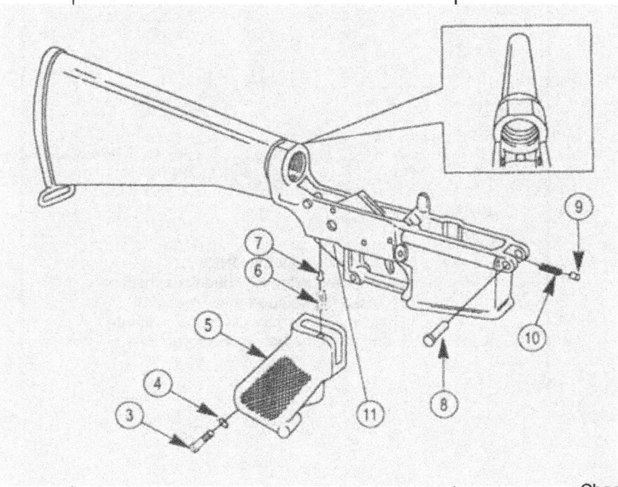

2-7. PREVENTIVE MAINTENANCE CHECKS AND SERVICES (CONT).

PREVENTIVE MAINTENANCE CHECKS AND SERVICES FOR M16A2 RIFLE (CONT)

Item No.	Interval	Item to be Checked or Serviced	Procedure	Not Fully Mission Capable if:
6	Quarterly (cont)	Lower receiver and buttstock assembly (serviceability check) (cont)	Lubricate helical compression spring and takedown pin detent (11) by placing one drop of lubricant on takedown pin detent and lowering the buttstock assembly (12) to vertical position. Allow the lubricant to work Its way around the helical compression spring and takedown pin detent (11). Check buttstock assembly(12) components for damage.	

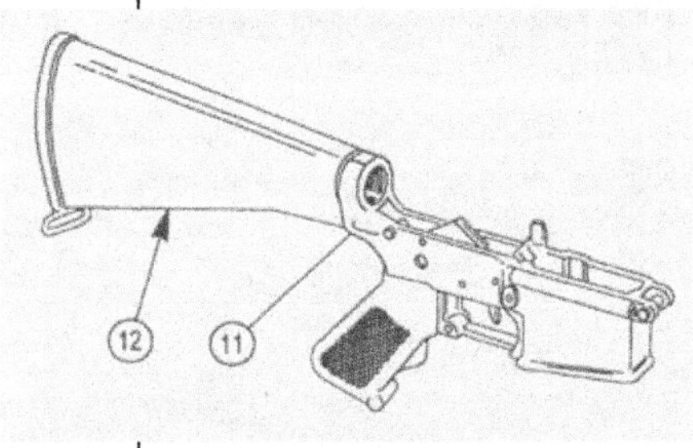

			RIFLE ONLY Under the following conditions, hairline cracks (no chipped away material allowed) originating from the buttplate end of the buttstock are acceptable.	Components are damaged.

PREVENTIVE MAINTENANCE CHECKS AND SERVICES FOR M16A2 RIFLE (CONT)

Item No.	Interval	Item to be Checked or Serviced	Procedure	Not Fully Mission Capable if:
6	Quarterly (cont)	Lower receiver and buttstock assembly (serviceability check) (cont)	**RIFLE ONLY** a. One hairline crack, not to exceed 1 in. (2.54 cm) in length, per side of buttstock	The buttstock is cracked in the critical area or does not meet the crack criteria.
			b. Two additional hairline cracks up to 0.25 in. (0.64 cm) in length, per side of buttstock.	
			c. A total of three cracks per side of the buttstock, originating from the buttplate end, are allowable.	
			Cracks in the critical area at the front end of the buttstock are not acceptable.	
			Check buttstock assembly (12) for forward to rear movement and/or a 1/32 in. (0.079 cm) gap between the buttstock assembly (12) and the lower receiver (13). If forward to rear movement and/or a 1/32 in. (0.079 cm) gap appears, tighten self-locking screw. If still not tight, remove buttstock assembly and check for loose lower receiver extension. If loose evacuate to support maintenance. If not loose, replace buttplate (14).	Lower receiver extension cannot be tightened.

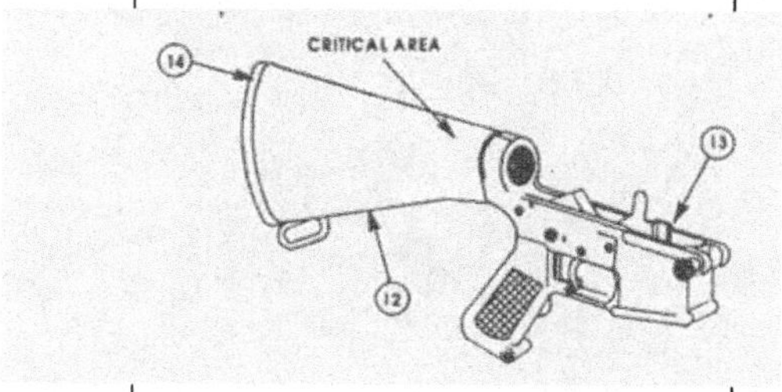

2-7. **PREVENTIVE MAINTENANCE CHECKS AND SERVICES (CONS).**

PREVENTIVE MAINTENANCE CHECKS AND SERVICES FOR M16A2 RIFLE (CONT)

Item No.	Interval	Item to be Checked or Serviced	Procedure	Not Fully Mission Capable if:
6	Quarterly (cont)	Lower receiver and buttstock assembly (serviceability check) (cont)	**RIFLE ONLY** Small amounts of side-to-side, up-and-down or rotational movement of the buttstock assembly is acceptable. (1) Cracks visible around the butt-plate mounting holes while screws are mounted SH - Cracks are visible around mounting holes when installed on rifle. (2) Cracks or separations around the door assembly are visible when the door assembly is closed SH - Cracks are visible when door assembly is closed. (3) If buttplate is cracked in excess of 0.25 in (O 64 cm) in length and extends through the buttplate (14). see pg. 2-64 for repair SH - If cracked in excess of 0 25 in. extends thru buttplate (4) The buttplate (14) should not be removed other than for repair or replacement of parts at which time a new self-locking screw, NSN 5305-01-147-8585, must be used.	

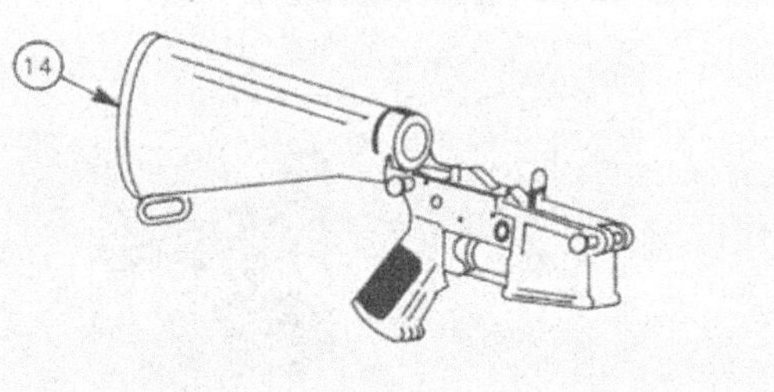

PREVENTIVE MAINTENANCE CHECKS AND SERVICES FOR M16A2 RIFLE (CONT)

Item No.	Interval	Item to be Checked or Serviced	Procedure	Not Fully Mission Capable if:
6	Quarterly (cont)	Lower receiver and buttstock assembly (serviceability check) (cont)	**NOTE** If a weapons lower receiver is missing one third or more of its exterior protective finish, resulting in an unprotected/light reflecting surface, it is candidate for overhaul. This missing finish will be considered a shortcoming. This shortcoming requires action to obtain a replacement weapon. Once a replacement has been received, evacuate the original weapon to depot for overhaul. If scratched or worn shiny in spots, repair in the same manner as outlined for upper receiver (see item 4 above).	
7	Quarterly	M16A2 Rifle	Assemble as in TM 9-1005-319-10 (operator's manual). Check sling for damage. If damaged, replace. Check for improperly assembled, broken, missing, or damaged parts. Check over all general appearance. Replace parts as required and authorize evacuation to support maintenance for repair.	

2-7. PREVENTIVE MAINTENANCE CHECKS AND SERVICES (CONT).

PREVENTIVE MAINTENANCE CHECKS AND SERVICES FOR M16A2 RIFLE (CONT)

Item No.	Interval	Item to be Checked or Serviced	Procedure	Not Fully Mission Capable if:
8	Quarterly	Annual DS safety and serviceability inspection and gaging	Check to ensure annual DS safety and serviceability inspection and gaging has been done and that the next gaging and inspection is scheduled. If annual gaging has not been performed within the last year, notify support maintenance.	Annual gaging has not been performed.

Section IV. TROUBLESHOOTING

2-8. GENERAL.

a. This section contains unit level troubleshooting information for locating and correcting most of the operating troubles which may develop in the M16A2 rifle, M4/M4A1 carbine. Each malfunction for the individual part or assembly is followed by a list of tests or inspections which will help you to determine the corrective actions in the order listed.

b. This manual cannot list all malfunctions that may occur, nor all tests or inspections and corrective actions. If a malfunction is not listed or is not corrected by listed corrective actions, see individual repair sections in the maintenance procedures on each major assembly.

2-9. TROUBLESHOOTING PROCEDURES. Refer to troubleshooting table for malfunctions, tests, and corrective actions. The symptom index is provided for a quick reference of the malfunctions covered in the table.

SYMPTOM INDEX

	Troubleshooting Procedures Page
Failure of magazine to lock in rifle	2-21
Failure to feed	2-22
Failure to chamber	2-22
Failure to lock	2-23
Failure to fire	2-24
Failure to unlock	2-26
Failure to extract	2-26
Failure to eject	2-27
Failure to cock	2-28
Short recoil	2-28
Rifle cannot be zeroed	2-30
Failure to cycle with selector lever set on BURST	2-31
Fires two rounds with one pull of trigger with selector lever set on SEMI (double firing)	2-31
Fires with selector lever on SAFE or when trigger is released with selector lever on SEMI	2-31
Bolt assembly falls to lock to rear after firing last round	2-32

TROUBLESHOOTING

MALFUNCTION
TEST OR INSPECTION
CORRECTIVE ACTION

1. FAILURE OF MAGAZINE TO LOCK IN RIFLE.

Step 1. Dirty or corroded magazine catch (1).

Disassemble and clean.

Step 2 Defective magazine catch spring (2)

Evacuate to support maintenance.

Step 3 Worn or broken magazine catch (1).

Evacuate to support maintenance

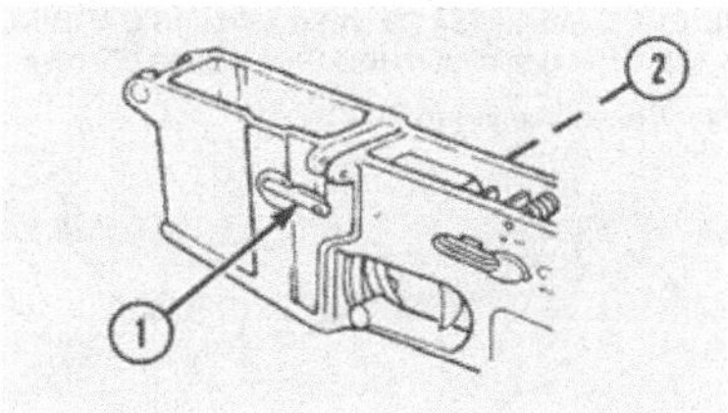

2-9. TROUBLESHOOTING PROCEDURES (CONT).

TROUBLESHOOTING (CONT)

MALFUNCTION
TEST OR INSPECTION
CORRECTIVE ACTION

2. FAILURE TO FEED

Step 1. Magazine catch spring weak or broken

Evacuate to support maintenance

Step 2. Magazine catch (1) defective.

Evacuate to support maintenance

Step 3. Magazine catch (1) out of adjustment (will not retain magazine)

Refer to operator's manual.

Step 4. Short recoil.

Refer to page 2-28

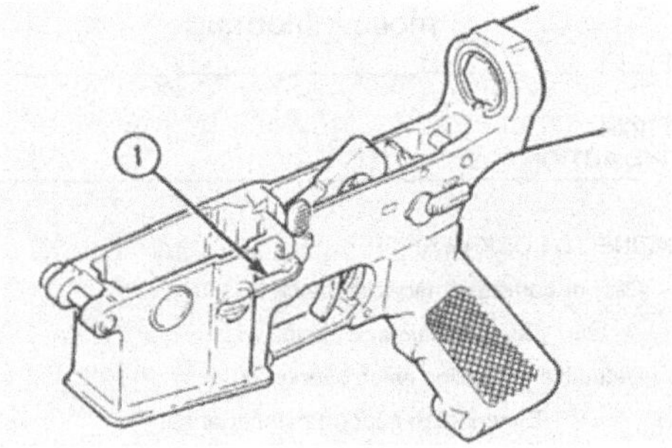

3. FAILURE TO CHAMBER.

Step 1. Weak or broken action spring (1), RIFLE ONLY (free length 11 3/4 inches (29.85 cm) minimum to 13 1.'2 Inches (34 29 cm) maximum) CARBINE ONLY (10 1 16 inches (2S 56 cm) minimum to 11 1,4 inches (28 58 cm) maximum)

Replace action spring (p 2-57)

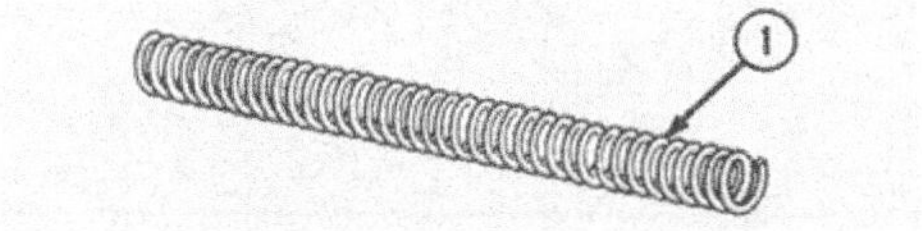

TROUBLESHOOTING (CONT)

MALFUNCTION
TEST OR INSPECTION
CORRECTIVE ACTION

Step 2. Short recoil

Refer to page 2-28

4. FAILURE TO LOCK

Step 1. Bolt cam pin (1) missing

Replace (p 2-35)

Step 2. Loose or damaged bolt carrier key (2).

a. Evacuate to support maintenance.

b. Dented bolt carrier key may be repaired (p 2 35).

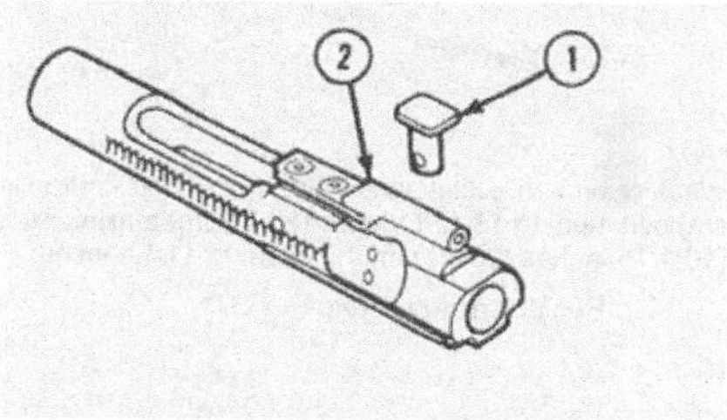

Step 3. Improperly assembled extractor spring assembly (3)

Assemble correctly (p 2-38).

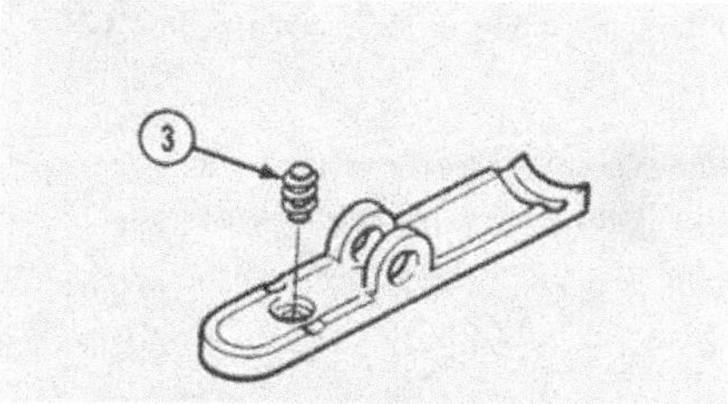

2-9. TROUBLESHOOTING PROCEDURES (CONT).

TROUBLESHOOTING (CONT)

MALFUNCTION
TEST OR INSPECTION
CORRECTIVE ACTION

4. FAILURE TO LOCK (CONT)

Step 4. Bent gas tube 14J

a Adjust to Its original configuration by bending in area of hand-guard assembly.

b If the gas tube cannot be returned to Its original configuration, evacuate the rifle to support maintenance

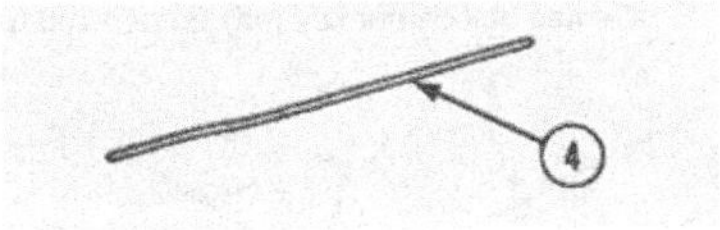

Step 5 Weak or broken action spring (5); RIFLE ONLY: (free length 11 3/4 inches (29,85 cm) minimum to 13 1,/2 Inches (34 29 cm) maximum). CARBINE ONLY (10 1/16 Inches (25 56 cm) minimum to 11 1,4 inches (28 58 cm) maximum

Replace action spring (p 2-571

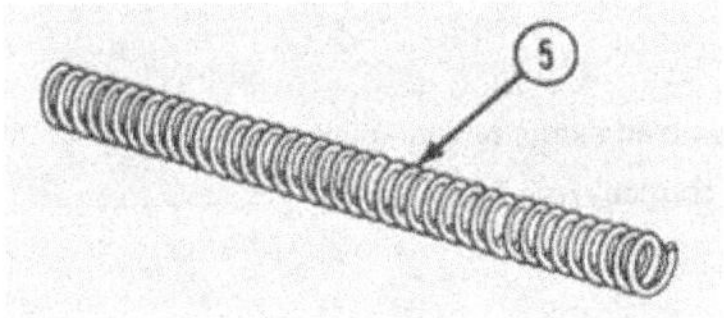

5. FAILURE TO FIRE.

Step 1 Broken or chipped firing pin (1).

Evacuate to support maintenance.

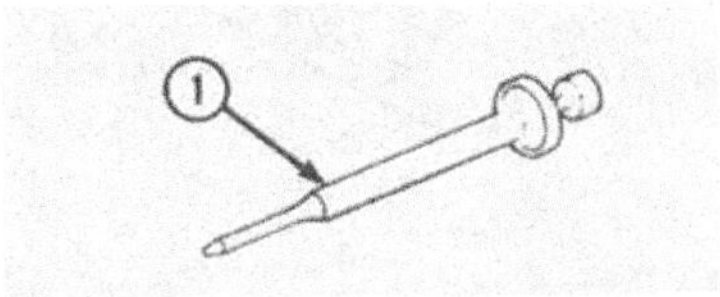

TROUBLESHOOTING (CONT)

MALFUNCTION
TEST OR INSPECTION
CORRECTIVE ACTION

Step 2 Carbon buildup in firing pin recess inside bolt assembly

Remove cartridge extractor and clean recess with pipe cleaner (Item 11, app D), refer to operator's manual.

Step 3. Firing mechanism (2) and or lower receiver assembly (31 improperly assembled or has worn, broken, or missing parts.

Evacuate to support maintenance

Step 4. Broken, defective, or missing firing pin retaining pin (4).

Replace (p 2-35)

Step 5. Selector lever (5) frozen on SAFE position.

Evacuate to support maintenance

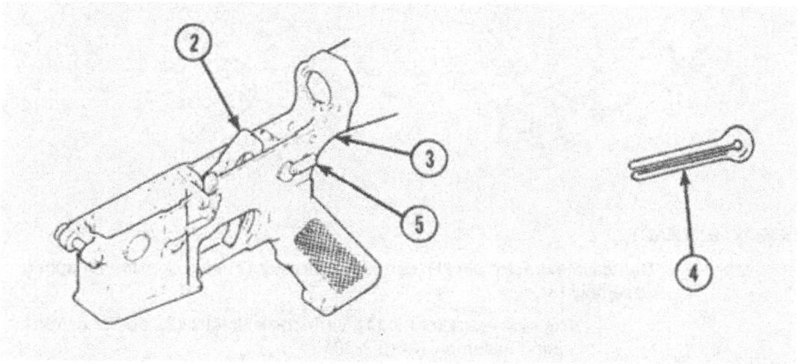

2-9. TROUBLESHOOTING PROCEDURES (CONT).

TROUBLESHOOTING (CONT)

MALFUNCTION
TEST OR INSPECTION
CORRECTIVE ACTION

6. FAILURE TO UNLOCK.

Step 1. Burred locking lugs (1) on bolt assembly.

Remove burrs.

Step 2. Burred lugs (2) on barrel extension

Remove burrs.

Step 3 Short recoil.

Refer to page 2-28.

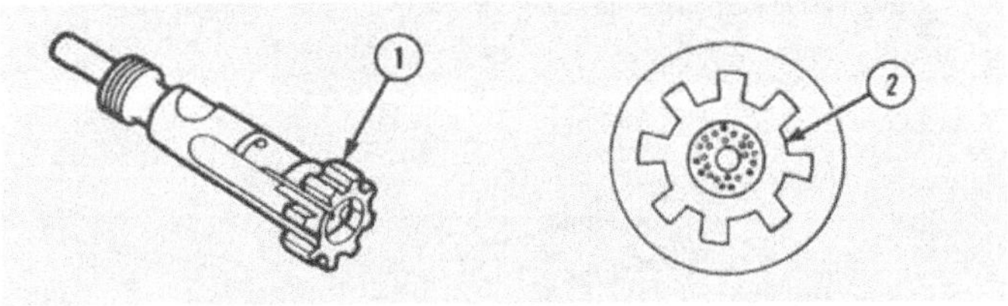

7. FAILURE TO EXTRACT.

Step 1. Defective extractor pin (1), cartridge extractor (2), and or extractor spring assembly (3).

Replace extractor pin (1), cartridge extractor (2), and or extra(:tor spring assembly (3) (p 2-381

Step 2. Short recoil

Refer to page 2-28

NOTE

Rubber insert and spring are an assembly Illustration shows Insert out of assembly for clarification only Do not remove the rubber Insert from the extractor spring assembly.

TROUBLESHOOTING (CONT)

MALFUNCTION
TEST OR INSPECTION
CORRECTIVE ACTION

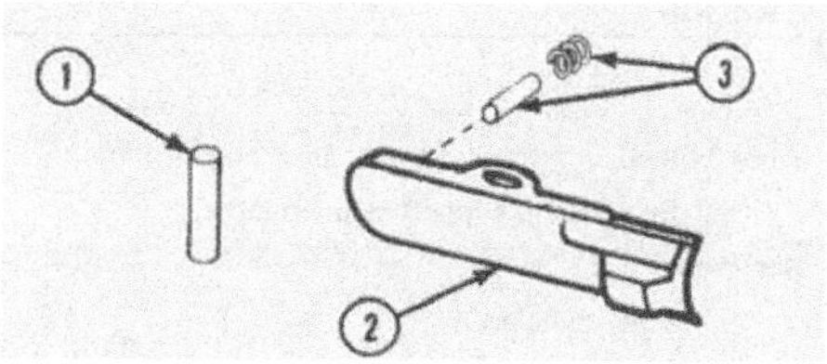

8. FAILURE TO EJECT.

Step 1. Broken cartridge ejector (1).

Replace (p 2-38).

Step 2. Cartridge ejector (1) stuck in bolt body (2).

Disassemble and clean (p 2-38).

Step 3. Weak or broken ejector spring (3).

Replace (p 2-38)

Step 4. Short recoil

Refer to page 2-28.

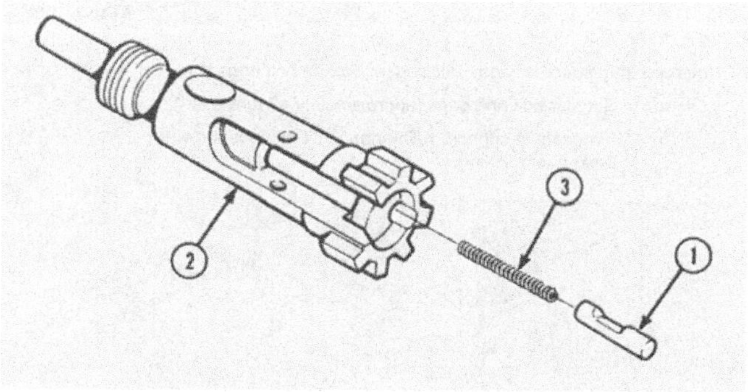

2-9. TROUBLESHOOTING PROCEDURES (CONT).

TROUBLESHOOTING (CONT)

MALFUNCTION
TEST OR INSPECTION
CORRECTIVE ACTION

9. FAILURE TO COCK

Step 1 Worn, broken, or missing parts of firing mechanism

Evacuate to support maintenance.

Step 2 Short recoil

Refer to below

10 SHORT RECOIL

Step 1 Broken or damaged action spring (1).

Replace action spring (p 2-57).

Step 2 Unlubricated or dirty action spring and receiver extension

Clean and lubricate

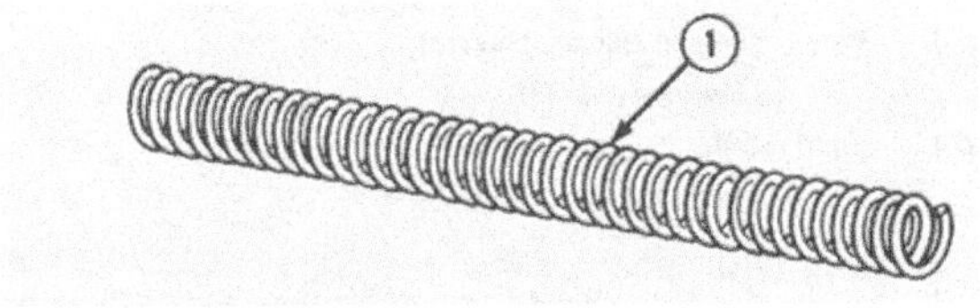

Step 3 Improper gap space or worn, missing, or broken bolt rings (2)

a. Stagger bolt ring gaps (approximately 13 turn apart}

b. Evacuate to support maintenance if bolt rings are worn, broken, or missing.

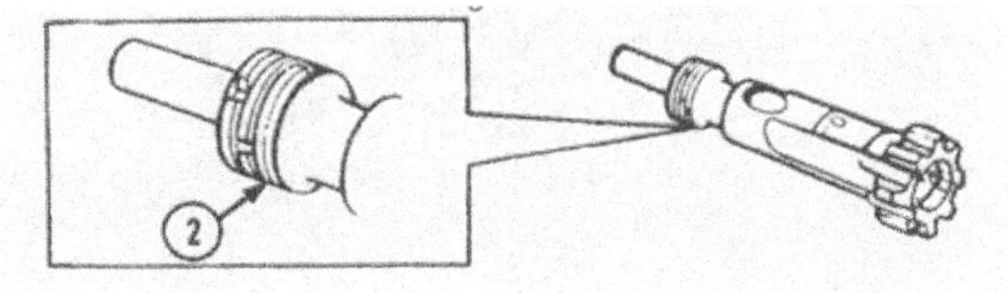

TROUBLESHOOTING (CONT)

MALFUNCTION
TEST OR INSPECTION
CORRECTIVE ACTION

Step 4. Carbon build-up or foreign matter In the narrow passage of the bolt carrier key (3).

Clean with CLP (Item 9, app D) and a pipe cleaner (Item 11, app D).

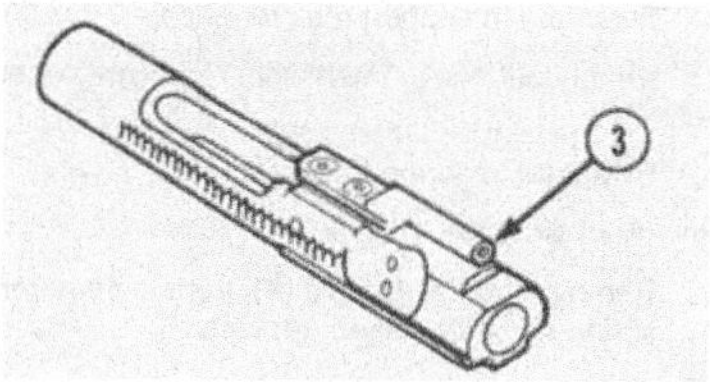

Step 5. Gas leakage caused by broken or loose gas tube (4) around front sight base.

Evacuate to support maintenance.

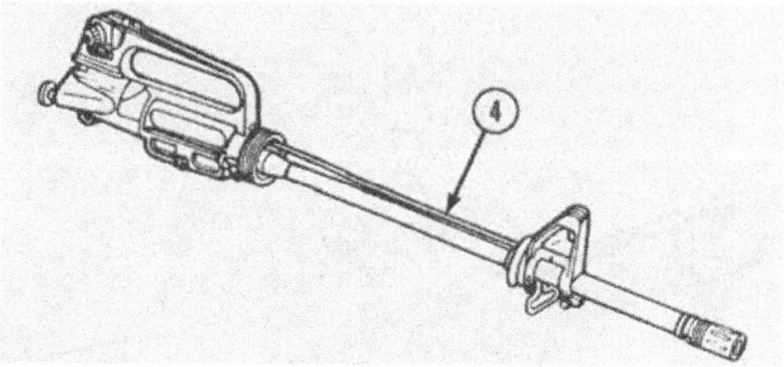

Step 6. Improper alignment of gas tube and carrier key.

a. Adjust gas tube alignment by bending in area of handguard assembly to Its original configuration

b. If gas tube cannot be returned to Its original configuration, evacuate the rifle to support maintenance.

2-9. TROUBLESHOOTING PROCEDURES (CONT).

TROUBLESHOOTING (CONT)

MALFUNCTION
TEST OR INSPECTION
CORRECTIVE ACTION

11. RIFLE CANNOT BE ZEROED.

Step 1. Defective barrel assembly (1).

Evacuate to support maintenance

Step 2. Barrel assembly out of alignment with rear sight assembly (2) on upper receiver.

Evacuate to support maintenance

Step 3. Defective front sight (3).

Remove front sight post (4), front sight detent (5), and helical spring (6). If damaged, replace.

Step 4. Defective -ear sight assembly (2).

Evacuate to support maintenance.

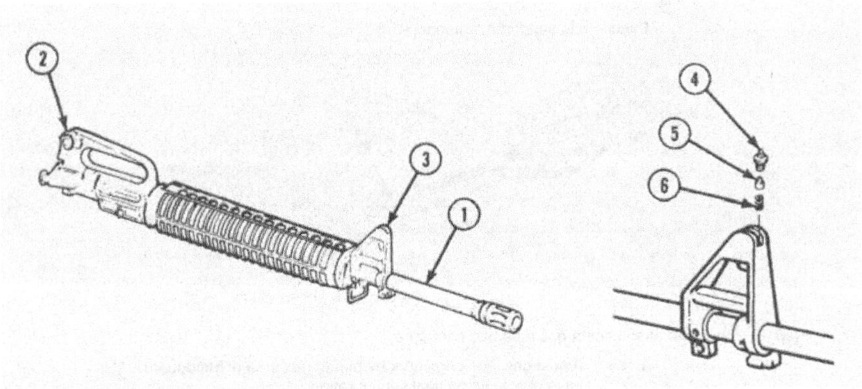

TROUBLESHOOTING (CONT)

MALFUNCTION
TEST OR INSPECTION
CORRECTIVE ACTION

12. FAILURE TO CYCLE WITH SELECTOR LEVER SET ON BURST (M16A2 AND M4).

Faulty selector lever or broken cam, cam clutch spring, or burst disconnector.

Evacuate to support maintenance.

12.1. FAILURE TO CYCLE WITH SELECTOR LEVER SET ON AUTO (M4A1).

Faulty selector lever.

Evacuate to support maintenance.

13. FIRES TWO ROUNDS WITH ONE PULL OF TRIGGER WITH SELECTOR LEVER SET ON SEMI (DOUBLE FIRING).

Perform function test.

If any part of function test (p 2-68) fails, evacuate to support maintenance.

14. FIRES WITH SELECTOR LEVER ON SAFE OR WHEN TRIGGER IS RELEASED WITH SELECTOR LEVER ON SEMI.

Worn, broken, or missing parts of firing mechanism.

Evacuate to support maintenance.

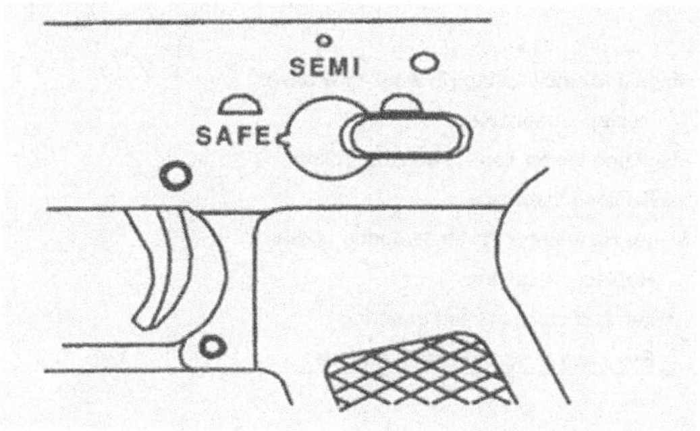

2-9. TROUBLESHOOTING PROCEDURES (CONT).

TROUBLESHOOTING (CONT)

MALFUNCTION
TEST OR INSPECTION
CORRECTIVE ACTION

15 BOLT ASSEMBLY FAILS TO LOCK TO REAR AFTER FIRING LAST ROUND.

Step 1. Magazine follower (1) worn or broken.

Replace magazine.

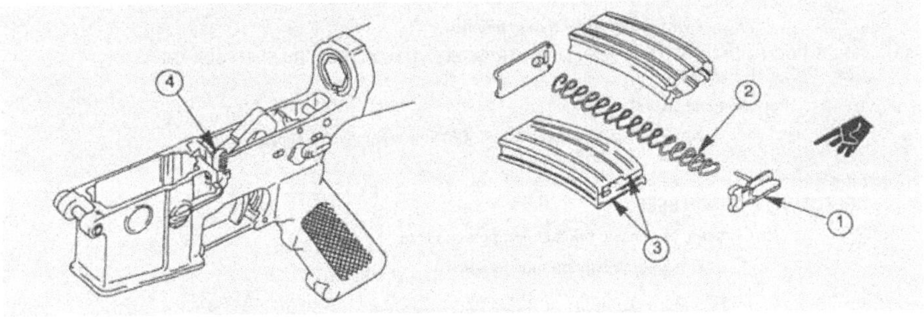

Step 2. Magazine catch spring (2) weak or broken.

Replace magazine.

Step 3. Magazine feeder lips (3) bent or broken.

Replace magazine.

Step 4. Magazine follower (1) binds during operation.

Replace magazine.

Step 5. Broken bolt catch (4) and/or spring.

Evacuate to support maintenance.

Section V. MAINTENANCE PROCEDURES

2-10. INITIAL SETUP. The following information will reduce the space required for the initial setup portion of the maintenance procedures.

a. Materials/Parts required are not listed unless they apply to the procedure

b. Personnel Required is listed only if the task requires more than one person If Personnel Required Is not listed, It means one person can do the job.

c. The normal standard equipment condition Is that the item is removed from the en(1 Item or next higher assembly and is in the assembled condition. Equipment Condition is not listed unless some other condition is required.

d. The approximate time required Is listed on the applicable Maintenance Allocation Chart (MAC).

e. When the term evacuate to support maintenance is used, the entire rifle must be evacuated.

2-11. LUBRICATION GENERAL.

a. Whenever the term "cleaner, lubricant, and preservative (CLP)" or the words "lubricant", "lube", "LSA", or "LAW" are cited In this TM, they are to be interpreted to mean CLP (Item 9, app D), weapons lubricating oil (LSA) (Item 23, app D), or weapons lubricating oil (LAW) (Item 22, app D) can be utilized as applicable The following constraints must be adhered to:

b. Under all but the coldest arctic conditions, LSA or CLP are the lubricants to use on the rifle. Either may be used at 100F (230C) and above. However, do not use both on the same rifle at the same time.

c. LAW is the lubricant to use during cold arctic conditions, + 100F (12C) and below

d. Any of the lubricants may be used from 10°F to + 100F (230C to 12°C)

e. ARMY ONLY: Do not mix lubricants on the same rifle The rifle must be thoroughly cleaned during change from one lubricant to another Dry Cleaning Solvent (SD) (Item 16, app D) is recommended for cleaning during change from one lubricant to another.

f. Rifle Bore Cleaner (RBC) (item 12, app D) may be used to remove carbon buildup In the bore and other portions of the rifle.

2-12. MAJOR COMPONENTS OF M16A2 RIFLE AND CARBINES.

This task covers disassembly

INITIAL SETUP:

References
TM 9-1005-319-10 (operator's manual)

Equipment Conditions
Rifle and Carbines assembled

General Safety Instructions
Before starting an inspection, be sure to clear the weapon. Do not keep live ammunition near the work area.

To avoid injury to your eyes, use care when removing and installing spring-loaded parts.

Below direct support maintenance, do not interchange bolt assemblies from one weapon to another. Doing so may result in injury to, or death of, personnel.

DISASSEMBLY

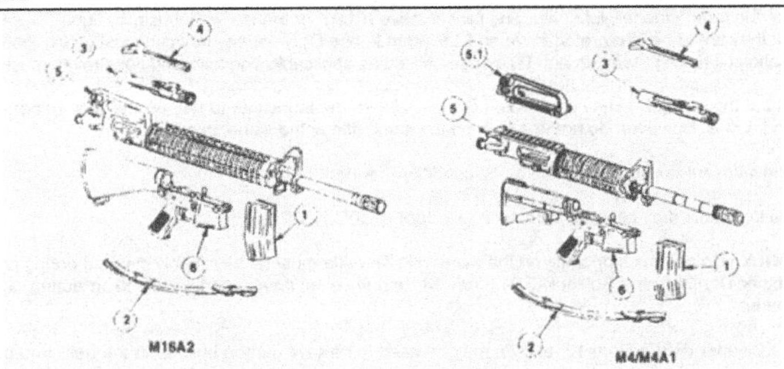

a. Refer to operator's manual.

b. Remove magazine (1), sling (2), bolt carrier assembly (3), charging handle assembly (4), and upper receiver and barrel assembly (5), carrying handle assembly (5.1), from lower receiver and buttstock assembly (6).

NOTE

Solid Film Lubricant (SFL) is the only authorized touchup for the MI 16A2 rifle and M4/M4A1 carbines and may be used on up to one third of the exterior finish of the weapon. **FOR ARMY CONUS USE ONLY AND AIR FORCE TRAINING WEAPONS ONLY**: SFL may be used as touchup without limitation on the upper receiver and barrel assembly. This is to say that units which DO NOT fall under the category of Divisional Combat Units or rapid deployment type units may have up to 100 percent of the exterior surface of the upper receiver and barrel assembly protected with SFL if necessary.

12-13. BOLT CARRIER ASSEMBLY.

This task covers:

a. **Disassembly**
b. **Cleaning**
c. **Inspection/Repair**
d. **Lubrication**
e. **Reassembly**

INITIAL SETUP:

Tools
(ARMY) Small Arms Repairman Tool Kit (item 3, app B)
Bolt Carrier Key Tool (item 11, figure C-17)

References
TM 9-1005-319-10 (operator's manual)

Equipment Conditions
2-34 Bolt carrier assembly removed

General Safety Instructions
Bolt cam pin must be installed or rifle will blow up while firing the first round. If the bolt cam pin is not installed, injury to, or death of, personnel may result.

Do not interchange bolt assemblies from one rifle to another. Doing so may result in injury to, or death of, personnel.

a. DISASSEMBLY

CAUTION

Do not spread or close legs of firing pin retaining pin (1).

1. Remove firing pin retaining pin (1). Tilt key and bolt carrier assembly (2) and catch firing pin (3) as it drops out.
2. Rotate bolt cam pin (4) one quarter turn and lift straight up to remove.
3. Remove bolt assembly (5) from key and bolt carrier assembly (2).

NOTE
For disassembly of bolt assembly (5), see page 2-38.

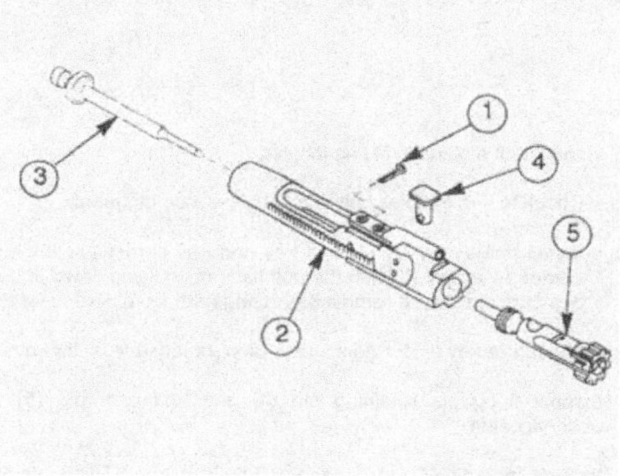

12-13. BOLT CARRIER ASSEMBLY (CONT).

b. CLEANING

Clean all items (operator's manual). Remove carbon deposits.

c. INSPECTION/REPAIR

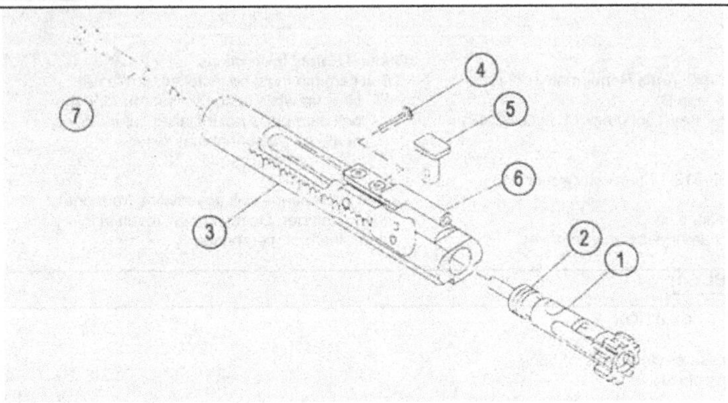

1. Inspect bolt assembly (1) as follows:

   a. Check to see that the bolt ring (2) gaps are staggered.

   b. Install bolt assembly (1) into key and bolt carrier assembly (3). Check for worn rings by holding the key and bolt carrier assembly (3) with the bolt assembly facing down. If bolt assembly falls out after firing pin retaining pin (4) and bolt cam pin (5) are removed, the rings are worn. Evacuate to support maintenance.

2.Inspect carrier key (6) for dents, distortion, or looseness. If dented or loose, evacuate to support maintenance.

3. Inspect firing pin retaining pin (4) and bolt cam pin (5) for cracks, damage, or excessive wear. Replace if unserviceable.

4. Inspect firing pin (7) for damage or If tip is chipped. Evacuate rifle to support maintenance if unserviceable

5. Inspect key and bolt carrier assembly (3) for damage or wear. If unserviceable, evacuate to support maintenance.

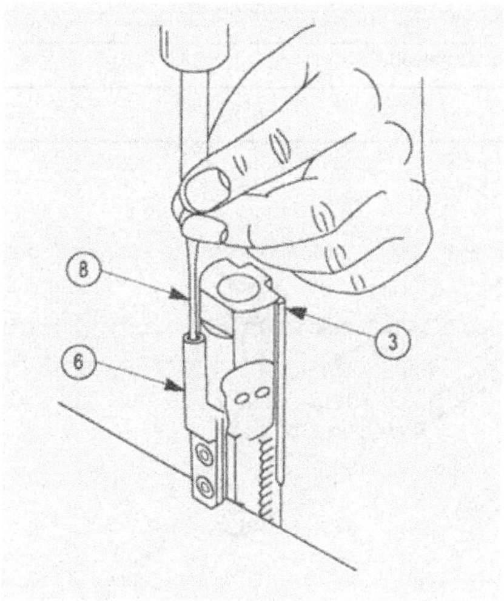

CAUTION

Extreme care must be exercised during the following procedure to assure that the striking force is not directed to the attaching screws and that the tube portion is not enlarged or flared beyond original requirement. Such enlargement would permit loss of gas pressure when the key and gas tube come together during functioning.

6. Repair small dents and/or distortions in carrier key (6) using bolt carrier key tool as follows:

a. Place the key and bolt carrier assembly (3) in a vertical position, supported In a manner that contact is made with the rear surface of the carrier key (6).

b. Insert the small end of the key tool (8) into the tube portion of the carrier key (6).

c. Strike the large end of the key tool (8) lightly with a 3 ounce, soft-brass hammer.

d. Repeat striking (gently) until carrier key (6) is reformed to original configuration.

e. If carrier key (6) cannot be reformed to original configuration, evacuate the weapon to support maintenance.

d. LUBRICATION

Lubricate all Items (p 2 33) (operator's manual)

e. REASSEMBLY

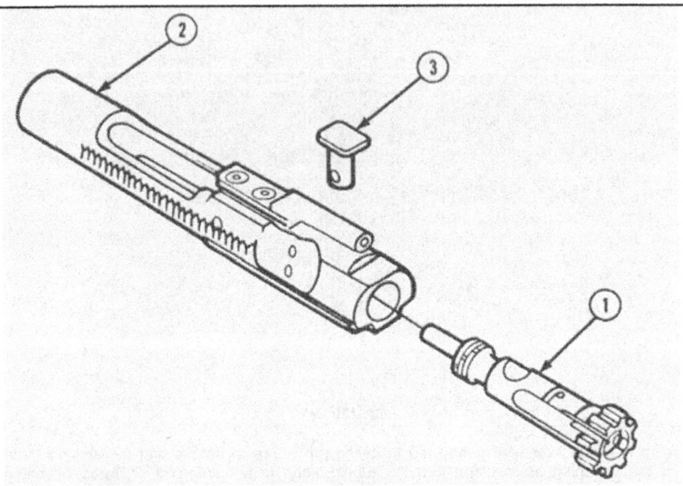

WARNING

Do not Interchange bolt assemblies from one rifle to another Doing so may result In Injury, or death of, personnel.

Bolt cam pin must be Installed or rifle will blow up while firing the first round. If the bolt cam pin is not Installed, Injury to, or death of, personnel may result

NOTE

Before installing bolt assembly, check to see that the bolt ring gaps are staggered to prevent loss of gas pressure

1. Install bolt assembly (1) In key and bolt carrier assembly (2)

2. Install bolt cam pin (3) and rotate one quarter turn

2-13. BOLT CARRIER ASSEMBLY (CONT).

e. REASSEMBLY (CONT)

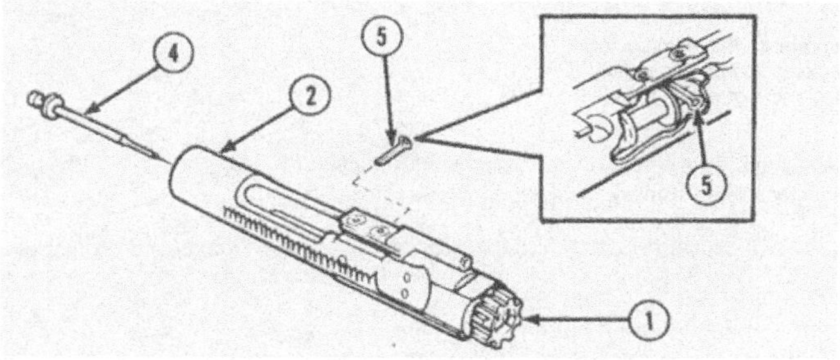

3. Hold key and bolt carrier assembly (2) with bolt assembly (1) down and drop in firing pin (4).

4. Install firing pin retaining pin (5) from left side only to ensure proper installation Check for proper installation by holding key and bolt carrier assembly (2) with bolt assembly (1) up, then attempt to shake out firing pin.

5. Reassemble rifle, refer to page 2-68.

2-14. BOLT ASSEMBLY.

This task covers:

- **a. Removal**
- **b. Cleaning**
- **c. Inspection**
- **d. Lubrication**
- **e. Reassembly**

INITIAL SETUP:

Tools
(ARMY) Small Arms Repairman Tool Kit (item 3, app B)

Equipment Conditions
2-35 Bolt assembly removed

General Safety Instructions
Do not Interchange bolt assemblies from one rifle to another. Doing so may result in injury to, or death of, personnel.

To avoid injury to your eyes, use care when removing and installing spring-loaded parts.

a. DISASSEMBLY

NOTE

Do not separate cartridge extractor and extractor spring assembly unless replacement of either or both is required

Do not remove the rubber Insert from the extractor spring assembly

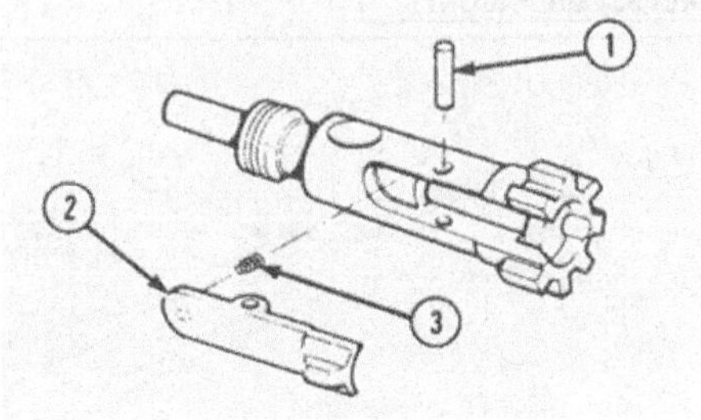

1. Push out extractor pin (1) and remove cartridge extractor (2) and extractor spring assembly (3) as a unit

2. If required, twist extractor spring assembly (3) counterclockwise to remove from cartridge extractor (2).

CAUTION

Be sure to use vise law protective caps

3. Hold bolt body In vise and remove spring pin (4) using 1 16 Inch punch and hammer

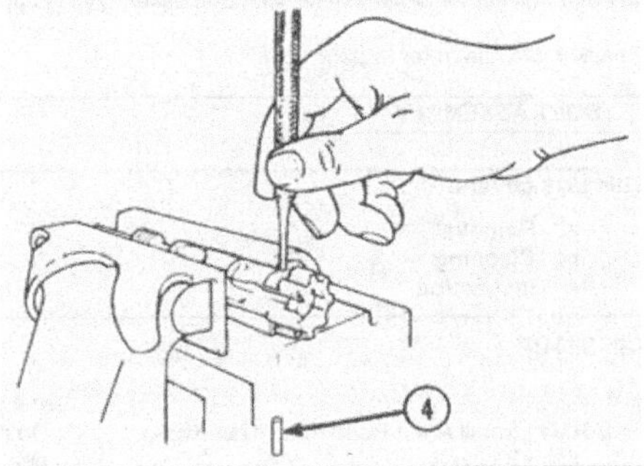

2-14. BOLT ASSEMBLY (CONT).

a. DISASSEMBLY (CONT)

WARNING

To avoid injury to your eyes, use care when removing and installing spring-loaded parts.

4. Remove punch, be sure to catch cartridge elector (5) and ejector spring (6) to prevent loss

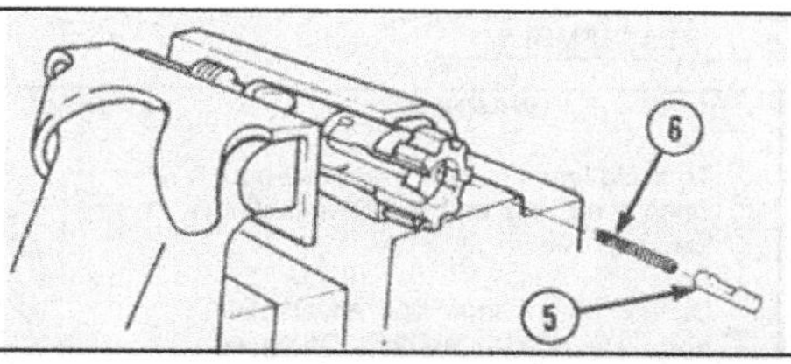

b. CLEANING

CAUTION

Do not distort extractor spring assembly during cleaning.

Clean all Items (operator's manual). Remove carbon deposits.

c. INSPECTION/REPAIR

1. Check bolt rings (1) for damage and proper stagger (gaps approximately 3 turn apart).

2. Inspect for cracks or damage, especially around bolt cam pin hole (2) and locking lugs (3). If cracked or damaged evacuate to support maintenance.
3. Inspect cartridge extractor (4), extractor spring assembly (5), and extractor pin (6) for cracks, breaks, chips, and other damage. Pay close attention to cartridge extractor lip (7). If damaged, replace.
4. Inspect cartridge ejector (8) and ejector spring (9) for cracks, breaks, and chips If damaged, replace.

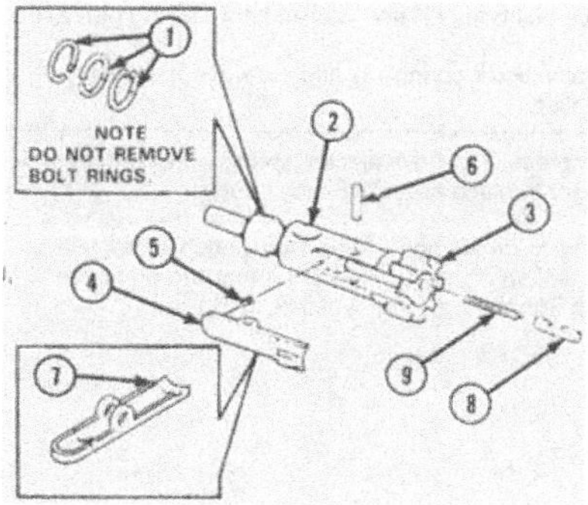

d. LUBRICATION

Lightly lubricate all items (p 2 33) (operator's manual)

e. REASSEMBLY

WARNING

To avoid Injury to your eyes, use care when removing and Installing spring loaded parts.

Do not Interchange bolt assemblies from one rifle to another Doing so may result In injury to, or death of, personnel.

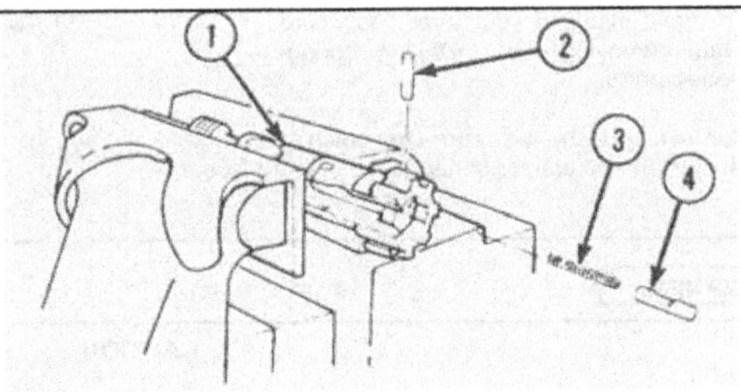

CAUTION

Be sure to use vise jaw protective caps

1. Place bolt body (1) in a vise and start spring pin (2) In hole.

2. Install ejector spring (3) and cartridge ejector (4). Align groove on cartridge ejector (4) so that spring pin 12) can be Installed

3. Compress and hold ejector spring and cartridge ejector In place with a 3 8 inch punch.

4. Using hammer and 1 16 Inch punch, complete Installation of spring pin (2) so that the ends are flush with the out side of bolt body (1).

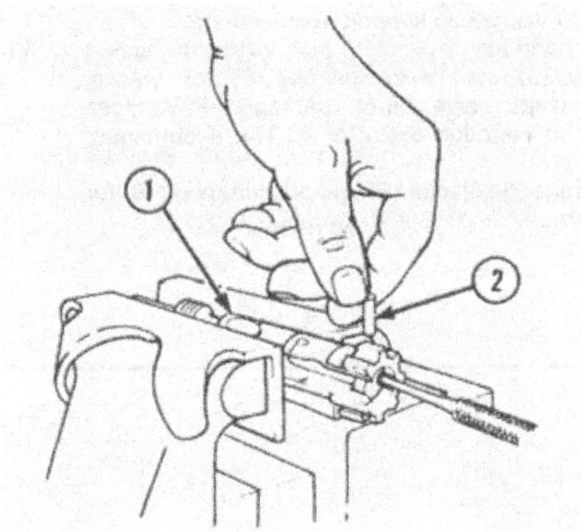

2-14. BOLT ASSEMBLY (CONT).

e. REASSEMBLY (CONT)

NOTE

Do not disassemble rubber Insert from extractor spring assembly.

5. If removed, Insert large end of extractor spring assembly (5) into cartridge extractor (6) and seat by pushing and turning clockwise.

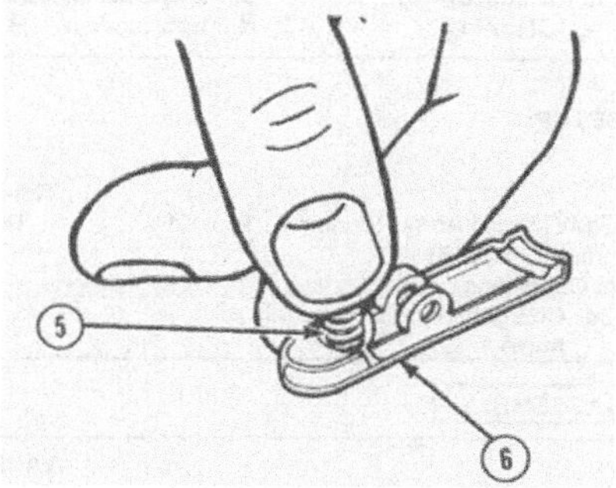

6 Position cartridge extractor (6) and ex tractor spring assembly (5) on bolt body (1).

7 Compress extractor spring assembly (5) and cartridge extractor (6) to align holes.

8. Install extractor pin (7) by hand.

9. Reassemble rifle, refer to page 2-68.

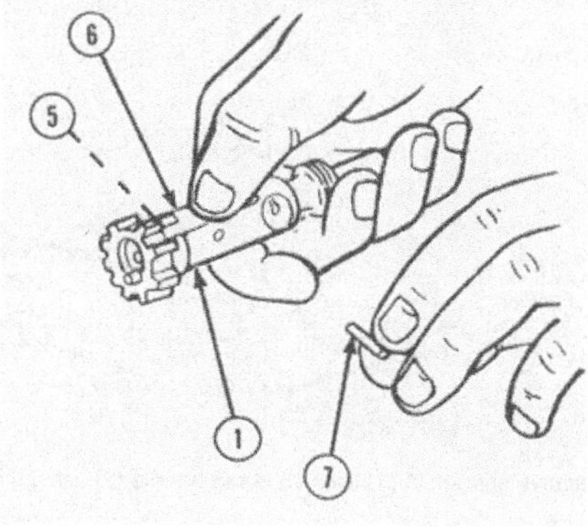

2-15. CHARGING HANDLE ASSEMBLY.

This task covers:

a. Disassembly **b. Cleaning** **c. Inspection/Repair** **d. Lubrication** **e. Reassembly**

INITIAL SETUP:

Tools
(ARMY) Small Arms Repairman Tool Kit (item 3, app B)

Equipment Conditions
2-34 Charging handle assembly removed

References
TM 9-1005-319-10 (operator's manual)

a. DISASSEMBLY

WARNING

To avoid injury to eyes, use care when removing and installing spring-loaded parts.

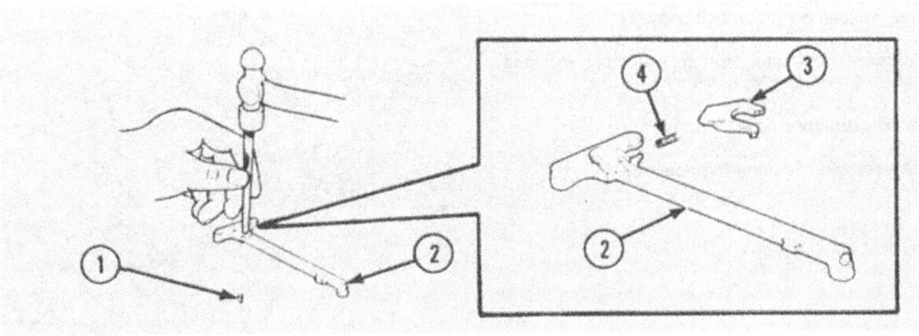

1. Remove spring pin (1) from charging handle (2) using a hammer and 1/16 inch punch.
2. As punch is withdrawn, catch charging handle latch (3) and helical spring (4) to prevent loss.

2-15. CHARGING HANDLE ASSEMBLY (CONT).

b. CLEANING

Clean all Items (operator's manual) Remove carbon deposits.

c. INSPECTION/REPAIR

Inspect all Items for breaks, cracks, or damage Replace all unserviceable Items

d. LUBRICATION

Lightly lubricate all Items (p 2-33) (operator's manual)

e. REASSEMBLY

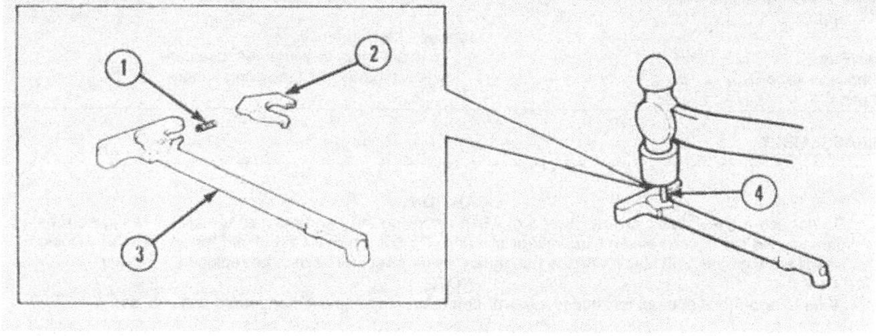

1. Position helical spring I1) and charging handle latch (2) in chdrgiin(J handle 13) Align holes and hold In position

2. Install spring pin (4) using a hammer Make sure spring pin is flush

3. Reassemble rifle, refer to page 2 68

2-16. UPPER RECEIVER AND BARREL ASSEMBLY, RIFLE BARREL ASSEMBLY, AND UPPER RECEIVER ASSEMBLY.

This task covers:

- **a.** Disassembly
- **b.** Cleaning
- **c.** Inspection
- **d.** Repair
- **e.** Lubrication
- **f.** Reassembly

INITIAL SETUP

Tools
(ARMY) Small Arms Repairman Tool Kit (item 3, app B)
Front sight post removal and installation tool (fig. E-2, app E)
Front sight detent depressor (fig. E-1, app E)

Materials/Parts
Lubricants (app D)

References
TM 9-1005-319-10 (operator's manual)

Equipment Conditions
2-34 Upper receiver and barrel assembly removed from lower receiver

General Safety Instructions
To avoid injury to your eyes, use care when removing and installing spring-loaded parts.

a. DISASSEMBLY

CAUTION

Do not use a screwdriver or any other tool when removing the handguard assemblies. Doing so may damage the handguard assemblies and/or slip ring. Do not remove heat shield for any reason. Doing so will damage the heat shield and the handguard assemblies will have to be replaced.

NOTE

Refer to operator's manual for "buddy system" procedure on removing handguard assemblies.

2-16. UPPER RECEIVER AND BARREL ASSEMBLY, RIFLE BARREL ASSEMBLY, AND UPPER RECEIVER ASSEMBLY (CONT).

a. DISASSEMBLY (CONT)

1. Push down on handguard slip ring (1) and lift upper handguard assembly (2) up and out.

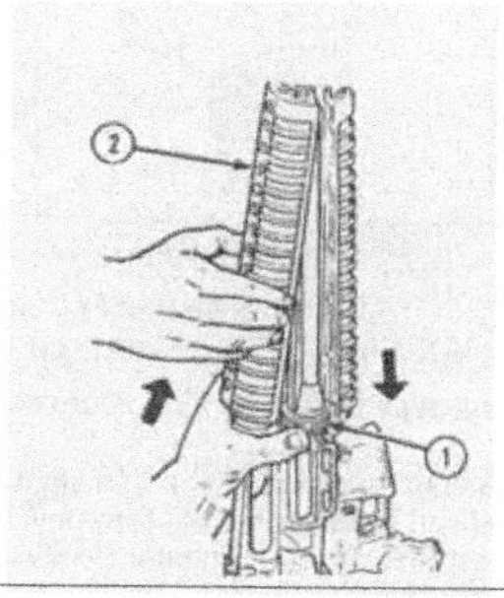

2. Push down on handguard slip ring (1) and lift the lower handguard assembly (3) up and out.

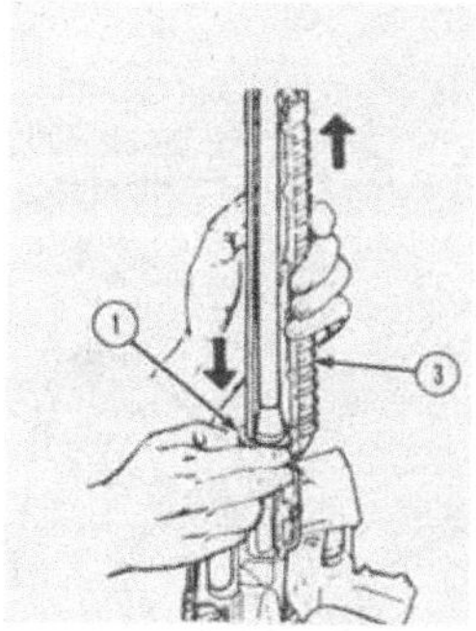

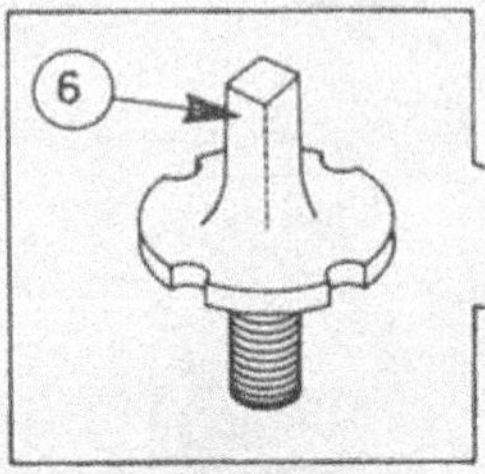

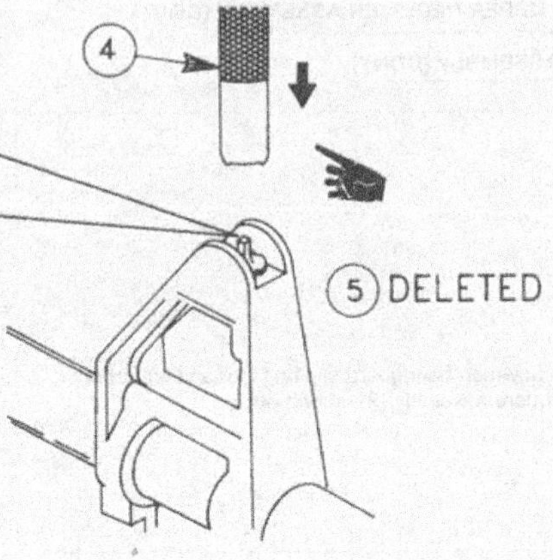

WARNING

To avoid injury to your eyes, use care when removing and installing spring-loaded parts.

NOTE

Remove front sight posts only if parts are damaged.

3. Using a dummy cartridge or a font sight post removal and installation tool (4).removed front sight post (6) by turning counter clockwise.

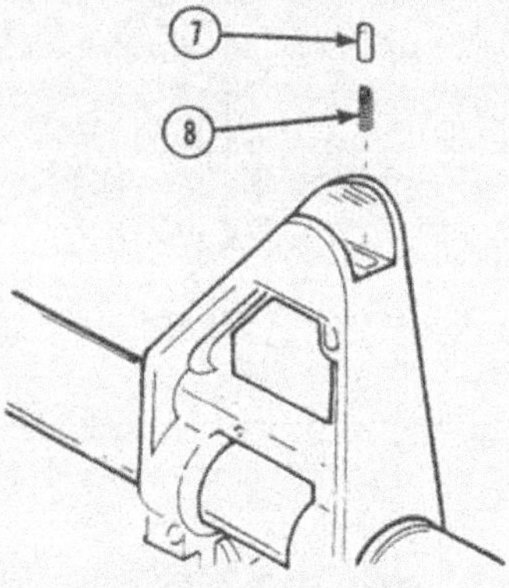

4. Catch front sight detent(7) and helical spring (8) to prevent loss.

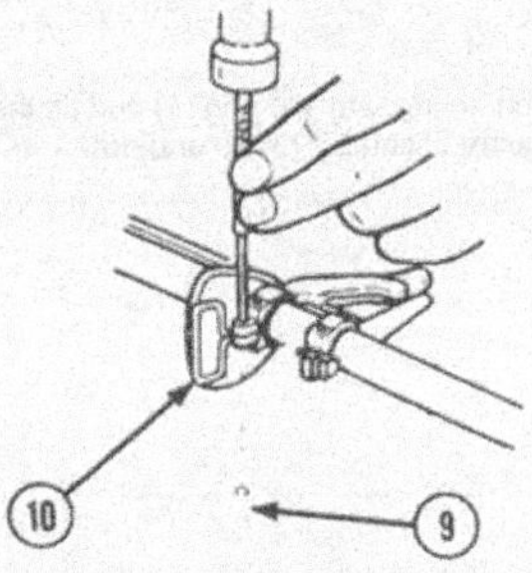

NOTE

RIFLE ONLY

Disassemble small sling swivel only if repair is necessary.

5. If necessary, knock out tubular rivet (9) with a hammer and punch and remove small sling swivel (10). Discard tubular rivet (9).

2-16. UPPER RECEIVE AND BARREL ASSEMBLY, RIFLE BARREL ASSEMBLY, AND UPPER RECEIVER ASSEMBLY (CONT).

a. DISASSEMBLY (CONT)

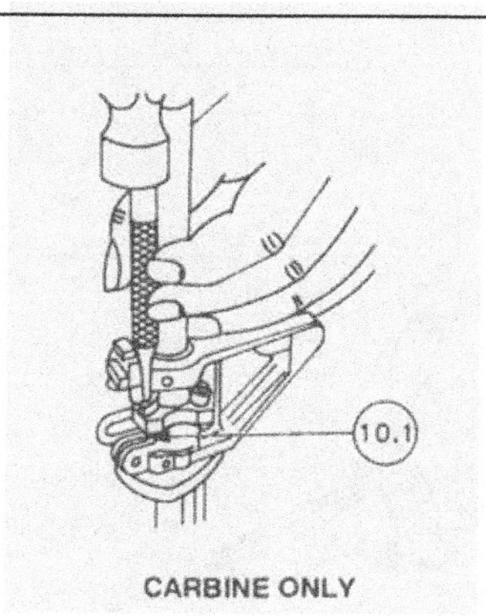

5A. Remove two spring pins (10 1)

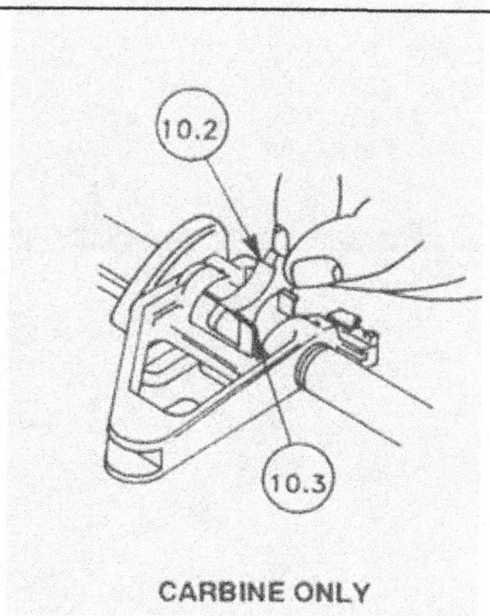

5B. Lift swivel locking bar (10 2) up and out of swivel mount (10 3)

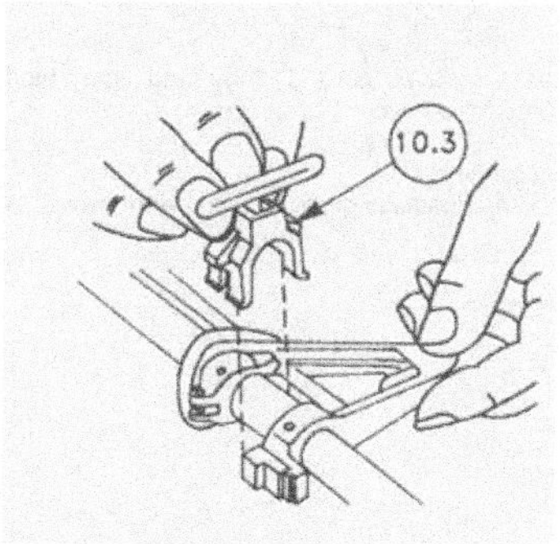

5C. Remove swivel mount (10.3).

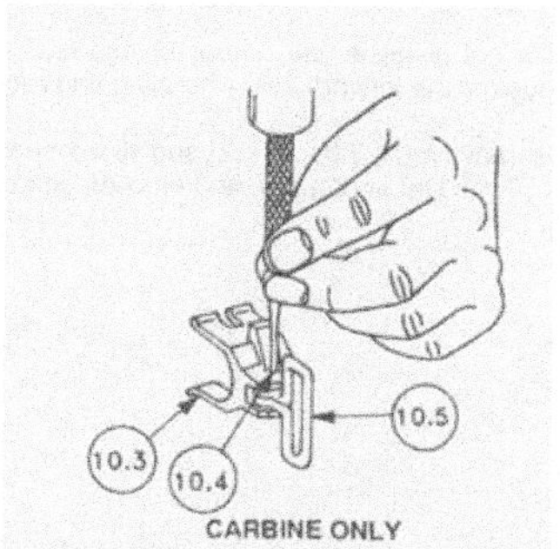

5D. Remove rivet (10 4) Separate front swivel (10 5) from swivel blank) mount (10 3)

2-16. UPPER RECEIVER AND BARREL ASSEMBLY, RIFLE BARREL ASSEMBLY, AND UPPER RECEIVER ASSEMBLY (CONT).

a. DISASSEMBLY (CONT)

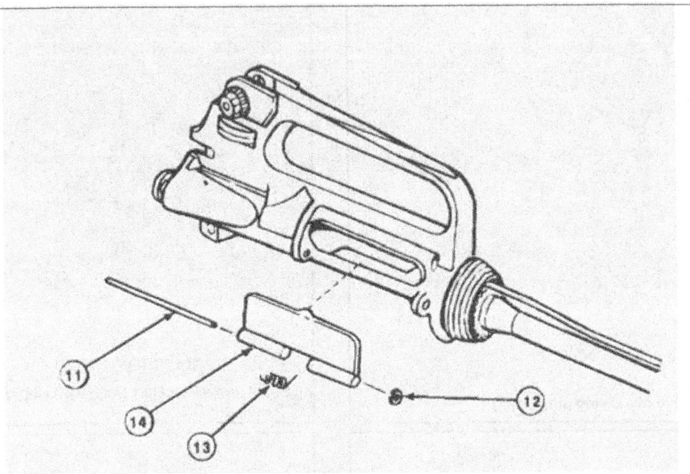

NOTE

Do not disassemble further unless repair is necessary. Headless grooved pin (11) may bind may bind against the forward assist housing and require some additional force to remove.

6. Remove retaining ring (12) and slide headles grooved pin (11) out to the rear.
7. Catch cover spring (13) and ejection port cover (14) to prevent loss as headless grooved pin is with-drawn.

M4/M4A1 ONLY

NOTE

Do not remove the carrying handle assembly nut (2 each) or clamp for normal maintenance. If carrying handle assembly nut or clamp are missing or the clamp is damaged to the point of being non-functional they may be replaced.

7.1. Disassemble as follows only if required to replace damaged or missing carrying handle assembly nut or clamp.

a. Observe the position of the clamp (15) in relation to the carrying handle (16).
b. Using care not to damage the surface finish, remove carrying handle assembly nut(s) (17) and clamp as required.
c. Inspect mating surfaces of the carrying handle (16) and the clamp (15) for damage. If damaged replace the damaged part.
d. Inspect for other damage, rust, etc. Repair or replace as required.
e. Apply a light coat of lubricant to all mating surfaces.
f. Install clamp (15) paying close attention to ensure that it is installed in its original position. The forward end of the clamp (15) must not extend past the forward end of the carrying handle (16).
g. Taking care not to cross thread, install handle assembly nut (17).

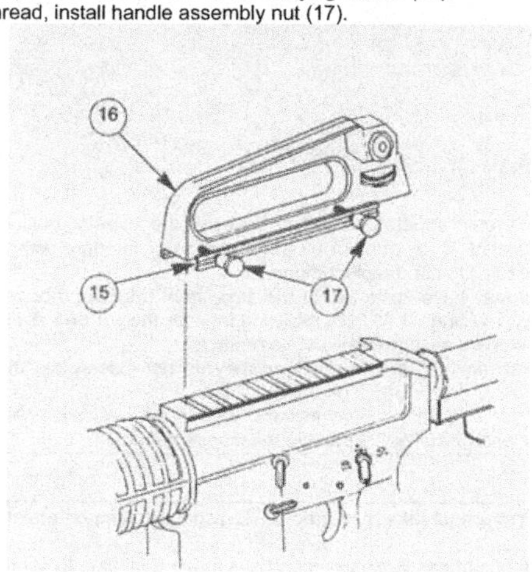

b. CLEANING

Clean all items (operator's manual).

c. INSPECTION

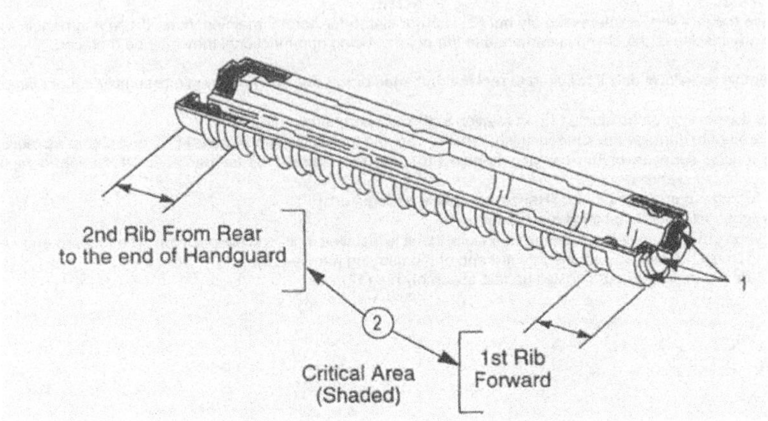

1. Inspect handguard assemblies for breaks separation, and cracks using the following guidelines:
   - **a.** Breaks and separations of material which prevent proper retention or interfere with functioning of the rifle will be cause for handguard assembly rejection and replacement.
   - **b.** M16A2 handguard assemblies may have up to two of the three front retaining tabs missing (1). M4/M4A1 may not have any front retaining tabs (1) missing. If all front retaining tabs for the M1 6A2 are missing, or any of the tabs for the M4/M4A1 are missing, handguard assemblies must be replaced.
   - **c.** Cracks up to one inch in length are acceptable provided they do not extend into the retaining flange (CRITICAL AREA) (2).
   - **d.** Replace severely cracked or damaged handguard assemblies. Handguard assemblies which have a heat shield which is loose enough to rattle when installed on the rifle must be replaced.

2. Inspect front sight assembly for chips, breaks, and cracks. Evacuate to support maintenance if broken, cracked, or bent.
3. Inspect front sight area for evidence of gas leakage around gas tube. Evacuate to support maintenance if short recoil results from gas leakage.
4. Inspect front sight post, front sight detent, and helical spring for damage. If damaged, replace.

2-16. UPPER RECEIVER AND BARREL ASSEMBLY, RIFLE BARREL ASSEMBLY, AND UPPER RECEIVER ASSEMBLY (CONT).

c. INSPECTION (CONT)

5. Inspect barrel for pits in bore, burrs, broken or worn locking lugs, and surface cracks and defects.
   a. Pits no wider than a land or groove and 3/8 inch (0.953 cm) or less in length are allowable in the bore.
   b. Uniformly fine pits in a densely pitted area of the bore are allowable.
   c. Lands that appear dark due to coating of gliding metal from projectiles are allowable.
   d. Striping of lands and grooves shall not be cause for rejection unless support maintenance determines by use of the barrel erosion gage.
   e. For pits other than mentioned above, broken or burred locking lugs, or surface cracks, evacuate to support maintenance.
6. Inspect bore for ringing. Definitely ringed bores or bores ringed sufficiently to bulge the outside surfaces of the barrel are causes for rejection. Evacuate to support maintenance.
7. Inspect chamber for pitting. Fine pits, or fine pits in a densely pitted area, are allowable. Pits 1/8 inch (0.318 cm) in length are cause for rejection. Evacuate to support maintenance.
8. Hand check compensator (3) for looseness on barrel. The third (middle) slot (4) must be straight up at top dead center (TDC). The alignment may vary as much as one half the width of the slot either direction. If loose or out of alignment, evacuate to support maintenance.
9. Inspect all items for serviceability in and tightness of latch assembly on ejection port cover. If items are damaged or nonfunctional, they are unserviceable.

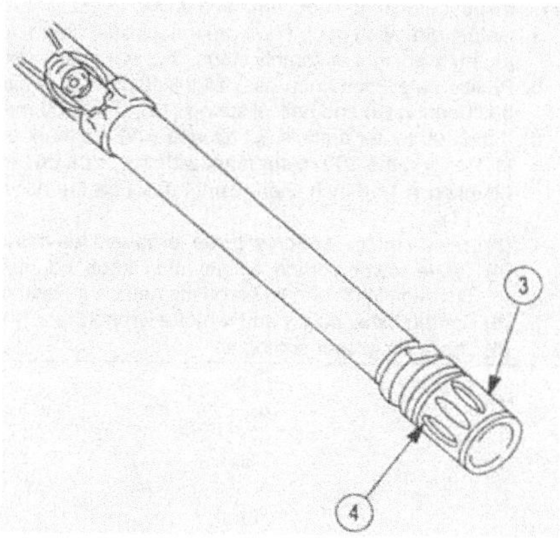

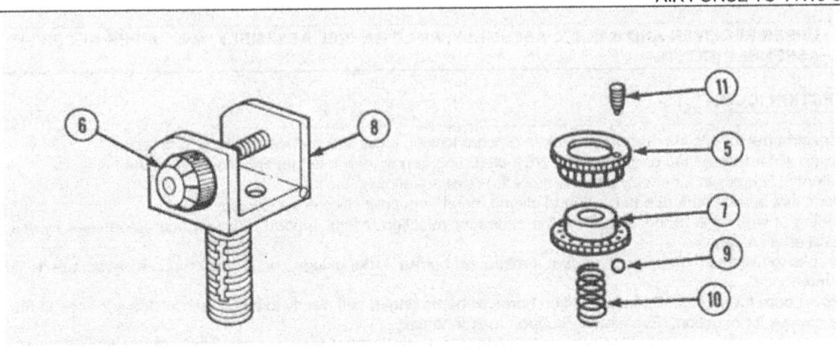

10. Rotate and test elevation index (5) and windage knob (6) for ease of functioning and legibility of markings.

11. Inspect elevation knob zero as follows:

a. Rotate elevation knob (7) counterclockwise until the rear sight assembly is all the way down. If a whole click is not felt as the rear sight assembly stops, the rear sight assembly has bottomed out and will not pivot freely.

b. Position elevation knob back slightly to Its last whole click so the rear sight assembly base (8) is under tension of the ball bearing (9) and helical spring (10). The 300 meter mark should align with the mark on the receiver.

c. If the 300 meter mark is not aligned with the mark on receiver, slip the range scale in the following manner:

(1) Position the 300 meter mark with the mark on the receiver.

(2) Insert a 1/16 inch allen wrench through the access hole of the rear sight assembly base and into the Index screw (11).

(3) Loosen the index screw three turns and leave the wrench in place.

(4) Rotate lower portion of elevation knob counterclockwise until it stops (range scale should not have moved). Elevation knob should be positioned on its last whole click.

(5) Tighten index screw and remove wrench.

(6) Check for proper setting.

2-16. UPPER RECEIVER AND BARREL ASSEMBLY, RIFLE BARREL ASSEMBLY, AND UPPER RECEIVER ASSEMBLY (CONT).

d. REPAIR

1. Replace all authorized unserviceable parts

2. Evacuate to support maintenance.

e. LUBRICATION

Lightly lubricate all Items (p 2 311 (oper ator's manual)

f. REASSEMBLY

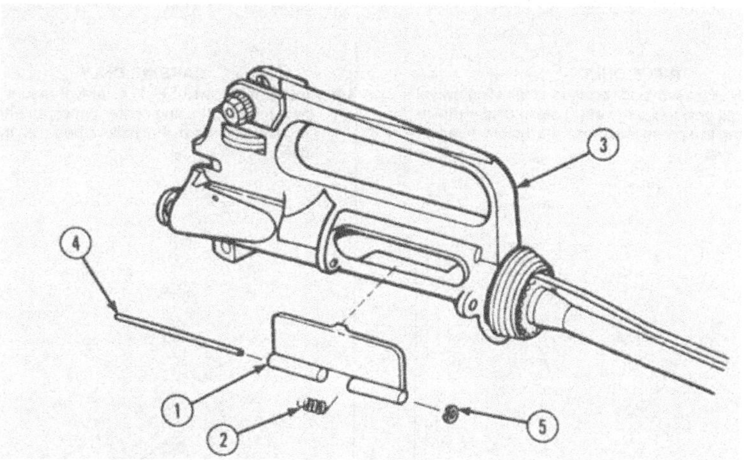

1. If previously disassembled, position election port cover 11) and helical spring (2) on Lip per receiver (3) with short leg of helical spring to the rear on Inside of ejection port cover

NOTE
Long legs of helical spring must be positioned and pretensioned before the headless grooved pin is installed

2. Hold helical spring (2) short leg In this position and turn long leg one half turn (180 degrees) with fingers of right hand

3. Position long leg of helical spring (21 against election port cover (1) Hold helical spring and election port cover in this position and Install headless grooved pin (4) Check for proper spring tension during installation of retaining ring (5)

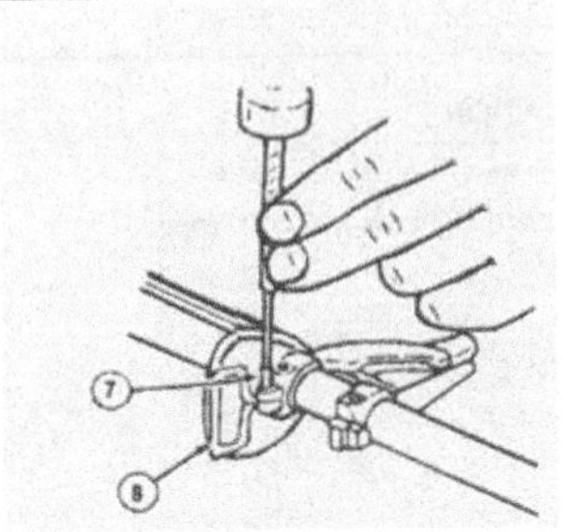

RIFLE ONLY

4. If previously disassembled, position small sling swivel (6) and install new tubular rivet (7) using center punch and hammer to spread and flare the hollow head of the tubular rivet.

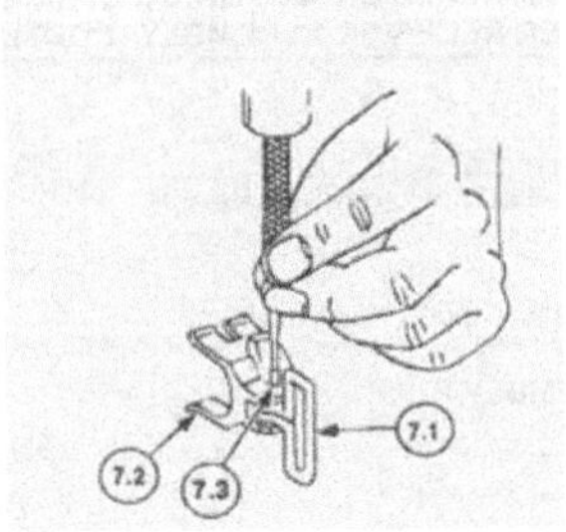

CARBINE ONLY

4A. Install front swivel (7.1) to swivel mount (7.2) with new rivet (7.3) using center punch and hammer to spread and flare the hollow head of the tubular rivet.

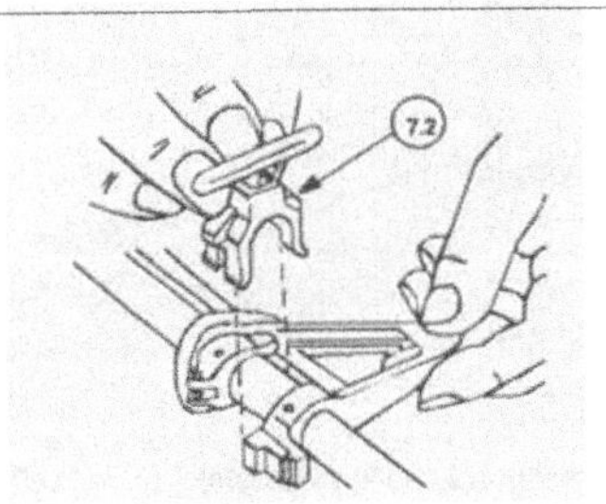

CARBINE ONLY

NOTE

It is recommended that the swivel mount be placed on the left side of the carbine for right handed operators and on the right side for left handed operators. This will keep the sling out of the way when the carbine is used.

4B. Install swivel mount (7.2).

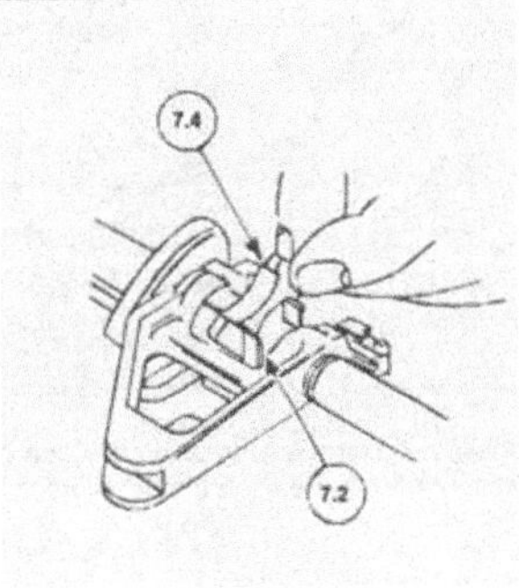

CARBINE ONLY

4C. Place swivel locking bar (7.4) in swivel mount (7.2).

2-16. UPPER RECEIVER AND BARREL ASSEMBLY, RIFLE BARREL ASSEMBLY, AND UPPER RECEIVER ASSEMBLY (CONT).

f. REASSEMBLY (CONT)

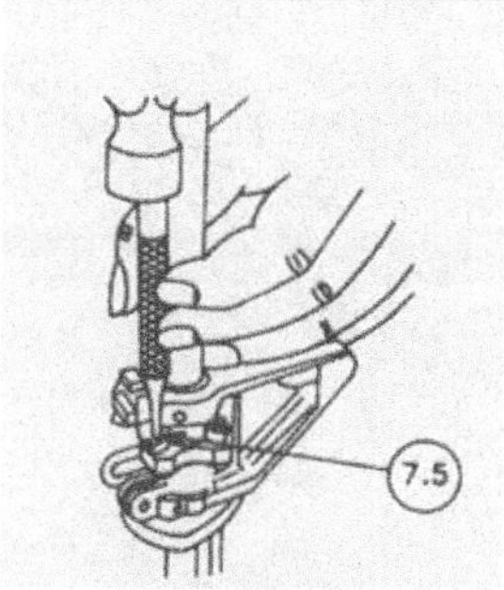

CARBINE ONLY

4D. Install two spring pins (7 5)

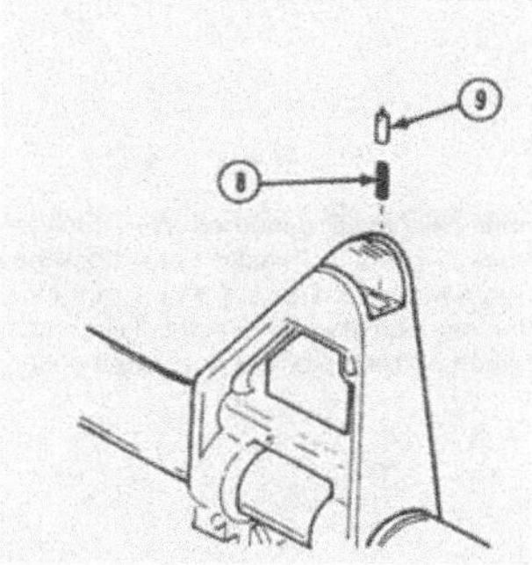

5. Position helical spring (8) and front sight detent (9).

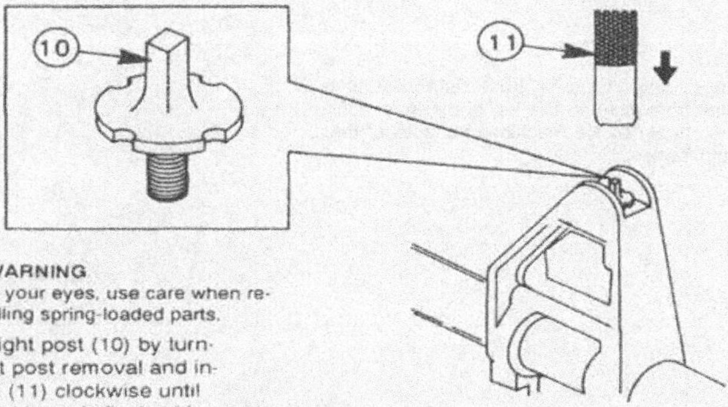

WARNING
To avoid injury to your eyes, use care when removing and installing spring-loaded parts.

6. Install front sight post (10) by turning front sight post removal and installation tool (11) clockwise until front sight post base is flush with or slightly below front sight frame.

2-16. UPPER RECEIVER AND BARREL ASSEMBLY, RIFLE BARREL ASSEMBLY, AND UPPER RECEIVER ASSEMBLY (CONT).

t. REASSEMBLY (CONT)

7. Mechanical Zero Procedures (A.F. Only).
   a. Mark a piece of plastic card stock or rigid paper with lines from 1 to 5mm in 1 mm increments Set the card on the front sight frame and check the height of the top of the front sight post.

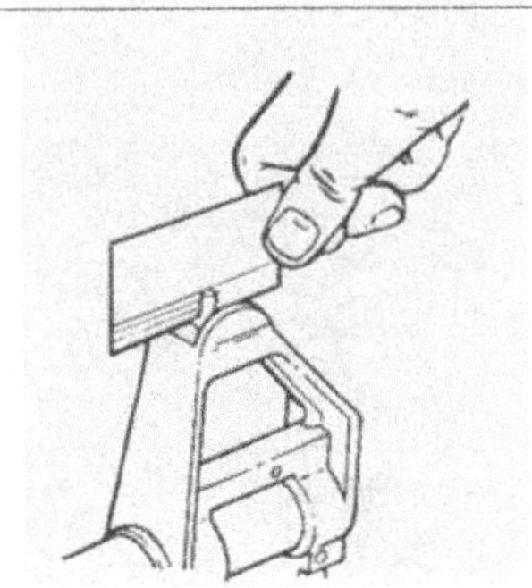

   b. Using a dummy round or front sight post tool adjust the front sight so the top of the front sight post is 5mm above the machined surfaces of the front sight frame.

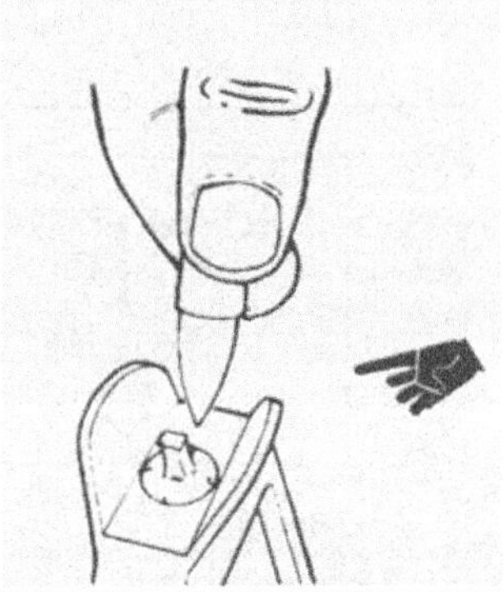

c. Visually check the front sight post top height by using the marked plastic or paper card. Card must set level on the machined surfaces of the front sight frame to obtain an accurate reading

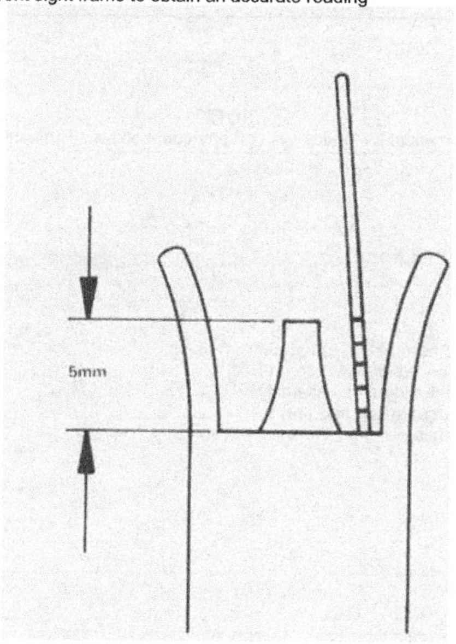

NOTE

This procedure, when used In conjunction with rear sight mechanical zero adjustment. will give an approximate battle sight zero to most M16A2 rifles The above steps can also be used before firing a new or newly assigned rifle Use the procedures to check rifles stored In preferred packaging during routine Inspections This will help ensure people armed with the rifles will stand a better chance of hitting an enemy If the rifles must be used before a live fire zero can be made Whenever possible, zeroing of the rifle should be accomplished using ball ammunition on a 25 meter zeroing target using the "L" aperture.

2-16. UPPER RECEIVER AND BARREL ASSEMBLY, RIFLE BARREL ASSEMBLY, AND UPPER RECEIVER ASSEMBLY (CONT).

f. REASSEMBLY (CONT)

NOTE

Refer to operator's manual for 'buddy system" procedure on installing hand guard assemblies

8. Install top of upper handguard assembly (12) In tube cap (13) while pushing down on handguard slip ring (14). Push bottom of upper handguard assembly (12) in place and release handguard slip ring (14) to lock handguard assembly In place

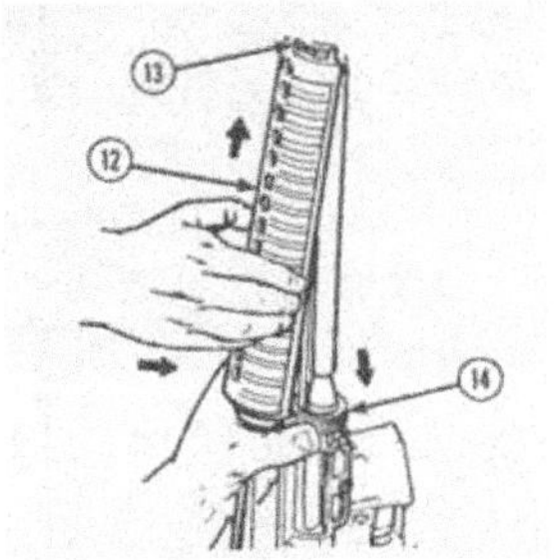

9. Install top of lower handguard assembly 1 5) In tube cap (13) while pushing down on handguard slip ring (14) Push bottom of lower handguard assembly (15) In place and release handguard slip ring {14) to lock both handguard assemblies In place

10. Reassemble rifle, refer to page 2-68

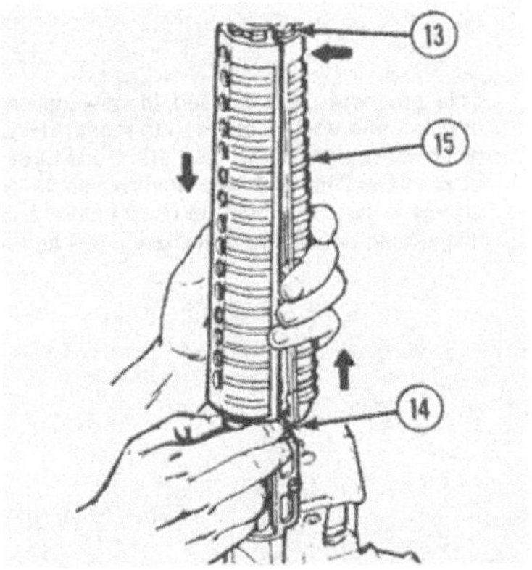

2-17. LOWER RECEIVER AND BUTTSTOCK ASSEMBLY.

This task covers:

a. Disassembly
b. Cleaning
c. Inspection
d. Repair
e. Lubrication
f. Reassembly

INITIAL SETUP

Tools
(ARMY) Small Arms Repairman Tol Kit (item 3, app B)
Pivot Pin Removal Tool (fig. E-3, app E)
Pivot Pin Installation Tool (fig. E-4, app E)

Materials/Parts
Lubricant, solid film (SFL) (item 21, app D)
Screw, self-locking (item 6, p C-1 1)

References
TM 9-1005-319-10 (operator's manual)

Equipment Conditions
2-34 Lower receiver and buttstock assembly removed

General Safety Instructions
To avoid injury to your eyes, use care when removing and installing spring-loaded parts.
When using solid film lubricant or dichloro-methane, be sure the area is well ventilated.

a. DISASSEMBLY

1. Remove screw (1) and lockwasher (2).

WARNING
To avoid injury to you eyes, use care when removing and installing spring-loaded parts.

2. Carefully remove pistol grip (3) and catch helical spring (4) and safety detent (5) to prevent loss.

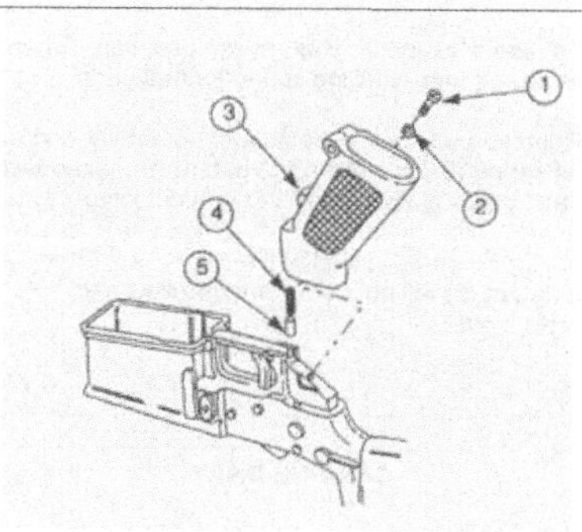

2-17. LOWER RECEIVER AND BUTTSTOCK ASSEMBLY (CONT).

a. DISASSEMBLY (CONT)

RIFLE ONLY

NOTE

If the self-locking screw is removed, it must be discarded and replaced with a new one

3. Remove self-locking screw (6).

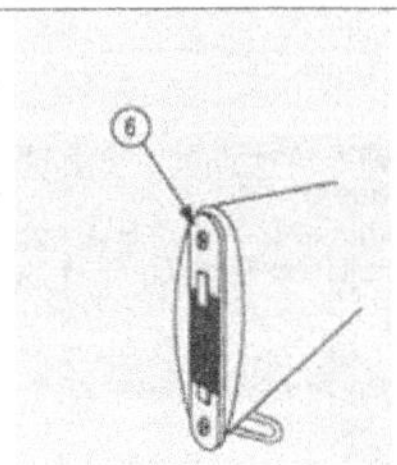

RIFLE ONLY

WARNING

To avoid injury to your eyes, use care when removing and installing spring-loaded parts.

4. Remove buttstock assembly (7) carefully and catch helical spring (8), takedown pin detent (9), takedown pin (10), and stepped spacer (11) to prevent loss.

NOTE

If detent (9) will not come out, use a wire to push it out.

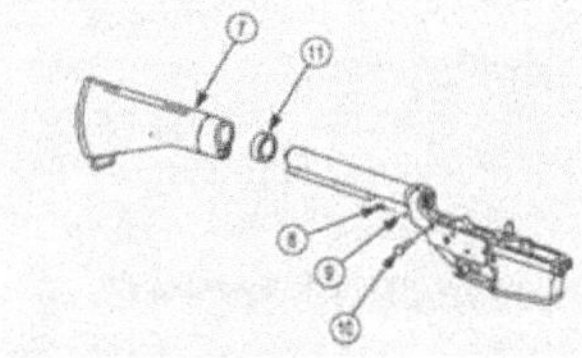

CARBINE ONLY

4A. Extend buttstock assembly (11.1).

4B Grasp the lock release lever (11.2) in the area of the retaining nut (11.3), pull down-ward, and slide buttstock to the rear to separate the buttstock assembly from the lower receiver extension

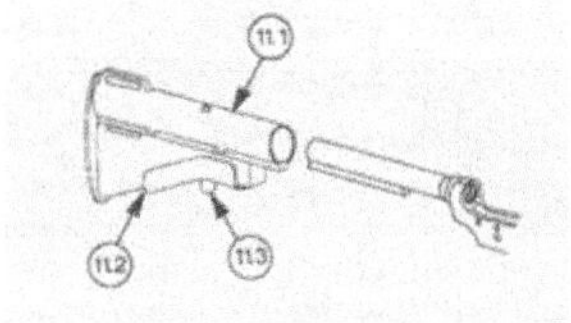

WARNING

To avoid injury to your eyes, use care when removing and installing spring-loaded parts

NOTE

Catch pivot pin detent and helical spring as pivot pin is removed (see step 6 on next page)

5. Depress pivot pin detent and spring and remove pin or Insert fabricated pivot pin removal tool (12) to compress pivot pin detent Turn pivot pin (13) a quarter turn Remove tool and pivot pin

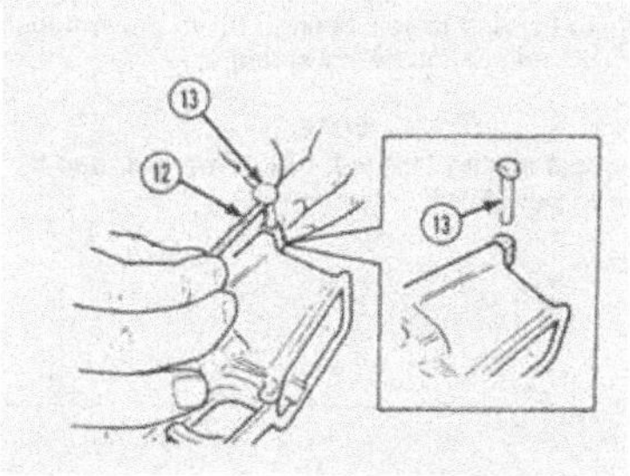

6. Be sure to hold cupped hand in front of pivot pin detent (14) and helical spring (15) to prevent loss of pivot pin detent and helical spring.

NOTE

If helical spring (15) will not come out, use a wire to pull it out.

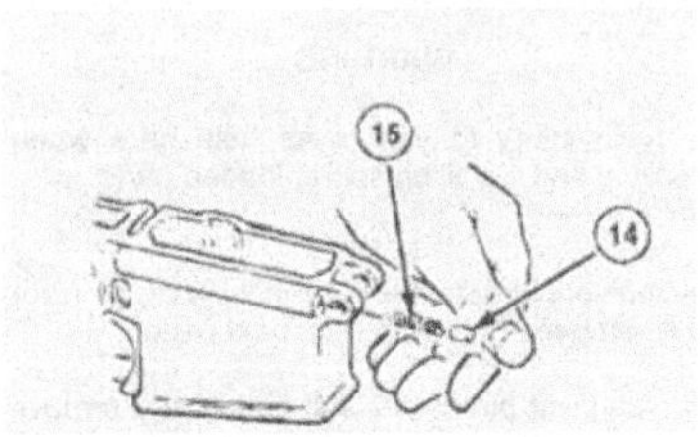

NOTE

Make sure the hammer is cocked and the selector lever is not set on BURST or AUTOMATIC before removing the buffer assembly.

7. Press buffer assembly (16) in about 1/4 inch (0.635 cm). Depress buffer retainer (17) and release buffer assembly (16) and action spring (18). Remove buffer assembly(16) and action spring (18) from receiver while depressing buffer retainer (17).

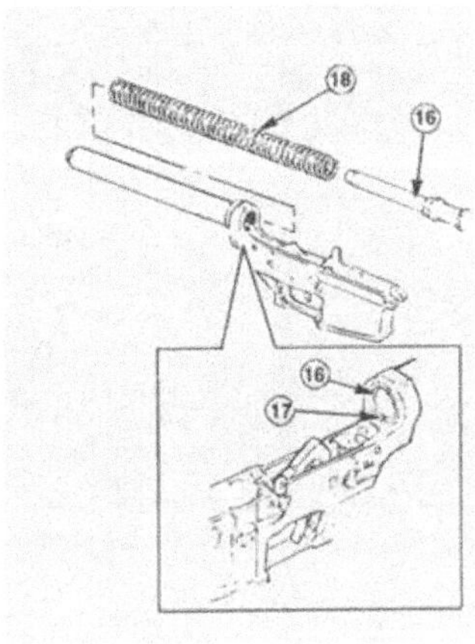

b. CLEANING

Clean all items (operator's manual). Remove carbon deposits

1 2-17. LOWER RECEIVER AND BUTTSTOCK ASSEMBLY (CONT).

c. INSPECTION I

1. Inspect buffer assembly (1, 2. 3, or 3 1) for cracks or damage

RIFLE ONLY

a. Some old buffers (1) have a hole with pin installed which protrudes equally on each side approximately 1/32 inch (O 08 cm)

b. Some buffers (2) have a hole In the housing but no pin

c. New buffers (3) do not have a hole In buffer body or a pin.

BOTH WEAPONS

2. If cracked or damaged, replace.

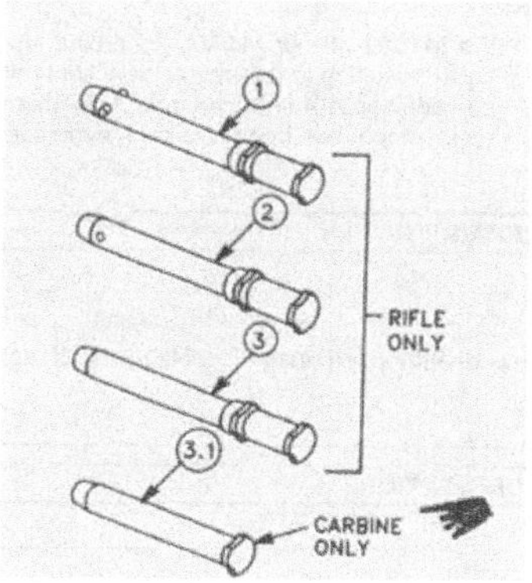

3. Check free length of action spring (4). The free length **FOR RIFLE ONLY** must be between 11 3/4 inches (29.85 cm) mini- mum and 13 1/2 inches (34 29 cm) maximum, **FOR CARBINE ONLY** must be between 10 1'/16 inches (25,56 cm) minimum and 11 114 Inches (28.58 cm) maximum, If not, replace Do not attempt to adjust the length by stretching the action spring

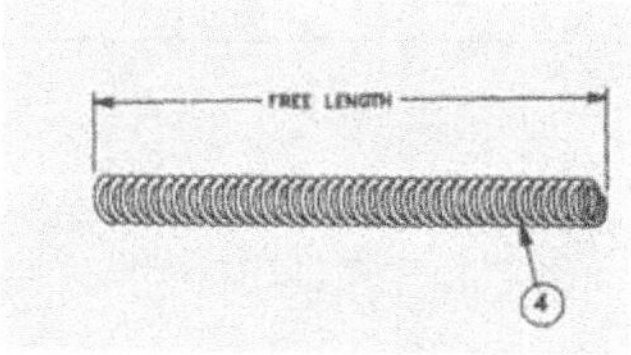

4. Inspect lower receiver (5) (without further disassembly) for legibility of serial number **ARMY ONLY**: If the serial number is hard to read, evacuate to direct support maintenance **AIR FORCE ONLY**: If the serial number is hard to read, evacuate to depot maintenance

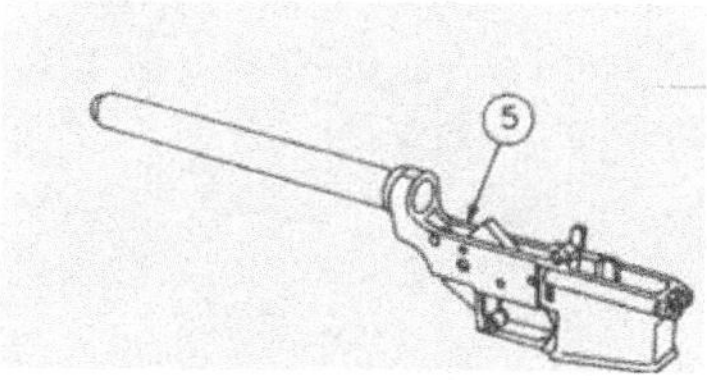

NOTE

AIR FORCE ONLY: Only depot maintenance is authorized to restamp the serial number.

5. Inspect for missing or damaged parts. Inspect finish of lower receiver for shiny spots. Touch up with solid film lubricant as required (p 2-34).

NOTE

If a M16A2 rifle or M4/M4A1 carbine lower receiver is missing one third or more of its exterior protective finish, resulting in an unprotected, light reflecting surface, it is candidate for overhaul. This missing finish will be considered a shortcoming. This shortcoming requires action to obtain a replacement weapon. Once a replacement has been received, evacuate the original weapon to depot maintenance for overhaul.

d. REPAIR

Replace all authorized unserviceable parts. If repair is not authorized at this level, evacuate to support maintenance.

e. LUBRICATION

Lightly lubricate all metal components (p 2-33) (operator's manual).

2-17. LOWER RECEIVER AND BUTTSTOCK ASSEMBLY (CONT).

f. REASSEMBLY

WARNING

To avoid Injury to your eyes, use care when removing and Installing spring loaded parts.

NOTE

Make sure the hammer Is cocked and the selector lever is not set on BURST before Installing the buffer assembly.

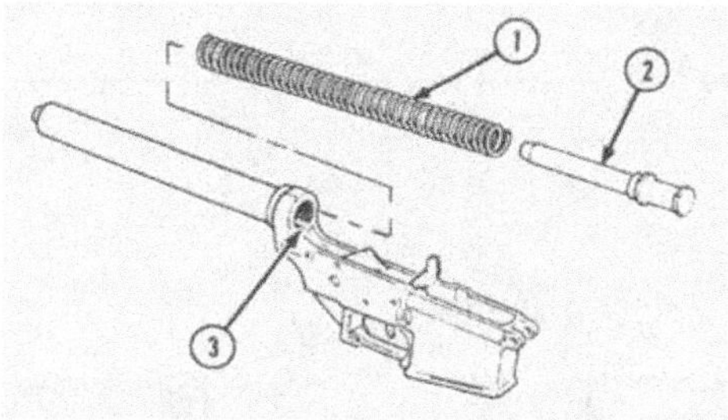

1. Press action spring (1) and buffer assembly (2) In until buffer retainer (3) snaps up and holds them In place.

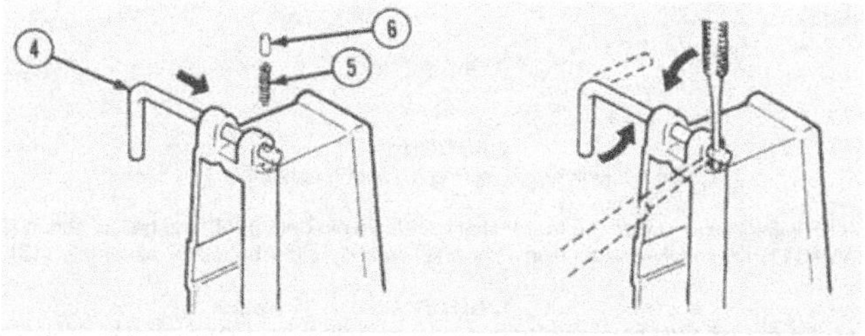

2. Install fabricated pivot pin installation tool (4) Insert helical spring 151 and pivot pin detent (6) Compress pivot pin detent in recess with 3 32 inch punch and rotate tool Remove 3'32 Inch punch

NOTE

Rounded end of pivot pin detent must be In the groove of the pivot pin (7) when assembly is complete.

3. Install pivot pin (7) while removing fabricated pivot pin Installation tool (4) Maintain pressure while sliding pivot pin (7) into hole Rotate pivot pin until pivot pin detent is inserted Into pivot pin groove

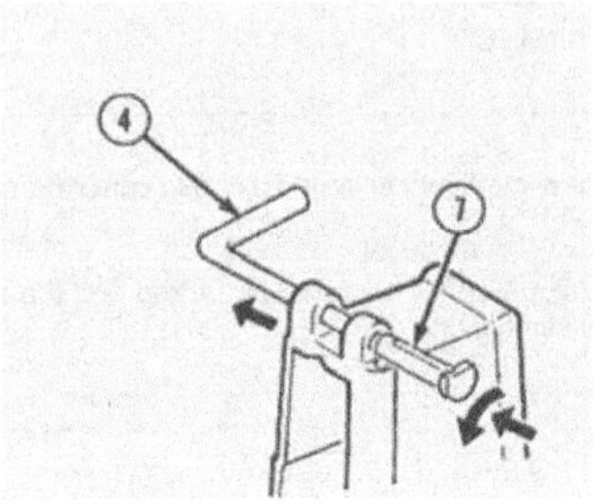

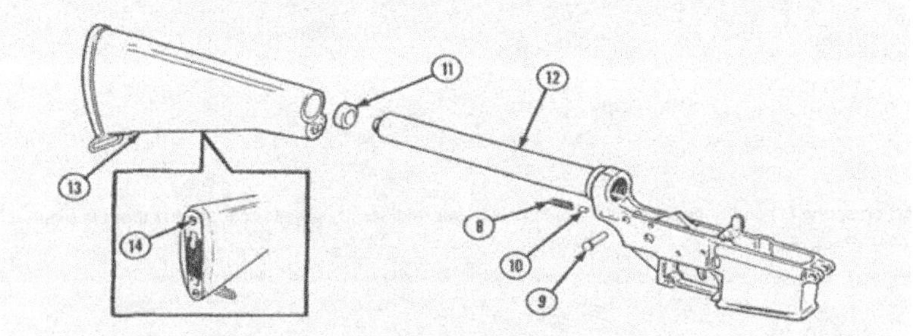

RIFLE ONLY

CAUTION

Do not kink helical spring (8) during assembly

4. Install takedown pin (9) with groove toward the rear Install takedown pin detent (10)and helical spring (8) from the rear
5. Install stepped spacer (11) on receiver extension (12) and carefully slide buttstock assembly (13) on to compress helical spring (8)

NOTE

Self-locking screw (14), If removed must be discarded and replaced with a new one 6. Install self-locking screw (14) to secure buttstock assembly (13)

6. Install self-locking screw (14) to secure buttstock assembly (13)

1 2-17. LOWER RECEIVER AND BUTTSTOCK ASSEMBLY (CONT).

f. REASSEMBLY (CONT)

CARBINE ONLY

6A. Grasp the lock release lever (14 1) In the area of the retaining nut (14 2) and pull to reinstall the buttstock assembly (14.3) onto the lower receiver extension (14 3) onto the lower receiver extension (14 4)

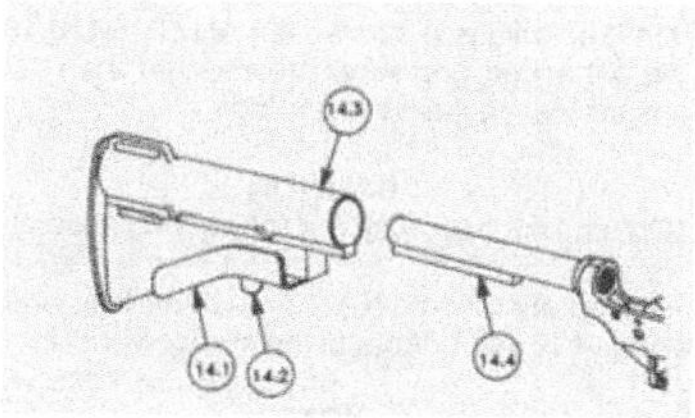

2-17. LOWER RECEIVER AND BUTTSTOCK ASSEMBLY (CONT).

f. REASSEMBLY (CONT)

WARNING

When utilizing the enhanced rifle grip (it has a bump between the second third finger for a better grip) PN 9349127, rifle grip screw. PN AN 501D416 18 (1-1/8 in.) or AN 501 D416 16 (1 in), is authorized to be(used with the enhanced grip Any screw longer than 1-1 8 In used with 9349127 could cause a hazardous situation Also, ensure the washer Is in 11place

CAUTION

Do not kink helical spring (15) during assembly

7. Install safety detent (16), pointed end first, and helical spring (1 5) into bottom of lower receiver (17)

NOTE

A portion of the helical spring will fit In a hole in the pistol grip

8. Carefully install pistol grip (18) to compress helical spring 115). Secure pistol grip 118) in place with lockwasher (19) and screw 120)

9. Reassemble rifle, refer to page 2 68

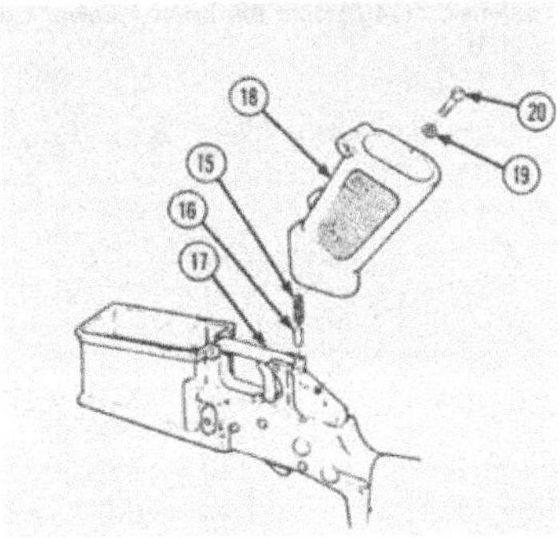

2-18. BUTTSTOCK ASSEMBLY.

This task covers:
a. Disassembly
b. Cleaning
c. Inspection
d. Repair
e. Lubrication
f. Reassembly

INITIAL SETUP

Tools
(ARMY) Small Arms Repairman Tool Kit
(item 3, app B)

Equipment Conditions
2-57 Buttstock assembly removed from lower receiver and buttstock assembly

a. DISASSEMBLY

RIFLE ONLY

1. Remove self-locking screw (1), small sling swivel (2), and buttplate group (3) from buttstock (4)

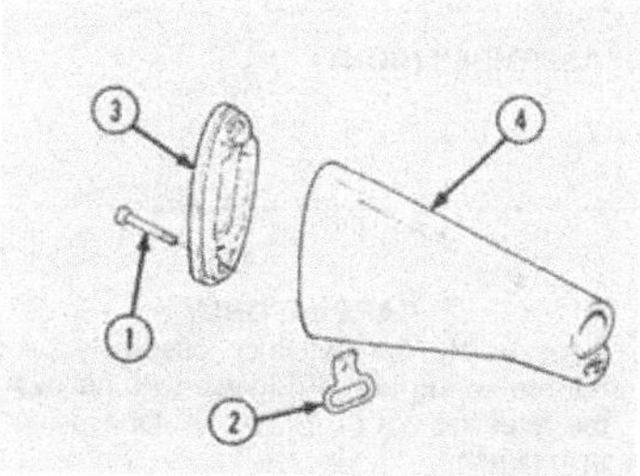

RIFLE ONLY

2. Push down on plunger (5) and lift door assembly (6) out of buttplate (7).

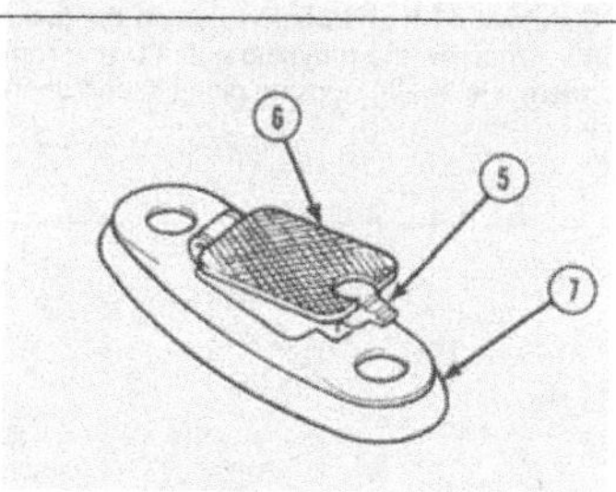

RIFLE ONLY

3. Remove straight pin 181 and separate hinge (9) and door assembly (61

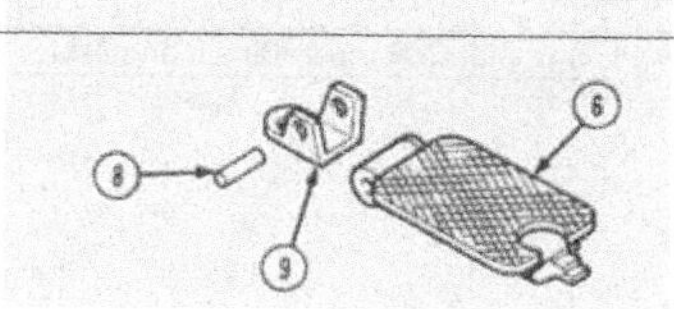

2-18. BUTTSTOCK ASSEMBLY (CONT).

a. DISASSEMBLY (CONT)

CARBINE ONLY

4. Disassemble the buttstock assembly by tapping out the spring pin (10) located in ,he oval slot of the retaining nut (1 1) This is done using a 1 16 Inch punch

5. Insert your Index finger into the forward end of the buttstock 112) and push down on the locking pin (13 Unscrew the retaining nut 11 and remove the release lever (14) locking pin (13) and spring (15)

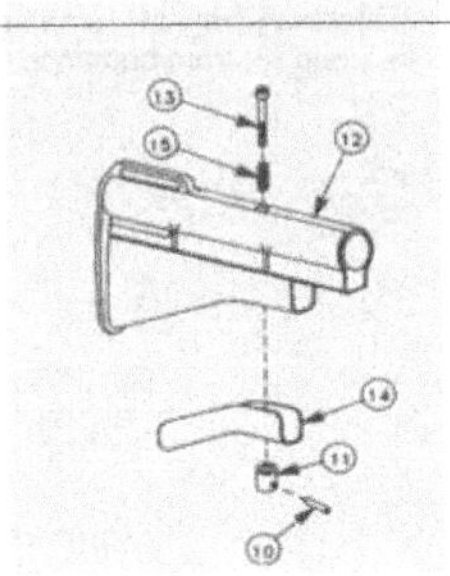

b. CLEANING (RIFLE ONLY)

Clean all parts with CLP (operator s manual) Use brush to clean knurled surface of door assembly

2-18. BUTTSTOCK ASSEMBLY (CONT).

c. INSPECTION

NOTE

M16A2 buttstocks, PN 9349121, with unauthorized markings may be used under the following conditions:

a. The only authorized markings are those which are temporary In nature, i e , paint, tape, etc

b. When marking a buttstock, only use temporary markings

c. Buttstocks with unauthorized markings that have been stamped into the surface of the buttstock will not be used

d. Unauthorized markings that have previously been scratched, etched, carved, etc may continue In use if the marks do not extend Into the fiber of the buttstock Cutting Into the fiber of the buttstock may weaken it

e. These marks may be at any location on the buttstock Unauthorized markings are not desirable. However, if previously applied, they will be allowed to continue In use due to the cost of the buttstock.

1. Inspect buttstock for cracks using the following guidelines:

   a. Under the following conditions, hairline cracks (no chipped away material allowed) originating from buttplate end of buttstock are acceptable.
      (1) One hairline crack, not to exceed 1 in. 12.54 cm) In length, per side of buttstock.
      (2) Two additional hairline cracks up to 0 25 in (0 64 cm) In length, per side of buttstock
      (3) A total of three cracks per side of buttstock, originating from buttplate end, are allowable.

   b. Cracks In the critical area at the front end of the buttstock are not acceptable and these buttstocks must be replaced.

2. While buttplate Is Installed on rifle, Inspect for cracks around the mounting holes Check for cracks In excess of 0 25 In. (O 64 cm) In length which extend through the buttplate Replace If cracked.

3. Inspect door assembly for cracks, corrosion, stuck plunger, separations on outer face, or other damage Replace If defective.

d. REPAIR

Replace all authorized unserviceable Items Unserviceable Items are those Items which are damaged.

e. LUBRICATION

Lubricate all metal components (p 2-33) (operator's manual)

f. REASSEMBLY

RIFLE ONLY

1. Position hinge (1) on door assembly (2) and Install straight pin (3)

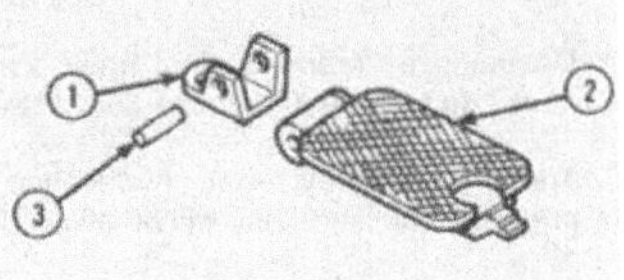

RIFLE ONLY

2. Install door assembly (2) into buttplate (4) and press plunger (5) to lock

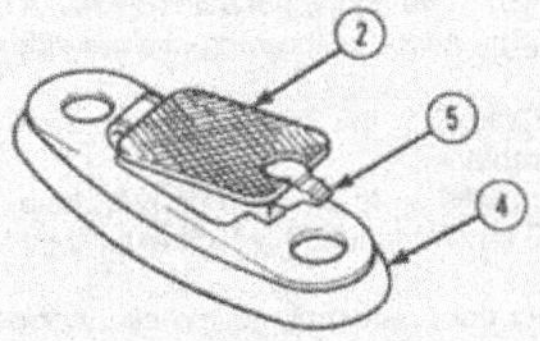

RIFLE ONLY

3. Position buttplate group (6) and small sling swivel (7) to the buttstock (8) and secure with self-locking screw (9).

NOTE

See page 2-62, reassembly, for reassembly of buttstock assembly to lower receiver

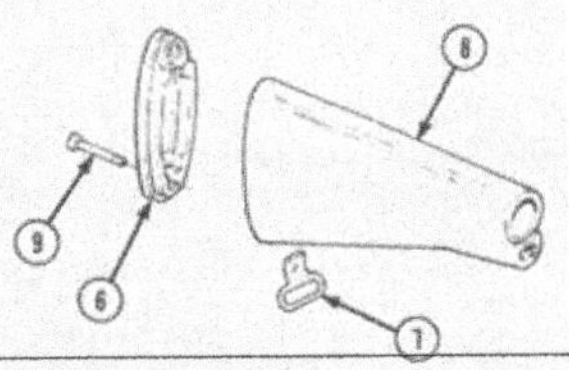

2-18. BUTTSTOCK ASSEMBLY (CONT).

f. REASSEMBLY (CONT)

CARBINE ONLY

4. Insert locking pin spring (10) onto locking pin (11)

5. Insert locking pin (11) and spring (10) Into the hole on top of the buttstock (12), threaded end first

6. Insert your index finger into the forward end of the buttstock (12) and push down on locking pin (11)

7. Install the release lever (13) onto the threaded portion of the locking pin (11) protruding through the bottom of the buttstock (12)

8. Screw on the locking nut (14) until flush with locking pin (11) Align the slot In locking nut (14) with the spring pin hole in the locking pin (11)

9. Lightly tap in spring pin (15) until flush on both sides of the retaining nut (14)

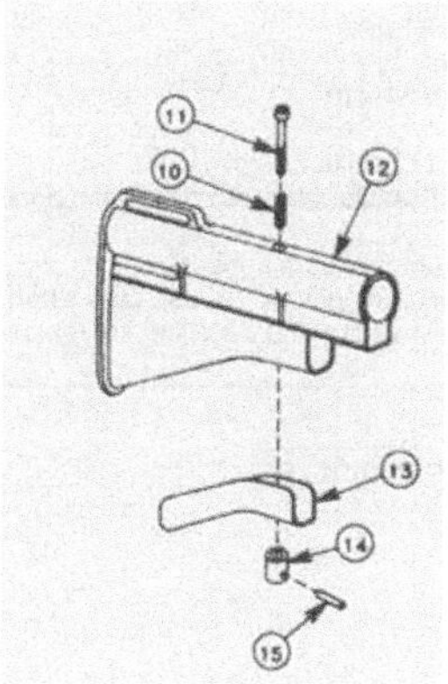

2-19. MAJOR COMPONENTS OF M16A2 RIFLE.

This task covers

a. Reassembly
b. Inspection
c. Stowage

INITIAL SETUP

References
TM 9 1005-319 10

Equipment Conditions
Rifle disassembled into major components

General Safety Instructions
To avoid injury to eyes, use care when removing and installing spring-loaded parts

Do not Interchange bolt assemblies from one rifle to another Doing so may result In Injury to, or death of. personnel
Do not keep live ammunition near the work area

a. REASSEMBLY

Refer to operator's manual and assemble lower receiver and buttstock assembly (1) upper receiver and barrel assembly (2). charging handle assembly (3), bolt carrier assembly (4), sling (5), and magazine(6).

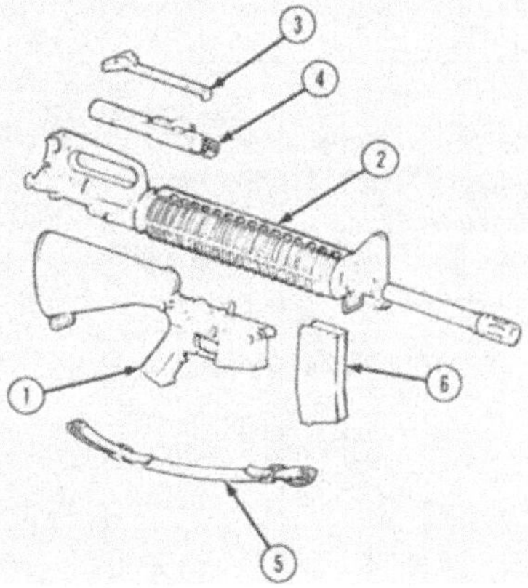

b. INSPECTION

Perform the following function checks on assembled weapon:

1. Remove magazine if installed. Pull charging handle assembly to rear. Check that chamber is clear. Let bolt and bolt carrier close. Do not pull trigger. Leave hammer in cocked position.

WARNING

If rifle fails any of the following tests, continued use of the rifle could result in injury to, or death of, personnel.

2. Place selector lever in SAFE position and pull trigger. Hammer should not fall.

3. Place selector lever in SEMI position. Pull trigger. Hammer should fall.

M16A2 and M4 ONLY
NOTE

For the purpose of the following check "SLOW" is defined as 1/4 to 1/2 the normal rate of trigger release.

4. Hold trigger to the rear, charge weapon, and release the trigger with a slow, smooth motion, without hesitations or stops, until the trigger is fully forward; an audible click should be heard. Hammer should not fall.

5. Repeat the SEMI position test five times. The weapon must not malfunction during any of these five tests. If the weapon malfunctions during any of these five tests, evacuate rifle to support maintenance for repair.

6. Place selector lever in BURST position. Charge weapon and pull trigger. Hammer should fall.

7. Hold trigger to the rear, pull charging handle assembly to rear and release three times. Release trigger. Hammer should not fall. The burst disconnector should have held the hammer to the rear while the trigger was in the pulled position.

8. Pull trigger. Hammer should fall. This should be the first round of a three round burst.

9. With hammer in forward position, attempt to place the selector lever in the SAFE position. If selector lever can be placed on SAFE, evacuate the weapon to support maintenance.

M4A1 ONLY

10. Place selector lever in AUTO position. Charge weapon and pull trigger. Hammer should fall.

11. Hold trigger to the rear, charge weapon and release trigger. Pull trigger. Hammer should not fall. AUTO sear should have released hammer while holding trigger in the squeezed position before releasing and resqueezing the trigger.

12. With hammer in forward position, attempt to place the selector lever in the SAFE position. If selector lever can be placed on SAFE, evacuate the weapon to support maintenance.

1 2-19. MAJOR COMPONENTS OF M16A2 RIFLE (CONT).

c. STOWAGE

Prior to stowing the weapon in arms room, perform the following procedures:

NOTE

Weapon passed into arms room issue window should be passed butt first with the bolt locked to the rear.

1. Clear. Refer to operator's manual.
2. Place selector lever in SEMI position.
3. Pull trigger. Hammer should fall.
4. Close ejection port (dust) cover.
5. Place weapon in rack.

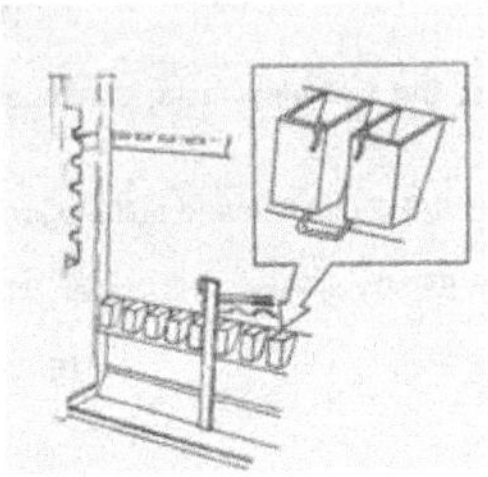

6. **M4/M4A1 Carbine Only, use M12 Arms Rack.**

a. The M12 arms rack is the correct arms rack in which to store the M4/M4A1 Carbine. The carbine must be stored with buttstock extended. When storing the M4/M4A1 Carbine in the M12 arms rack a mounting bracket, NSN 5340-01-230-3181, (app D, item 23.1) may be used for each M4/M4A1 Carbine being stored. This option is for the convenience of the person who opens and closes the arms rack to store the carbines.

b. To install the mounting bracket on the M12 arms rack, for use with the M4/M4A1 Carbine, install the bracket with the hooks of the bracket facing toward the carbine, so that the lower receiver extension will contact the bent end of the bracket. The bent end of the bracket will hold the carbine upright when the arms rack is opened. The bracket can be turned around when not in use for the carbine to allow storage of the M16A2 rifle.

c. When storing the M4/M4A1 Carbine in the M12 arms rack, an adapter bar (fig E-6, app E), MUST be used for security reasons. To install the adapter bar to the M12 arms rack:

(1) Remove all weapons from the rack and position the rack to gain access to the back.

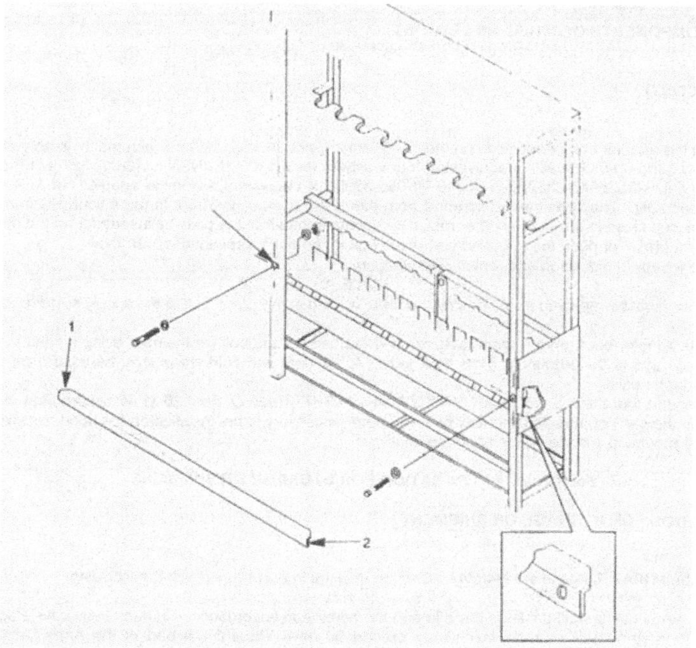

(2) The side of the adapter bar with the corners (1) cut off is the top and the side with square corners (2) is the back. The portion of the bar must be placed so the cut off corners (1) face the front of the rack.

(3) Holding the adapter bar at an angle, place one end into position inside the rear leg of the arms rack. Lower the other end of the bar into position. Allow the adapter bar to rest on the rack.

(4) Clamp both ends of the adapter bar into position. Mark the center line of the leg and adapter bar where they meet (see illustration) Using a center punch, mark the location of the holes to be drilled where the center lines cross. The holes must be centered on both the leg and the adapter bar. Drill a 1/8 inch pilot hole through both arms rack legs and the ends of the adapter bar. Drill a 3/8 inch hole through both arms rack legs and the ends of the adapter bar using the pilot hole as a guide. Remove the adapter bar. File the edges of all holes smooth Paint all bare metal surfaces with olive drab enamel paint, NSN 8010-01-350-5249, or equivalent.

2-19. MAJOR COMPONENTS OF M16A2 RIFLE (CONT).

c. STOWAGE (CONT)

(5) Reinsert the adapter bar into position on the arms rack. Using two 3/8 inch x 2 inch machine screws (MS35206-315, NSN 5305-00-984-5695) or equivalent, four washers (MS21306-1C, NSN 5310-01- 327-9713) or equivalent and two nuts (MS35649-2382, NSN 5310-00-056-3395) or equivalent assemble adapter bar to arms rack and tighten securely. The bolts can be inserted from either the back or the front to meet your requirements. If the rack is placed close to a wall or another rack It is recommended that the bolts be inserted from the back

(6) Tack weld, braze, or peen the threaded end of the bolt to the nut to prevent easy removal

(7) Place rack back into position and replace the weapons

NOTE

Adapter bar must be removed from arms rack prior to turning in the arms rack to the supply system.

d. To remove the adapter bar from the arms rack, remove the head of the bolt (or the nut), using a hand held grinder or file to avoid damage to the aluminum arms rack legs. A hammer and cold chisel may be used If no other way to remove the adapter exist

e. It is recommended that the rail protector NSN 1005-01-394-7677, (app D, item 26.1) be used during storage of the carbines when the carrying handle assembly or some other accessory is not installed on the upper receiver to prevent damage to the mounting surface on the upper receiver.

Section VI. PREPARATION FOR STORAGE OR SHIPMENT

2-20. PREPARATION FOR STORAGE OR SHIPMENT.

a. Packaging of the M16A2 Rifle and the M4/M4A1 Carbine shall be in accordance with the following

ARMY ONLY: Army users shall package the rifle and the carbine in accordance with each respective Packaging Data Sheet (PDS) for shipment or storage which may exceed 90 days The PDS is part of the Army Master Data File Retrieval Microform System (ARMS) Packaging
File

AIR FORCE ONLY: Air Force users shall package the rifle in accordance with each respective Special Packaging Instruction (SPI) 00-856-6885 for shipment or storage which may exceed 90 days The SPIs are part of the Army Master Data File Retrieval Microform System (ARMS) Packaging File

b. Packaging, if required, for shipping/storage which will not exceed 90 days shall be as follows

(1) Clean in accordance with operator's manual

(2) Wrap with MIL-B-121 waterproof material.

(3) Place in barrier bag MIL-B-117, Type I, Class C, or wrap with MIL-B-121, Type I, Grade A, and seal with tape, PPP-T-76.

(4) Place one or more of item in minimum size container. Block and brace in accordance with MIL-STD-1186. Cushion the M16 and similar weight items with PPP-C-843, and use PPP-F-320 as filler, to create a tight pack.

(a) Fiber board containers shall be in accordance with PPP-B-636 and may be Class Domestic. Gross weight and size of material shall determine grade of fiberboard container. PPP-B-640 may also be used.

(b) Wood containers shall be in accordance with PPP-B-601 or PPP-B-621.

(5) Equivalent materials may be used.

c. NSNs are not assigned to all the specified material. If it is necessary to specify an NSN in the TMs, the packing materials will have to be spared and part numbers and NSNs assigned.

d. The specifications used are:

(1)	MIL-B-121	Barrier Material, Greaseproofed, Waterproofed, Flexible (NSN 8135-00-753-4661)
(2)	MIL-B-117	Bag, Sleeve and Tubing -- Interior Packaging (NSN 8135-00-543-6574)
(3)	PPP-B-636	Boxes, Shipping, Fiberboard
(4)	PPP-B-601	Boxes, Wood, Cleated Plywood
(5)	MIL-STD-129	Marking for Shipment and Storage
(6)	PPP-T-76	Tape, Packaging, Paper
(7)	MIL-STD-1186	Cushioning, Anchoring, Bracing, Blocking and Waterproofing; with Appropriated Test Methods
(8)	PPP-C-843	Cushioning Material, Cellulosic
(9)	PPP-F-320	Fiberboard, Corrugated and Solid Sheet Stock (Container Grade), and Cut Shapes
(10)	PPP-B-640	Boxes, Fiberboard, Corrugated, Triple-Wall
(11)	PPP-B-621	Boxes, Wood, Nailed and Locked -- Corner

CHAPTER 3

DIRECT SUPPORT MAINTENANCE INSTRUCTIONS

CHAPTER OVERVIEW

This chapter provides information and instructions to keep the weapon in good repair and contains the following sections:

a. Repair Parts and Special Tools
b. Direct Support Troubleshooting
c. Direct Support Maintenance Procedures for the M1 6A2 Rifle and M4/M4A1 Carbine
d. Preparation for Storage or Shipment
e. Preembarkation Inspection of Material in Units Slated for Overseas Movement

Section I. REPAIR PARTS, SPECIAL TOOLS, TMDE, AND SUPPORT EQUIPMENT

3-1. COMMON TOOLS AND EQUIPMENT. For authorized common tools and equipment, refer to the Modified Table of Organization and Equipment (MTOE) applicable to your unit. Air Force users must maintain the following common tools:

Flat tip screwdriver
Socket wrench handle and socket head screw socket wrench
Vise jaw caps
Machinist's vise
Solid center punch
Hammer
Combination wrench
Torque wrench
Retaining ring pliers
(Two) 8-inch adjustable wrenches
Flat file
Ball-peen hammer
Trigger pull test fixture rod and weights
5/64-inch drive pin punch
1/8-inch drive pin punch
1/16-inch drive pin punch
3/32-inch drive pin punch

3-2. SPECIAL TOOLS, TMDE, AD SUPPORT EQUIPMENT. Special tools required for direct support maintenance are listed in appendixes B and C. Fabricated tools are listed and illustrated in appendix E.

3-3. REPAIR PARTS. Repair parts are listed and illustrated in appendix C of this manual.

NOTE

Bolt assemblies, and/or barrel assemblies may be interchanged , at the Direct Support Maintenance level, from one rifle to another, under the provisions of the note on page C-3: If these parts are interchanged the rifle must be checked/inspected as depicted on pages 3-17, 3-21, and 3-33. While performing these checks and inspections, pay special attention to the headspace requirements on page 3-47.

Section II. TROUBLESHOOTING

3-4. GENERAL.

a. This section contains direct support level troubleshooting information for locating and correcting most of the operating troubles which may develop in the M16A2 rifle and M4/M4A1 carbine. Each malfunction for the individual part or assembly is followed by a list of tests or inspections which will help you to determine the corrective action to take. You should perform the tests/inspections and corrective actions in the order listed.

b. This manual cannot list all malfunctions that may occur, nor all tests or inspections and corrective actions. If a malfunction is not listed or is not corrected by listed corrective actions, see individual repair sections in the maintenance procedures on each major assembly.

3-5. TROUBLESHOOTING PROCEDURES. Refer to troubleshooting table for malfunctions, tests, and corrective actions. The symptom index is provided for a quick reference of the malfunctions covered in the table.

SYMPTOM INDEX

	Troubleshooting Procedures Page
Failure of magazine to lock in rifle	3-3
Failure to feed	3-3
Failure to chamber	3-3
Failure to lock	3-4
Failure to fire	3-5
Failure to unlock	3-6
Failure to extract	3-6
Failure to eject	3-6
Failure to cock	3-7
Short recoil	3-9
Rifle cannot be zeroed	3-10
Failure to cycle with selector lever set o n BURST	3-11
Fires two rounds with one pull of trigger with selector lever set on SEMI (double firing)	3-13
Fires with selector lever on SAFE on when trigger is released with selector lever on SEMI	3-13
Hammer pin "walks"	3-14
Bolt assembly fails to lock to rear after firing last round	3-14

TROUBLESHOOTING

MALFUNCTION
TEST OR INSPECTION
CORRECTIVE ACTION

1. FAILURE OF MAGAZINE TO LOCK IN RIFLE.
   Step 1. Dirty or corroded magazine catch (1).
   Disassemble and clean.
   Step 2. Defective magazine catch spring (2).
   Replace magazine catch spring (2) (p 3-62).
   Step 3. Worn or broken magazine catch (1).
   Replace magazine catch (1) (p 3-62).

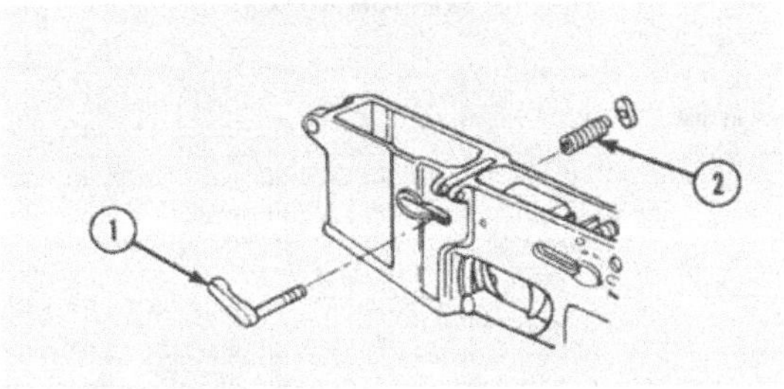

2 FAILURE TO FEED
   Step 1 Magazine catch spring (1) weak or broken.
   Replace magazine catch spring I1) (p 3-62)

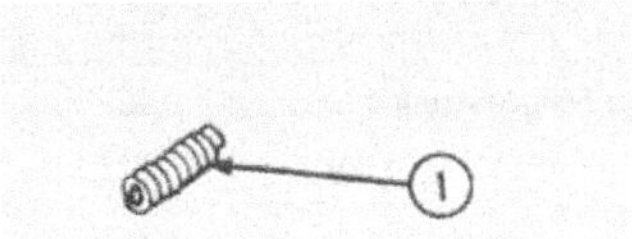

   Step 2 Short recoil
   Refer to page 3-9

3 FAILURE TO CHAMBER
   Short recoil.
   Refer to page 3-9.

3-5. TROUBLESHOOTING PROCEDURES (CONT).

TROUBLESHOOTING (CONT)

MALFUNCTION
TEST OR INSPECTION
CORRECTIVE ACTION

4. FAILURE TO LOCK.

Step 1. Damaged bolt carrier key (1).

Repair or replace bolt carrier key (1) (p 3-25)

Step 2 Loose screws (2) on bolt carrier key (1).

a. Disassemble and repair (p 3-25)
b. Reassemble using new screws.

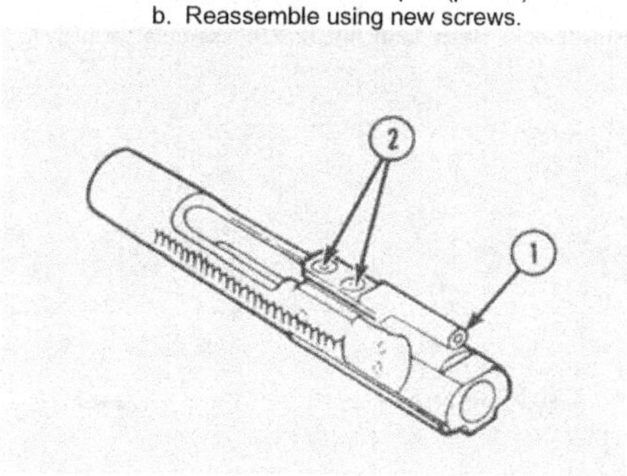

Step 3 Bent gas tube (3).

a Adjust by bending gas tube 13) In area of handguards
b Replace gas tube (3) and check alignment (p 3-29)

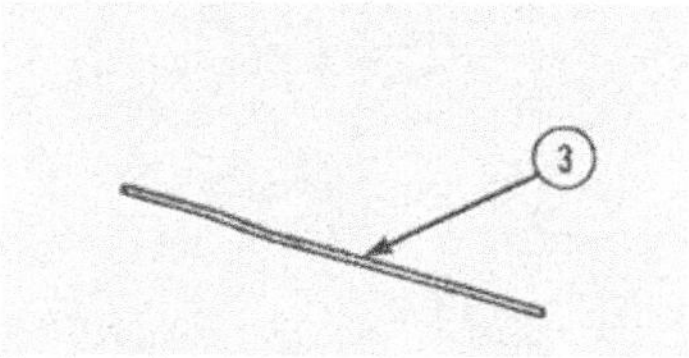

Step 4. Short recoil

Refer to page 3 9

TROUBLESHOOTING (CONT)

MALFUNCTION
TEST OR INSPECTION
CORRECTIVE ACTION

5. FAILURE TO FIRE.

Step 1. Broken hammer (1).
Replace hammer (1) (p 3-73).

Step 2. Weak or broken hammer spring (2).
Replace hammer spring (2) (p 3-73).

Step 3. Hammer spring (2) improperly assembled.
Assemble properly.

Step 4. **(M16A2 and M4 ONLY)** Burst cam (3) and/or cam spring (4) frozen or improperly assembled
Disassemble, clean, lubricate, and reassemble correctly (p 3-73).

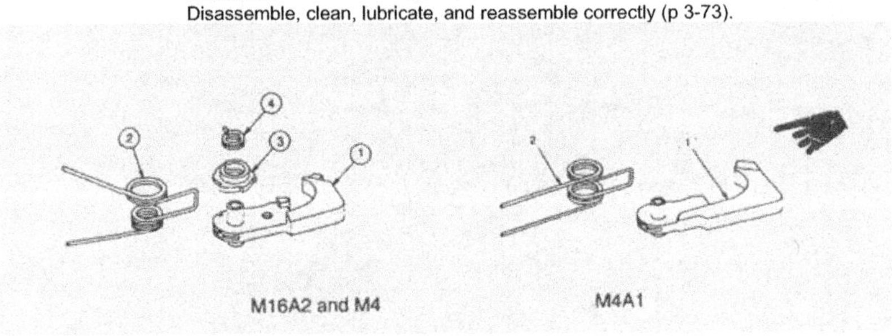

Step 5. Selector lever (5) frozen on SAFE position.
Disassemble and clean (p 3-62)

Step 6. Broken firing pin (6) or firing pin does not meet gage protrusion requirement.
Replace firing pin (6) (p 3-16).

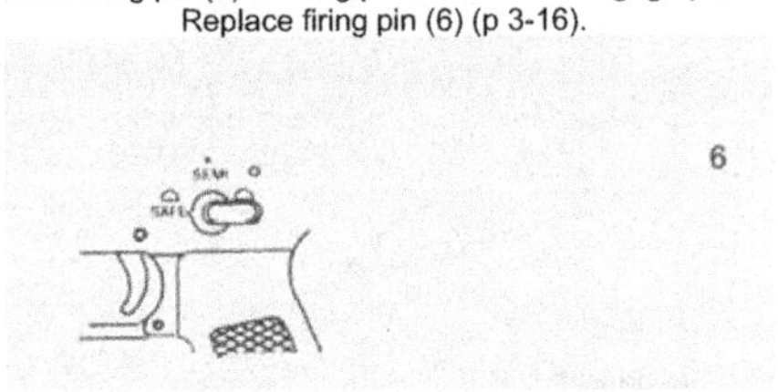

3-5. TROUBLESHOOTING PROCEDURES (CONT).

TROUBLESHOOTING (CONT)

MALFUNCTION
TEST OR INSPECTION
CORRECTIVE ACTION

6. FAILURE TO UNLOCK.

Step 1 Burred locking lugs (1) on bolt assembly.
Remove burrs.

Step 2 Burred lugs (2) on barrel extension.
Remove burrs.

Step 3. Short recoil.
Refer to page 3-9.

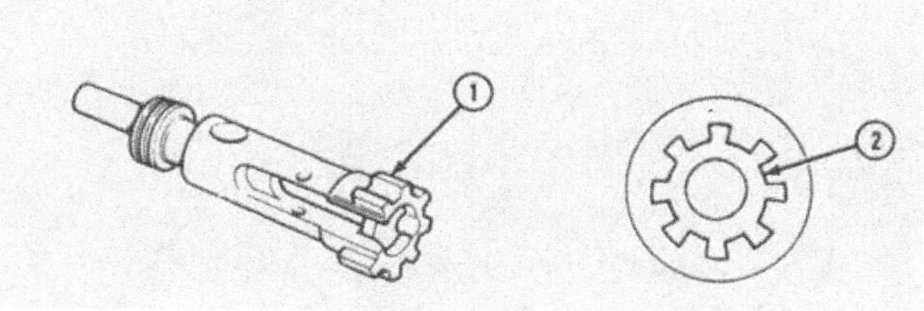

7. FAILURE TO EXTRACT

Step 1. Inspect cartridge extractor and extractor spring assembly.
Replace if cracked or broken (p 2-35).

Step 2. Inspect badly pitted chamber with reflector tool (Item 2, fig. C-1 7).
Replace barrel assembly If chamber Is badly pitted (p 3-29).

8 FAILURE TO EJECT.

Short recoil.
Refer to page 3-9.

TROUBLESHOOTING (CONT)

MALFUNCTION
TEST OR INSPECTION
CORRECTIVE ACTION

9. FAILURE TO COCK.

Step 1. Worn or broken trigger nose (1) or trigger spring (2).
Replace trigger (3) or defective trigger spring (2) (p 3-62).

Step 2. Worn or broken hammer trigger notch (4).
Replace hammer (5) (p 3-62).

Step 3. Worn or broken hammer disconnector hook (6).
Replace hammer (5) (p 3-62).

Step 4. Worn or broken hammer automatic sear hook (7).
Replace hammer (5) (p 3-62).

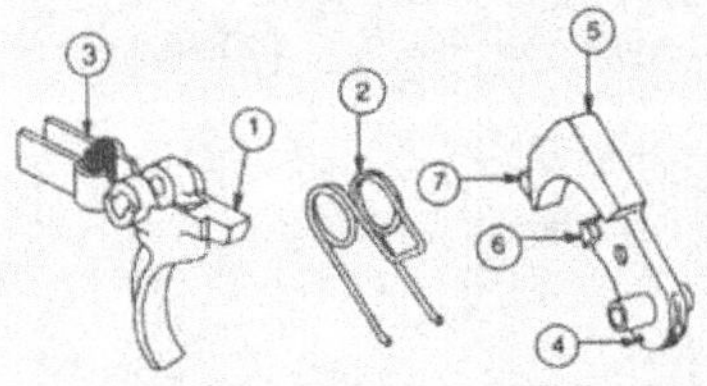

M16A2 and M4 ONLY

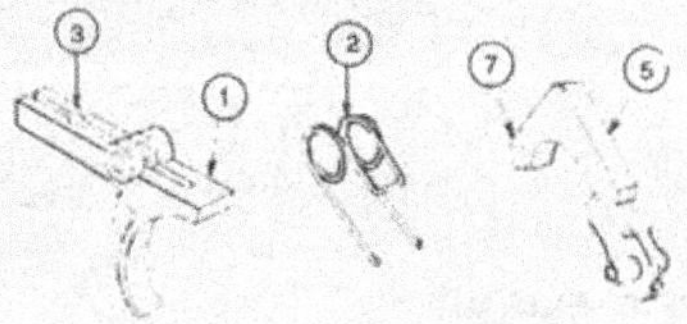

M4A1 ONLY

3-5. TROUBLESHOOTING PROCEDURES (CONT).

TROUBLESHOOTING (CONT)

MALFUNCTION
TEST OR INSPECTION
CORRECTIVE ACTION

Step 5 **(M4A1 has only 1 disconnector)** Worn or broken disconnector hooks (8).
Replace defective disconnectors (9) (p3-62).

Step 6 **(M4A1 has only 1 spring)** Weak, broken, or missing disconnector springs (10).
Replace disconnector springs (10) (p 3-62).

Step 7. Worn, broken, or missing automatic sear (11).
Replace automatic sear (11) (p 3-62).

Step 8 Weak or broken automatic sear spring (12).
Replace automatic sear (11) (p 3-62).

Step 9. Long leg (13) of automatic sear spring incorrectly assembled in receiver.
Remove automatic sear assembly (11) and install correctly (p 3-62).

Step 10. **(M16A2 and M4 ONLY)** Burst cam (14) or clutch spring (15) frozen or improperly assembled.
Disassemble, inspect, clean, lubricate, or replace as required (p 3-73).

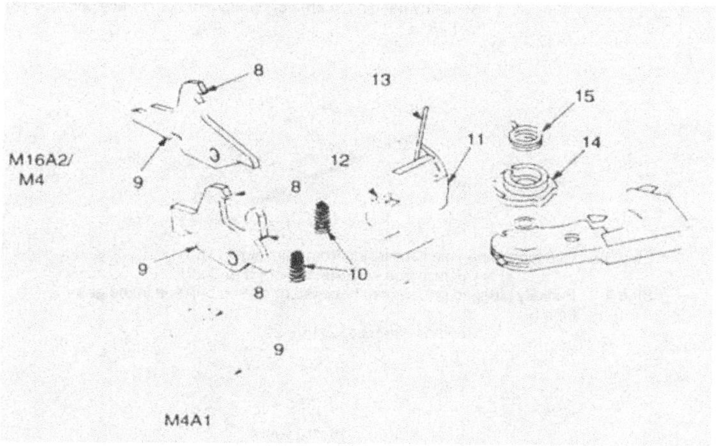

TROUBLESHOOTING (CONT)

MALFUNCTION
TEST OR INSPECTION
CORRECTIVE ACTION

10. SHORT RECOIL.

Step 1. Improper gap space or worn, missing, or broken bolt rings (1).

a. Stagger bolt ring gaps (p 3-21).

b. Replace bolt rings (1) and stagger gaps (p 3-21).

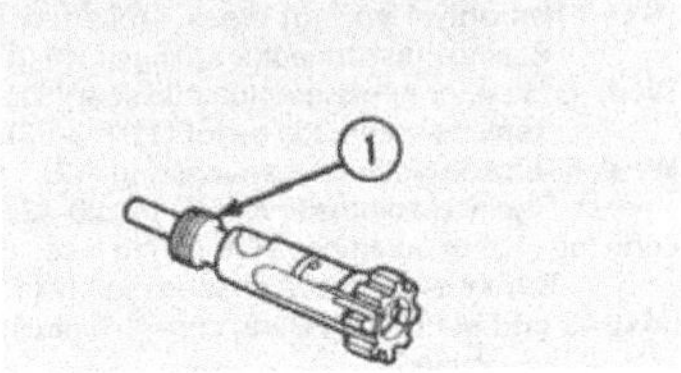

Step 2. Broken or bent gas tube (2).

Adjust by bending in area of handguards or replace gas tube (2) (p 3-29).

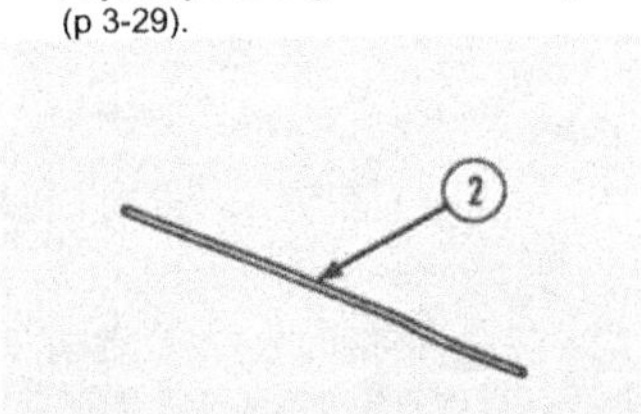

Step 3. Gas tube spring pin (3) missing from front sight (4).

Replace gas tube spring pin (3) (p 3-29).

Step 4. Partially plugged gas system because of carbon build-up in the gas tube (2).

Replace gas tube (2) (p 3-29).

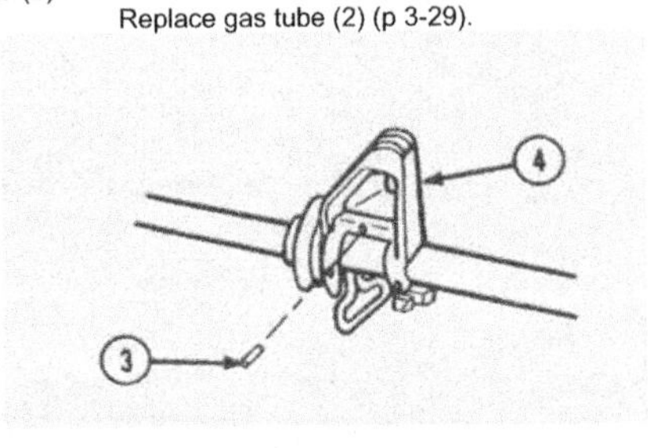

3-5. TROUBLESHOOTING PROCEDURES (CONT).

TROUBLESHOOTING (CONT)

MALFUNCTION
TEST OR INSPECTION
CORRECTIVE ACTION

10 SHORT RECOIL (CONT).

WARNING

When using carbon removing compound (Item 8, app D), avoid skin contact. If it comes in contact with the skin, wash off thoroughly with running water. The use of a good lanolin base cream after exposure to compound Is helpful. The use of gloves and protective equipment Is required. Step 5. Carbon build-up in barrel gas port (5). Remove carbon build-up by soaking barrel In carbon removing compound (item 8, app D). Use rubber gloves (Item 18, app D) with carbon removing compound. Use a bore small arms cleaning brush (Item 4, app D).

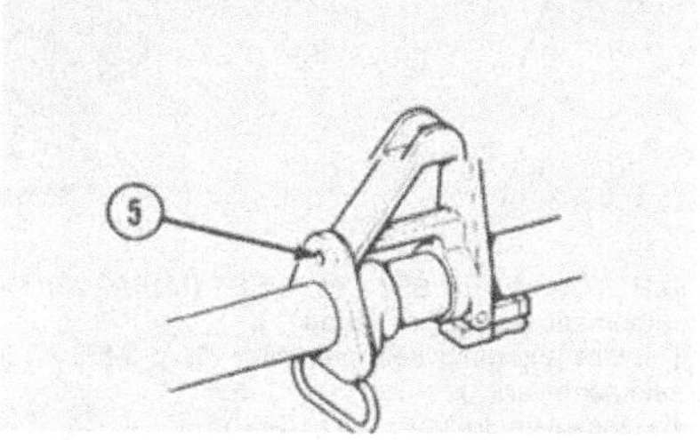

11. RIFLE CANNOT BE ZEROED.

Step 1 Inspect for defective or bent barrel assembly (1) (p 3-29).
Replace barrel assembly (1) (p 3-29).

Step 2 (For windage) barrel assembly (1) out of alignment with rear sight on up per receiver.
Align barrel assembly (1) and upper receiver (p 3-291

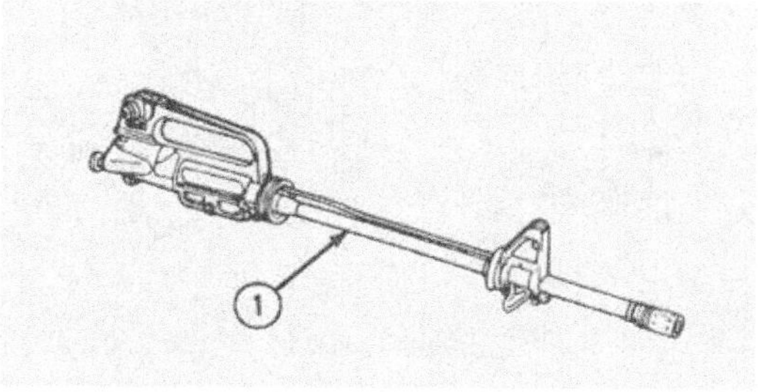

TROUBLESHOOTING (CONT)

MALFUNCTION
TEST OR INSPECTION
CORRECTIVE ACTION

Step 3. (For elevation) defective front sight (2) or rear sight (3).
Repair as required (p 3-48).

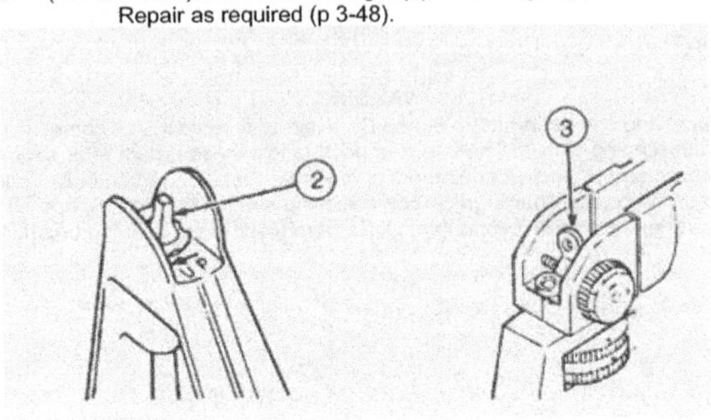

12. FAILURE TO CYCLE WITH SELECTOR LEVER SET ON BURST (M16A2 and M4 ONLY).

Step 1. Broken automatic sear (1) or spring (2).
Replace automatic sear assembly (1) (p 3-62).

Step 2. Faulty selector lever (3).
Replace selector lever (3) (p 3-62).

Step 3. Broken tooth on burst cam (4).
Replace burst cam (4) (p 3-73).

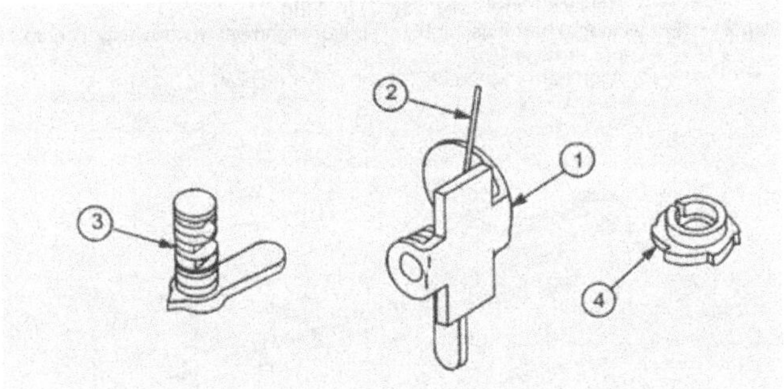

M16A2 and M4

3-5. TROUBLESHOOTING PROCEDURES (CONT).

TROUBLESHOOTING (CONT)

MALFUNCTION
TEST OR INSPECTION
CORRECTIVE ACTION

12 FAILURE TO CYCLE WITH SELECTOR LEVER SET ON BURST (M16A2 and M4 ONLY) (CONT).

Step 4 Broken cam clutch spring (5) Cam clutch spring should be bent and properly formed without any sharp edges or corners

Inspect and replace If required.

Step 5 The bend In the cam clutch spring (5) installed backwards (toward outside).

Install cam clutch spring (5) properly with the bend to the inside (p 3-73).

NOTE

When hammer Is rotated back to cocked position, cam should rotate to allow the burst disconnector to latch in the next notch.

Step 6. Cam clutch spring (5) fails to "clutch" and burst cam (4) fails to rotate back with hammer (6).

Replace cam clutch spring (5) (p 3-73). If problem continues, replace hammer (6) and cam (4) (p 3-73).

Step 7. Broken front hook (7) on burst disconnector (8).

Replace burst disconnector (8) (p 3-62).

Step 8. Short recoil.

Refer to page 3-9

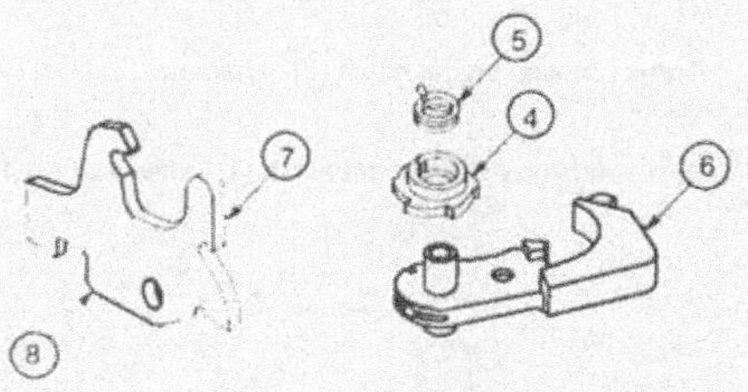

M16A2 and M4

TROUBLESHOOTING (CONT)

MALFUNCTION
TEST OR INSPECTION
CORRECTIVE ACTION

12.1. FAILURE TO CYCLE WITH SELECTOR LEVER SET ON AUTO (M4A1 ONLY)

Step 1. Broken automatic sear (1) or spring (2).
Replace automatic sear assembly (1).

Step 2. Faulty selector lever (3).
Replace selector lever (3).

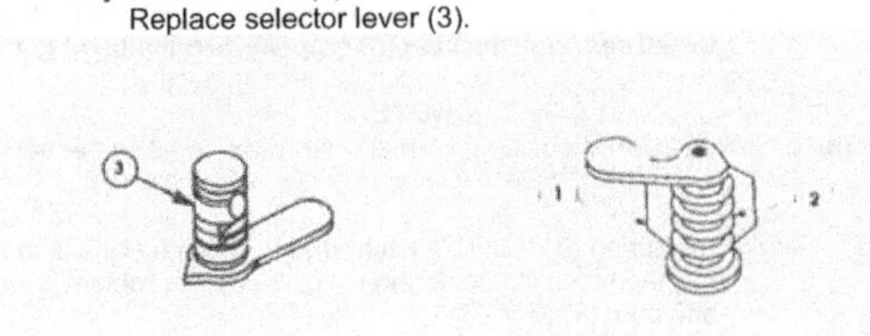

Step 3. Short recoil.
Refer to page 3-9.

13. FIRES TWO ROUNDS WITH ONE PULL OF TRIGGER WITH SELECTOR LEVER SET ON SEMI (DOUBLE FIRING).

Step 1. Defective semiautomatic disconnector (1). Replace semiautomatic disconnector (1) (p 3-62).

Step 2 Worn or broken trigger notch (2) of hammer (3) (searing portion). Replace hammer (3) (p 3-62).

Step 3. Worn or broken disconnector notch (4) of hammer (3). Replace hammer (3) (p 3-62).

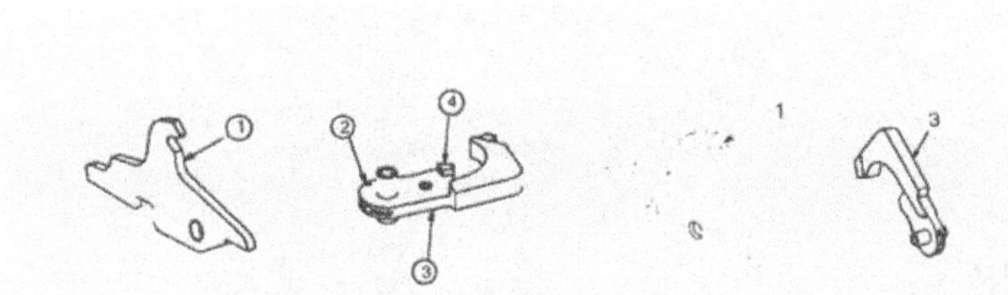

M16A2 and M4 M4A1

3-5. TROUBLESHOOTING PROCEDURES (CONT).

TROUBLESHOOTING (CONT)

MALFUNCTION
TEST OR INSPECTION
CORRECTIVE ACTION

13 FIRES TWO ROUNDS WITH ONE PULL OF TRIGGER WITH SELECTOR LEVER SET ON SEMI (DOUBLE FIRING) (CONT).

Step 4 Worn or broken trigger (5) (searing portion).
Replace trigger (5) (p 3-62).

Step 5 Worn trigger or hammer pin hole (6).
Gage trigger pin hole (7) and hammer pin hole (6) (p 3-62). If test falls, replace weapon.

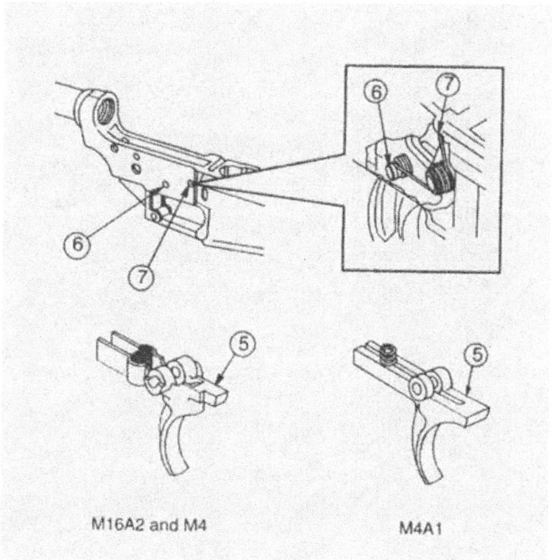

TROUBLESHOOTING (CONT.)

MALFUNCTION

TEST OR INSPECTION

CORRECTIVE ACTION

14. FIRES WITH SELECTOR LEVER ON SAFE OR WHEN TRIGGER IS RELEASED WITH SELECTOR LEVER ON SEMI.

Step 1. Defective selector lever (1).

Replace selector lever (1) (p 3-62).

Step 2. Worn or broken trigger (rear portion) (2).

Replace trigger (3) (p 3-62).

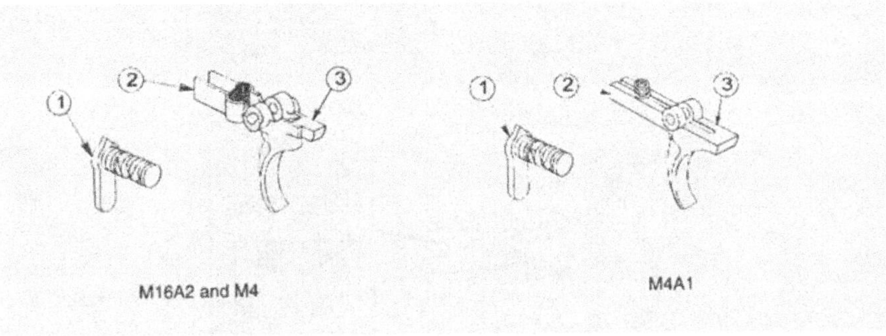

3-5. TROUBLESHOOTING PROCEDURES (CONT.).

TROUBLESHOOTING (CONT.)

MALFUNCTION

TEST OR INSPECTION

CORRECTIVE ACTION

15. HAMMER PIN "WALKS".

Hammer pin (1) "walks" or works loose during firing or hammer pin Is very easy to push out of receiver when hammer Is Installed.

Replace hammer assembly (p 3-62).

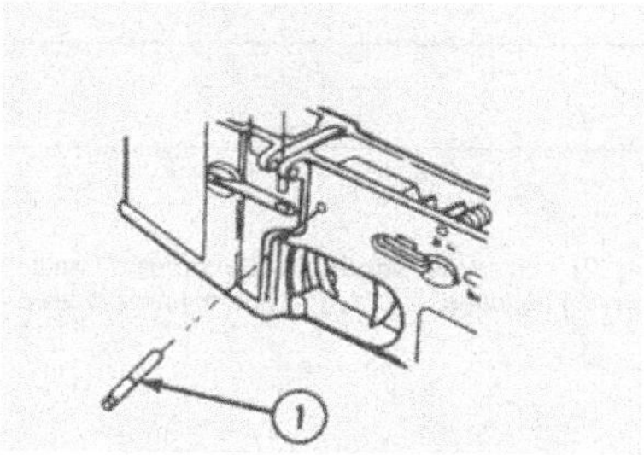

16. BOLT ASSEMBLY FAILS TO LOCK TO REAR AFTER FIRING LAST ROUND.

Step 1. Broken bolt catch (1).

Replace bolt catch (1) (p 3 62,.

Step 2. Weak or broken bolt catch spring (2).

Replace bolt catch spring (2) (p 3-62)

Step 3 Restricted movement of bolt catch (1).

Disassemble and clean.

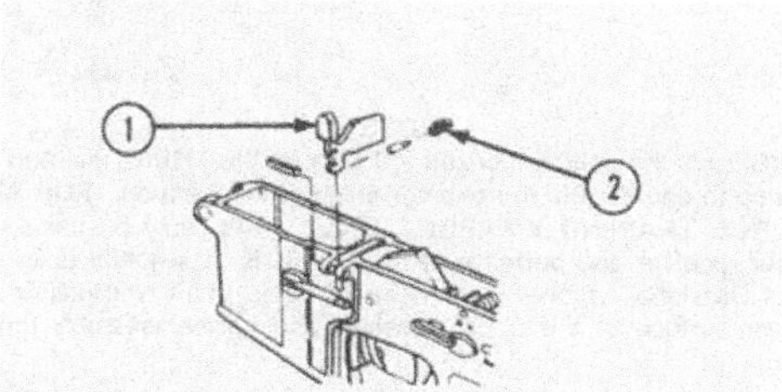

Section III. MAINTENANCE PROCEDURES FOR THE M16A2 RIFLE AND M4/M4A1 CARBINES

3-6. MAJOR COMPONENTS OF M16A2 RIFLE AND M4/M4A1 CARBINES.

This task covers disassembly.

INITIAL SETUP

Tools
(ARMY) Small Arms Repairman Tool Kit (item 3, app B)

References
TM 9-1005-319-10 (operator's manual)

Equipment Conditions
Weapon assembled

General Safety Instructions
Before starting an inspection, be sure to clear the weapon Do not pull the trigger until the weapon has been cleared.
Inspect the chamber and receiver to ensure no ammunition is present Do not keep live ammunition near work area.

DISASSEMBLY

a. Refer to operator's manual.

b. Remove magazine (1), sling (2), bolt carrier assembly (3), charging handle assembly (4), and upper receiver and barrel assembly (5), carrying handle assembly (5.1), from lower receiver and buttstock assembly (6).

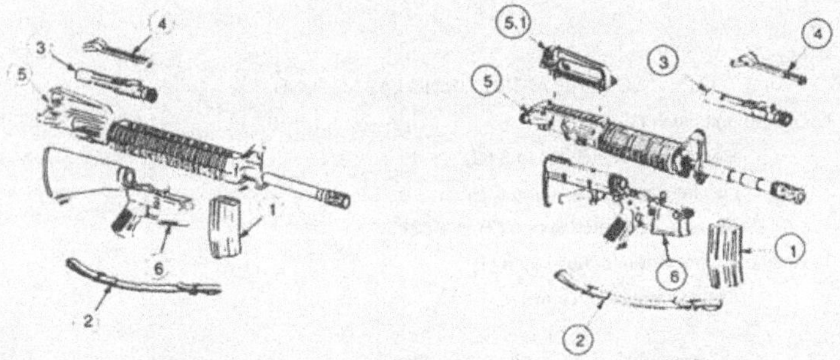

M16A2 M4/M4A1

NOTE

Solid Film Lubricant (SFL) is the only authorized touchup for the M16A2 rifle and M4/M4A1 carbines and may be used on up to one third of the exterior finish of the weapon. **FOR ARMY CONUS USE ONLY AND AIR FORCE TRAINING WEAPONS ONLY**: SFL may be used as touchup without limitation on the upper receiver and barrel assembly. This is to say that units which DO NOT fall under the category of Divisional Combat Units or rapid deployment type units may have up to 100 percent of the exterior surface of the upper receiver and barrel assembly protected with SFL if necessary.

3-7. BOLT CARRIER ASSEMBLY.

This task covers:

a	Disassembly	**d**	Test
b	Cleaning	**e**	Repair
c	Inspection	**f**	Reassembly

INITIAL SETUP

Test Equipment
Tool and Gage Set (item 2, app B)

Tools
(ARMY) Small Arms Repairman Tool Kit (item 3, app B)

Materials/Parts
Cleaner, lubricant, and preservative (CLP) (item 9, app D)
Cleaning compound, rifle bore (RBC) (item 12, app D)
Pipe cleaner (item 11, app D)

Equipment Conditions
3-1 5 Bolt carrier assembly removed

a. DISASSEMBLY

1. Remove firing pin retaining pin (1).
2. Tip key and bolt carrier assembly (2) allowing firing pin (3) to drop out. Catch the firing pin.
3. Rotate bolt cam pin (4) one quarter turn and lift straight up to remove.
4. Remove bolt assembly (5) from key and bolt carrier assembly (2)

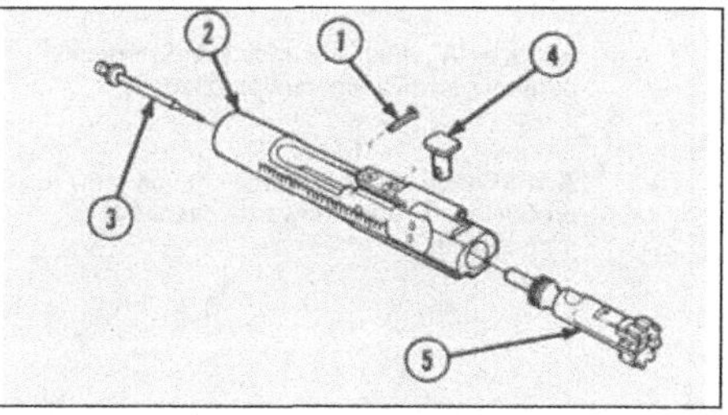

b. CLEANING

Clean carrier key and gas relief ports using CLP or RBC and pipe cleaner

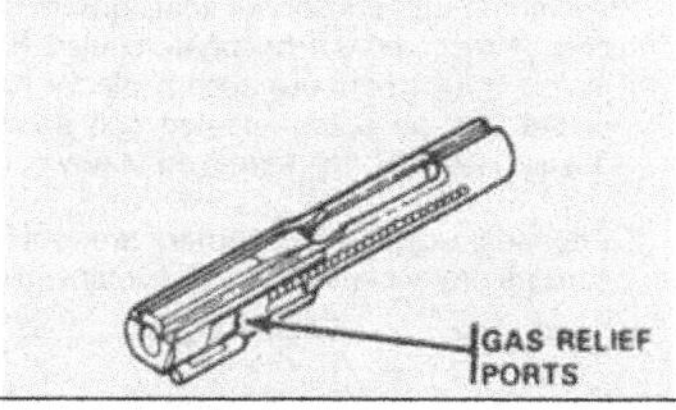

c. INSPECTION

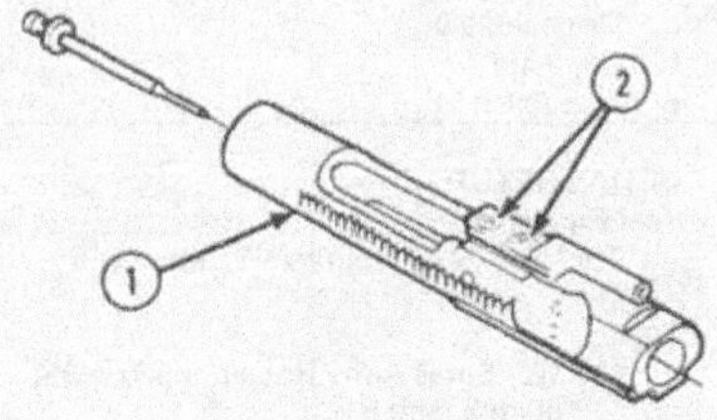

1. Inspect bolt carrier assembly using the following guidelines.

   (a) Inspect bolt carrier assembly (1) for burrs, cracks, wear, and evidence of gas loss.

   (b) Visually inspect the carrier and key screws (2) for looseness and proper staking as shown below.

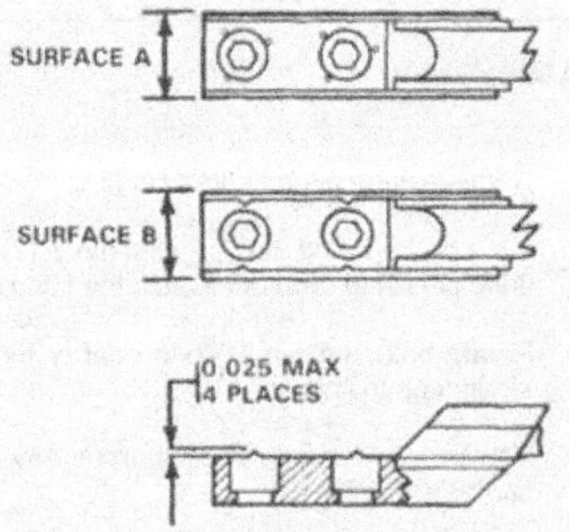

NOTE

Do not attempt to retorque if there is no loosening of the screws indicated by the staking marks.

Surface "A" must not Indicate distortion or damage which Impairs parallelism.

NOTE

A maximum of 0.025 in. (0.064 cm) protrusion In an upward direction is permissible.

NOTE

There are bolts and bolt carriers on fielded rifles, some with chrome-plated exterior surface finishes and some with phosphate coating Both finishes are acceptable under certain operational requirements and or restrictions Phosphate-coated bolt carriers are required for divisional combat units Chrome plated bolt carriers are acceptable for divisional noncombat units and training center units. Chrome-plated and phosphate-coated bolt assemblies, bolt carrier assemblies, and repair parts for these assemblies may be intermixed In any combination, with the following exception:

Phosphate-coated bolt carriers are required for all deployable and deploying units Chrome-plated bolt carriers are acceptable for nondeployable and training center units.

3-7. BOLT CARRIER ASSEMBLY (CONT.).

c. INSPECTION (CONT.)

2. Inspect firing pin 131 tip for proper con tour Inspect for pitting, wear, and burrs Pits or wear in area (4) Is permissible Replace firing pin If defective

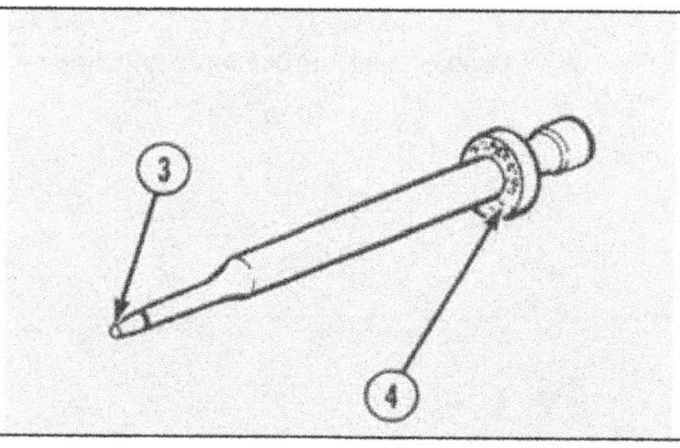

3. Prior to reassembly, insert bolt assembly (5) Into key and bolt carrier assembly (6) (do not insert bolt cam pin) and exercise bolt assembly in and out of key and bolt carrier assembly Check for binding

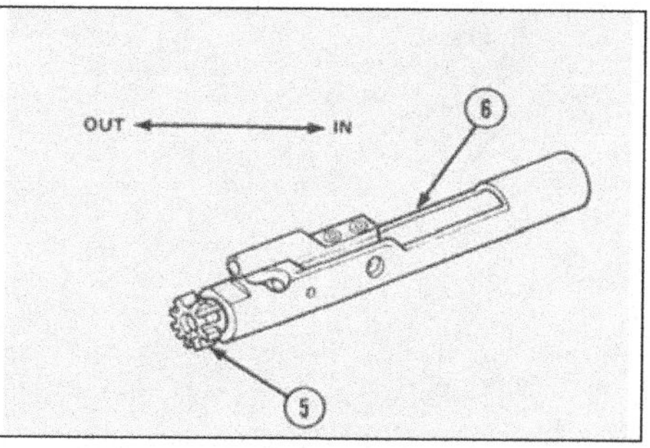

4. Check bolt assembly (5) for proper fit with bolt cam pin removed Turn key and bolt carrier assembly (61 and suspend so the bolt assembly is pointed down

NOTE

The bolt assembly must not drop out If weight of bolt assembly allows It to drop out of key and bolt carrier assembly, replace bolt rings (p 3-21)

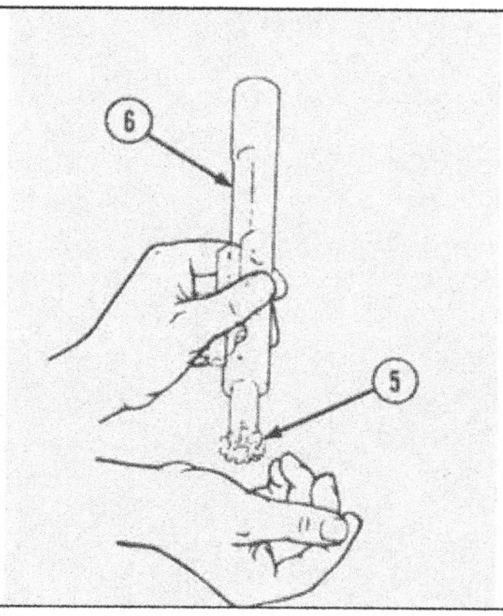

This page is blank. Procedures and figure were moved to page 2-36.

3-7. BOLT CARRIER ASSEMBLY (CONT.).

d. TEST

1. Insert firing pin (1) through bolt (2).

2. Position firing pin protrusion gage (3) PN 7799735 to check for proper firing pin (1) protrusion (minimum 0.028 in. (0 07 cm)-maximum 0 036 In. (0.09 cm))

   NOTE

   Firing pin should touch the gage on minimum but should not touch on maximum.

3. Replace a defective firing pin.

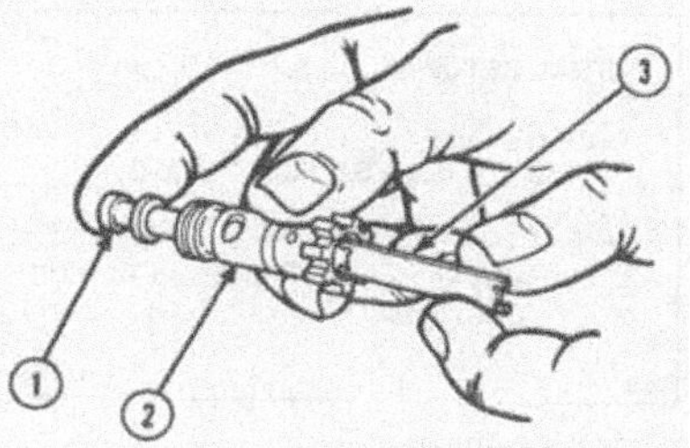

e. REPAIR

Replace all authorized unserviceable items. Retest all replaced parts.

f. REASSEMBLY

1. Install bolt assembly (1) into key and bolt carrier assembly (2).

2 Install bolt cam pin (3) and rotate one quarter turn to secure bolt assembly (1)

3. Hold key and bolt carrier assembly (2) with bolt assembly (1) down and drop In firing(pin (4)

4. Install firing pin retaining pin (5) from left side only to ensure proper Installation Check Installation by attempting to shake out firing pin

5. Reassemble rifle, refer to page 3-77

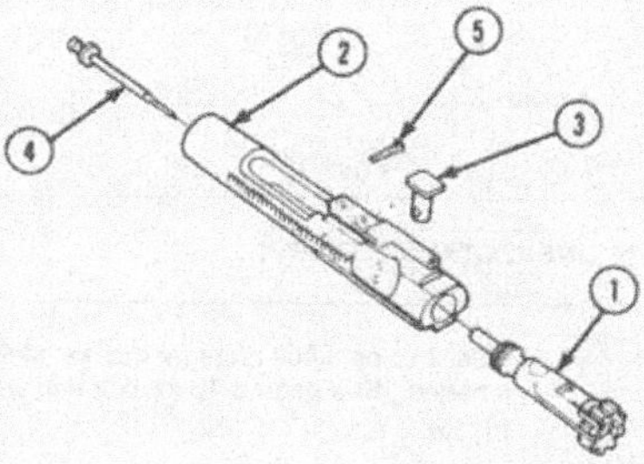

3-8. BOLT ASSEMBLY.

This task covers:
a. Disassembly
b. Inspection/Repair
c. Test
d. Reassembly

INITIAL SETUP

Test Equipment
Tool and Gage Set (Item 2, app B)

Tools
(ARMY) Small Arms Repairman Tool Kit (Item 3, app B)

Materials/Parts
Penetrant kit (item 25, app D)
Rag, wiping (item 26, app D)

Equipment Conditions
3-16 Bolt assembly removed

a. DISASSEMBLY

NOTE
Do not remove bolt rings unless they require replacement and three new replacement bolt rings are on hand.

Using small flat tip jeweler's screwdriver, remove the three bolt rings (1) from the bolt (2).

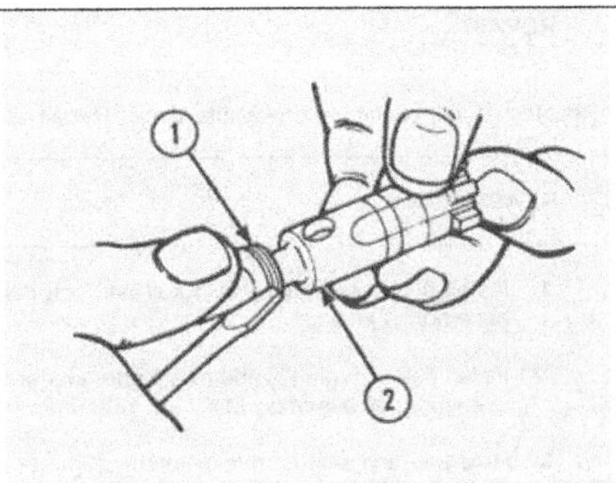

b. INSPECTION/REPAIR

1. Visually Inspect bolt rings for cracks, kinks, and bends. Replace all three bolt rings if one or more bolt rings is damaged. See page 3-16 for bolt ring wear check.

3-8. BOLT ASSEMBLY (CONT.).

b. INSPECTION/REPAIR (CONT.)

2. Inspect bolt for pits, burrs, and wear as follows.

(a) Bolt faces with a cluster of pits which are touching or tightly grouped, covering an area measuring approximately 1,8 Inch across, will be rejected and replaced.

(b) Bolts which contain individual pits or a scattered pattern will not be cause for rejection.

(c) Bolts that contain pits extending Into the firing pin hole will not be rejected unless firing pin hole gaging check determines excess wear.

(d) Rings on the bolt face (machine tool marks), grooves, or ridges less than approximately 0.010 inch will not be cause for rejection.

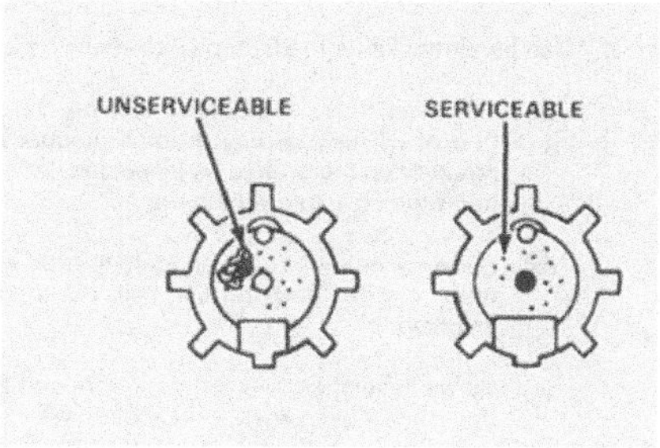

3. Inspect bolt for cracks In the locking lugs and the bolt cam pin hole area. Use black light If available; otherwise, use a glass of no more than 3X magnification or use a penetrant kit (item 25, app D). Pay close attention to the area where the locking lugs meet the body. Replace bolt assembly if bolt is defective.

WARNING

Dry cleaning solvent is flammable and toxic and should be used in a well-ventilated area. The use of rubber gloves is necessary to protect the skin when washing rifle parts.

4. Use penetrant kit (Item 25, app D) to check for cracks in bolt as follows:

(a) The area to be inspected must be clean, free of oil, etc. Spray a small amount of remover on the area to be inspected, let dry and wipe off with a wiping rag.

(b) Spray penetrant (only enough to wet the area) on the area of the bolt (1) to be inspected.

(c) Spray developer over the penetrant and let the developer work. Cracks will be indicated by a change in color where there is a crack. If there are cracks, the component 1' unserviceable.

(d) Pay close attention to the area where the locking lugs (2) meet the body.

(e) If there are no cracks, spray remover on the area; let dry and wipe off with a wiping rag. Oil the area to prevent corrosion.

(f) Replace bolt assembly If bolt (1) is defective.

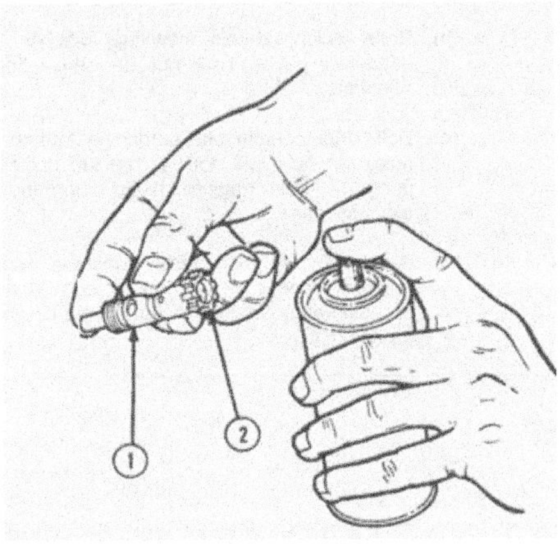

NOTE

Replacement of the bolt assembly will require that the headspace be tested (p 3-45, TEST).

3-8. BOLT ASSEMBLY (CONT.).

c. TEST

Test bolt (1) for elongated or oversized firing pin hole using special no-go plug gage (2) PN 12620101.

NOTE

Bolts with firing pin holes which permit the special no-go plug gage to fully penetrate at any position on the circumference will be rejected and replaced.

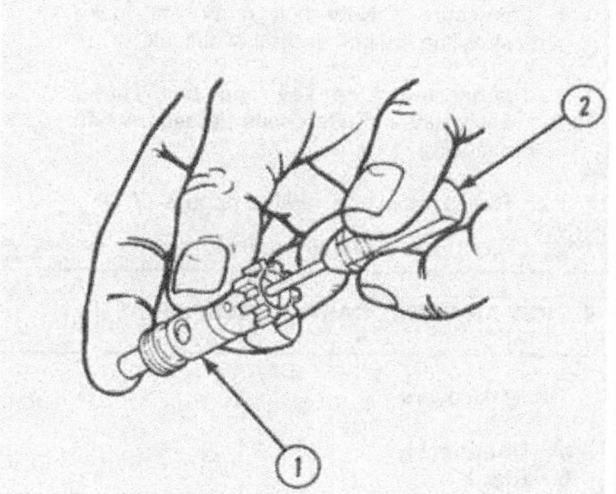

d. REASSEMBLY

NOTE

To install a bolt ring, carefully place one end in the bolt ring groove and hold In place with the thumb of one hand With the index finger of the other hand, gently guide and push the rest of the bolt ring into the groove a little bit at a time until the entire bolt ring is in place.

1. Install the three bolt rings (1) one at a time onto the bolt (2) using care not to bend or "spring" new bolt rings Stagger the bolt ring gaps (approximately ' turn apart).

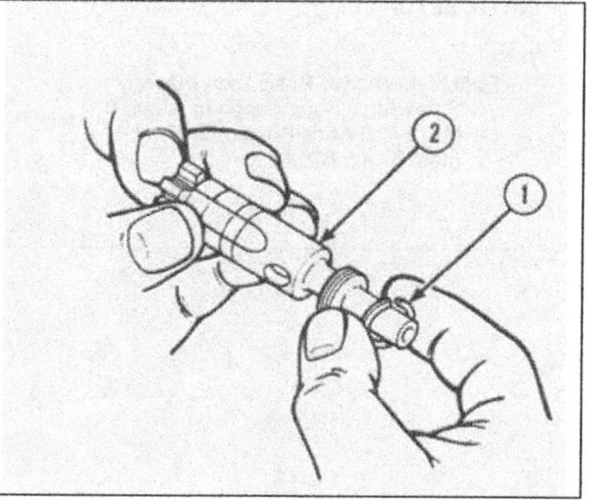

NOTE

Make certain bolt ring gaps are staggered to prevent loss of gas pressure. New bolt rings will make installing the bolt assembly difficult.

Lubricate inside key and bolt carrier assembly and use gentle pressure when installing.

2. Reassemble rifle, refer to page 3-77.

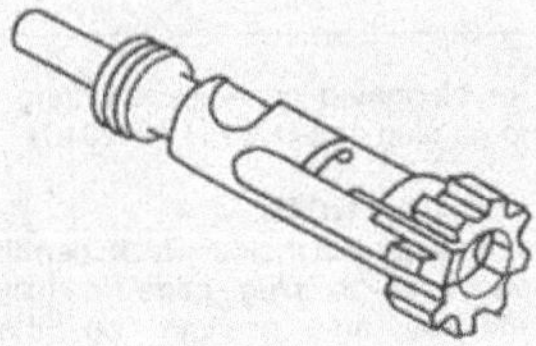

3-9. KEY AND BOLT CARRIER ASSEMBLY.

This task covers:

a Disassembly
b Repair
c Reassembly

INITIAL SETUP

Tools
- Field Maintenance Basic Less Power Small Arms Shop Set (item 1, app B)
- (ARMY) Small Arms Repairman Tool Kit (item 3, app B)

Materials/Parts
- Carrier and key screws (2) (8448508)

Equipment Conditions
- 3-16 Key and bolt carrier assembly removed

3-9. KEY AND BOLT CARRIER ASSEMBLY (CONT.).

a. DISASSEMBLY

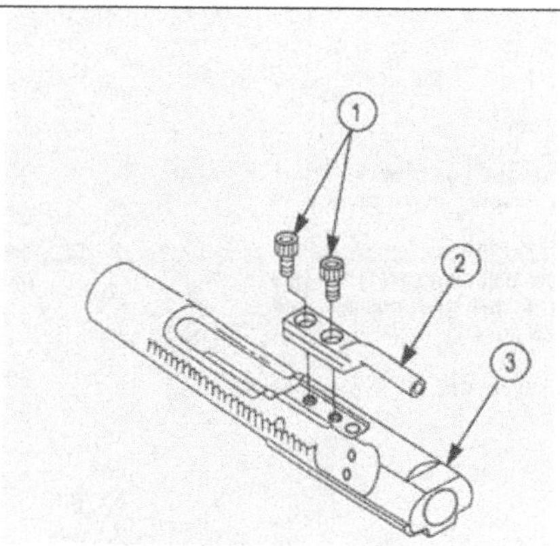

NOTE
Do not disassemble the key and bolt carrier assembly unless the bolt carrier key is defective as determined by inspection procedures described in section III, preceding.

1. Using socket wrench handle and tight fitting 1/8 inch socket head screw socket wrench attachment, remove two carrier and key screws (1).

NOTE
The heads and part of the bolt carrier key may be ground off in order to remove bolt carrier key from bolt carrier if carrier and key screws cannot otherwise be removed.

2. Remove bolt carrier key (2) from bolt carrier (3).

b. REPAIR

NOTE
Do not retorque carrier and key screws if staking marks do not indicate loosening screws.

Repair by replacing, torquing, and restaking carrier and key screws. Refer to the following reassembly procedures.

c. REASSEMBLY

NOTE

Do not reuse old carrier and key screws New carrier and key screws must be used at assembly.

1. If disassembled, place bolt carrier (1) In vise using vise law caps Install and position bolt carrier key (2) on bolt carrier (1).

2. Install two carrier and key screws (3). Always use new screws.

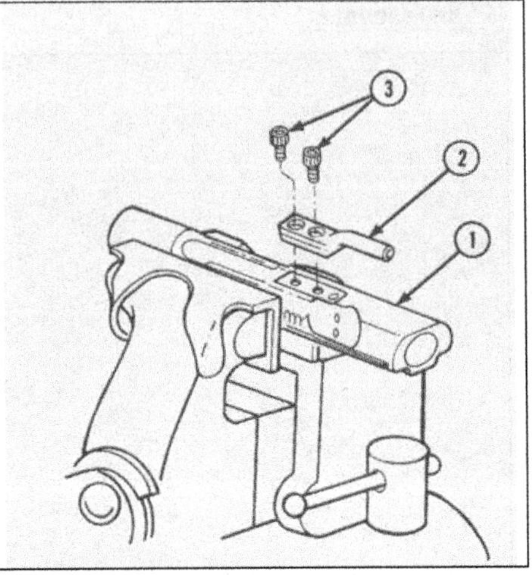

3. Use a tight-fitting 1 8 inch socket head screw wrench attachment and an Inch-pound torque wrench to torque the carrier and key screws (3) to 35 to 40 Inch-pounds (3 95 to 4 52 N m)

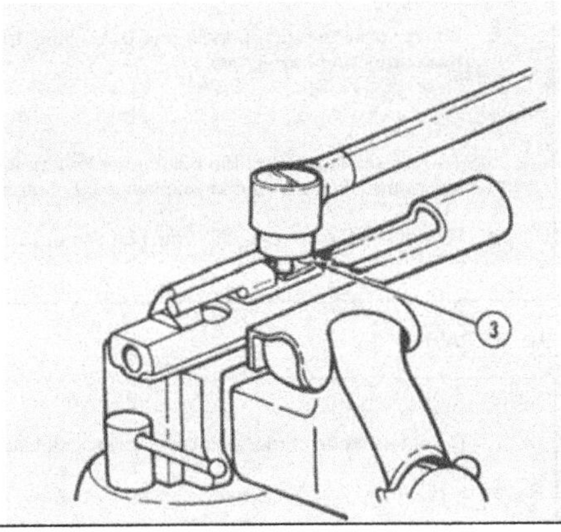

3-9. KEY AND BOLT CARRIER ASSEMBLY (CONT.).

c. REASSEMBLY (CONT.)

NOTE
Field staking method will be used by field units.

4. Use solid center punch and hand hammer to stake the two carrier and key screws (3) In three places.

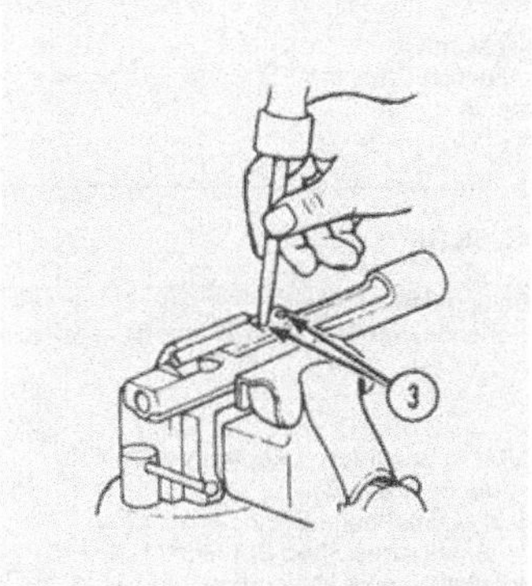

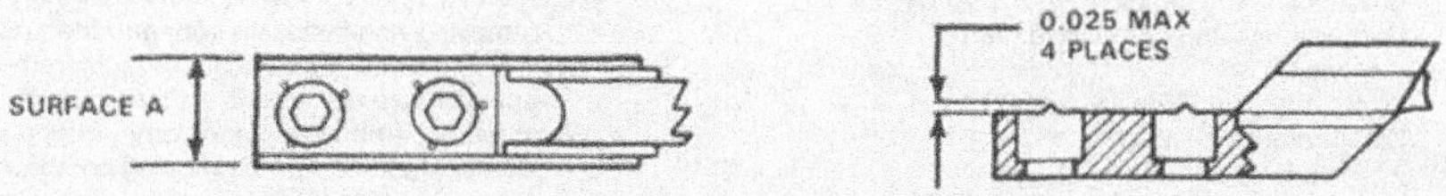

FIELD REPLACEMENT STAKING

5. Reassemble rifle, refer to page 3-77

CAUTION
If blanks are used, blank firing attachment (BFA) must be attached

NOTE
If the bolt carrier key is replaced, three to eight rounds of blank or ball ammunition must be fired to ensure a seal is created. Manual operation of the rifle may be required. If blank ammunition is utilized, M16A2 Blank Firing Attachment must be adapted

3-10. UPPER RECEIVER AND BARREL ASSEMBLY.

This task covers:

a Disassembly
b Inspection/Cleaning
c Repair
d Reassembly
e Test

INITIAL SETUP

Test Equipment
Tool and Gage Set (item 2, app B)

Tools
(ARMY) Small Arms Repairman Tool Kit (item 3, app B)
Field Maintenance Basic Less Power Small Arms Shop Set (item 1, app B)
Modified Needle Nose Pliers (fig E-7, app E)

Materials/Parts
Brush, cleaning, small (item 3, app D)
Carbon removing compound (item 8, app D)
Cloth, abrasive (item 13, app D)
Dichloromethane, technical (item 15, app D)
Dry cleaning solvent (item 16, app D)
Gloves, chemical and oil protective (item 18, app D)
Grease, molybdenum disulfide (item 19, app D)
Lubricant, solid film (item 21, app D)
Pan, wash (item 24, app D)
Polyethylene (item 32, app D)
Sealing compound (item 28, app D)
Target (item 31, app D)

References
FM 23-9
TM 9-1005-319-10

Equipment Conditions
3-15 Upper receiver and barrel assembly removed
2-43 Handguard assemblies removed

General Safety Instructions
To avoid injury to your eyes, use care when removing and installing spring-loaded parts.
When using solid film lubricant or dichloromethane, be sure the area is well ventilated.
When using carbon removing compound, avoid skin contact If carbon removing compound comes in contact with the skin, wash thoroughly with running water Using a good lanolin base cream after exposure to the compound is helpful Using gloves and protective equipment is required.

3-10. UPPER RECEIVER AND BARREL ASSEMBLY (CONT.).

a. DISASSEMBLY

1. Using ball-peen hammer and 5/64 inch diameter drive pin punch, drive spring pin (1) (which retains the gas tube) out of front sight assembly (2).

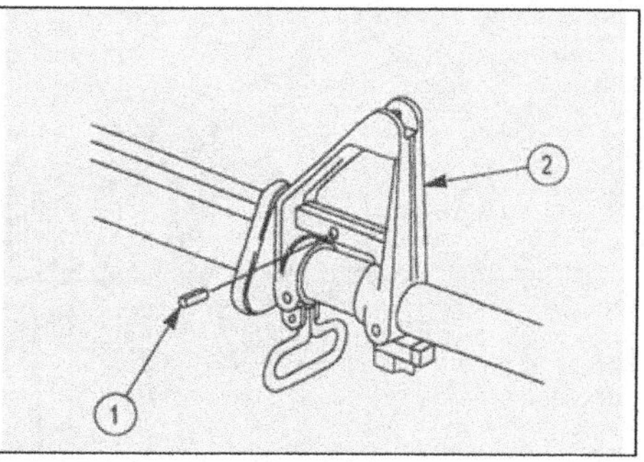

2. Slide gas tube (3) back into upper receiver assembly (4) to clear front sight assembly (2). Then lift slightly, pull forward, and remove gas tube (3).

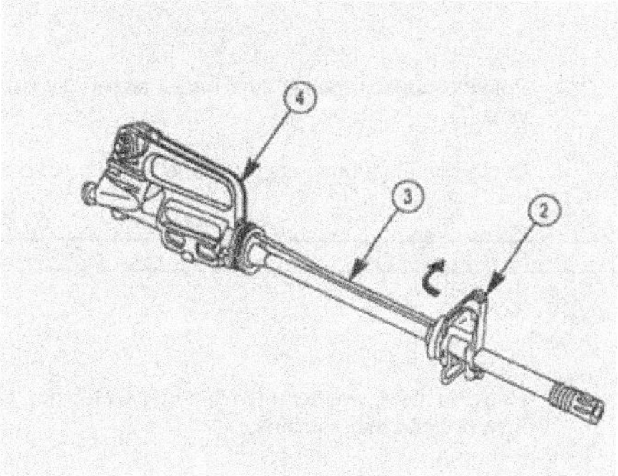

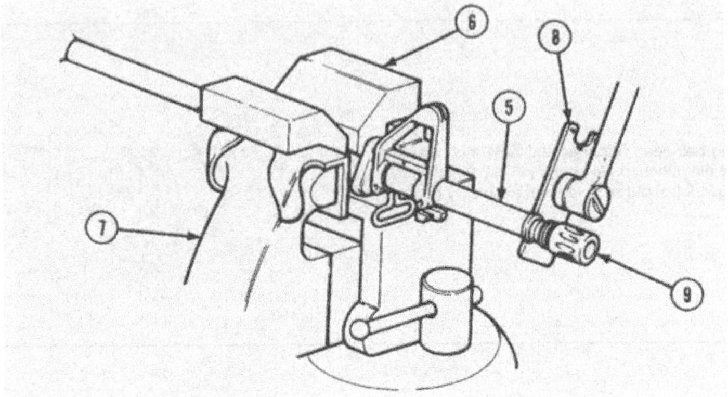

3. Position upper receiver and barrel assembly (5) in barrel removal fixture (6) and secure both In machinist's vise (7).

4. Using combination wrench (8) and 12 inch drive handle, remove compensator (9).

5. Remove peel washer (10) being careful not to lose or bend thin sections.

6. Remove upper receiver and barrel assembly (5) from barrel removal fixture and machinist's vise.

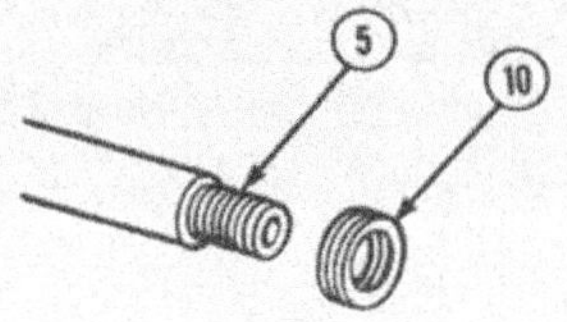

3-10. UPPER RECEIVER AND BARREL ASSEMBLY (CONT.).

a. DISASSEMBLY (CONT.)

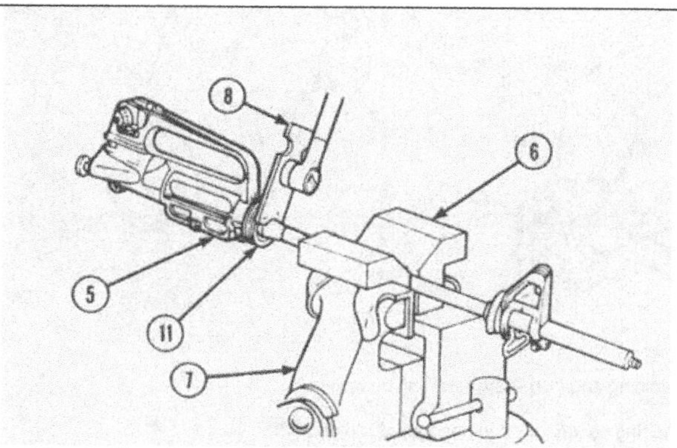

7. Place upper receiver and barrel assembly (5) Into barrel removal fixture (6) and clamp Into machinist's vise (7).

NOTE

Be sure all three drive pins on combination wrench are fully engaged with barrel nut assembly. Wrench must be pushed toward upper receiver assembly to compress the slip ring spring In barrel nut assembly. Do not use a torque wrench to loosen the barrel nut assembly.

8. Using '2 Inch drive handle and combination wrench (8), loosen barrel nut assembly (11)

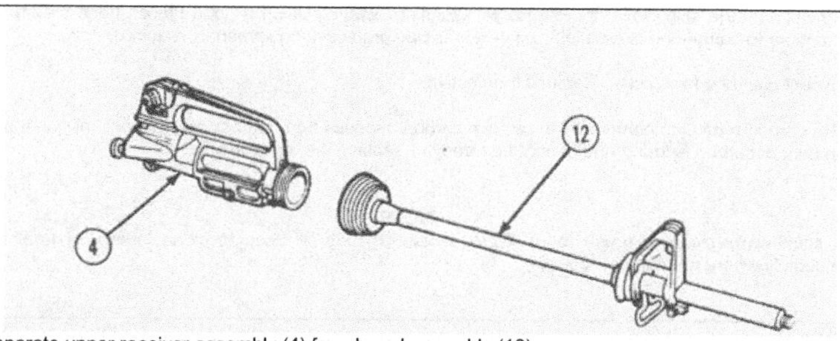

9. Separate upper receiver assembly (4) from barrel assembly (12)

10. Remove barrel assembly (12) from machinist's vise and barrel removal fixture

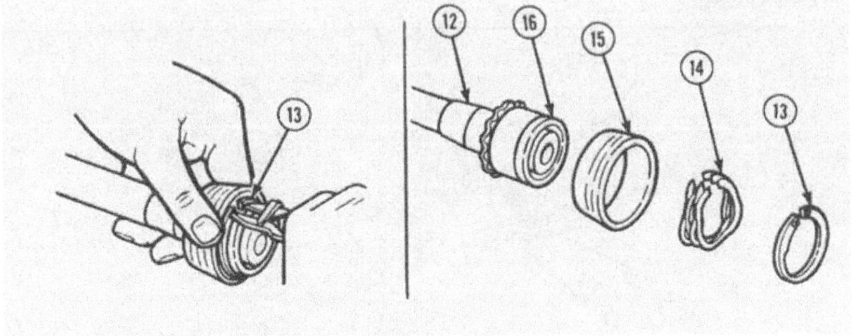

11. Remove retaining ring (13) using retaining ring pliers.

12. Remove slip ring spring (14) and handguard slip ring (15).

13. Do not remove barrel nut (16) from barrel assembly (12).

b. INSPECTION/CLEANING

WARNING

When using carbon removing compound, avoid skin contact. If carbon removing compound comes In contact with the skin, wash thoroughly with running water. Using a good lanolin base cream after exposure to compound is help ful. Using gloves and protective equipment is required.

1. Inspect gas tube for cracks. Replace if defective

2. Use carbon removing compound to remove carbon deposits from interior and exterior of gas tube. If a large amount of carbon is found and cannot be removed, replace the gas tube

NOTE

A small arms cleaning brush (bore) (item 4, app D) may be used to clean Interior of front sight assembly where gas tube is secured.

3-10. UPPER RECEIVER AND BARREL ASSEMBLY (CONT.).

b. INSPECTION/REPAIR (CONT.)

3. Inspect bore for burrs, cracks, rust, bulges, and pits using the following guidelines

   (a) Pits no wider than a land or groove and no longer than 3'8 inch (0.95 cm) are al lowed in the bore.

   (b) Lands that appear dark blue due to coating of gliding metal from projectiles are allowable.

   (c) Definitely ringed bores or bores ringed sufficiently to bulge the outside surface of the barrel are cause for rejection. Replace barrel assembly if defective

4. If the upper receiver is separated from the barrel assembly, Inspect chamber for pits utilizing a flashlight. Pits 1'8 inch 10 32 cm) In length are cause for rejection Replace rifle barrel assembly If defective

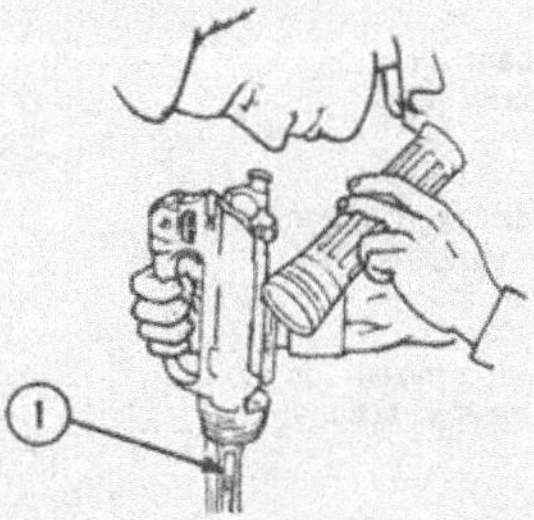

5. If upper receiver and barrel assembly is assembled, Inspect chamber using reflector tool (1) and flashlight. Pits 1 '8 inch (O 32 cm) In length are cause for rejection. Replace rifle barrel assembly If defective. If rifle barrel assembly Is replaced, Inspect headspace (p 3-45).

6. Inspect upper receiver assembly for cracks, corrosion, wear, or damage

   (a) Small dents or gouges that do not affect functioning will not be cause for rejection.

   (b) If upper receiver assembly contains cracks or holes, the upper receiver assembly will be replaced

7. Inspect all parts for damage and wear. Replace all defective parts.

NOTE

Damaged or missing teeth of the barrel nut is not cause for rejection provided the proper torque value can be obtained during installation using the tools depicted. If removal of the barrel is not possible with the combination tool, a pipe wrench or other such tool may be used during removal.

8. Inspect front sight guards for bends, if bent see page 3-37 for repair procedures.

c. REPAIR

1. Repair corroded upper receiver assembly surfaces as follows:

   (a) Sand corroded area with abrasive cloth and make sure all corrosion has been removed.

WARNING

When using solid film lubricant or dichloromethane, be sure the area is well ventilated

   (b) Wash area with technical dichloromethane (methylenechloride) to re move all dirt, grease, and foreign material

   (c) Apply sealing compound, mixed In accordance with manufacturer's directions, to areas to be filled

   (d) Spread sealing compound as smoothly as possible Into defective area using a putty knife or similar tool

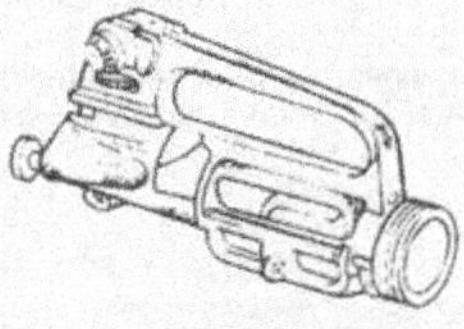

CORRODED
(REPARABLE)

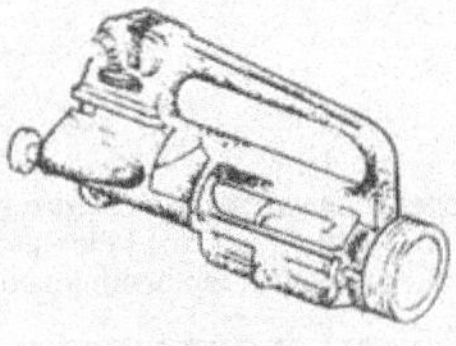

CORRODED
(NONREPARABLE)

3-10. UPPER RECEIVER AND BARREL ASSEMBLY (CONT.).

c. REPAIR (CONT.)

NOTE
Do not feather edges

(e) Place a sheet of polyethylene (Item 32, app D), cut to size, over filled area Rub by hand to smooth.

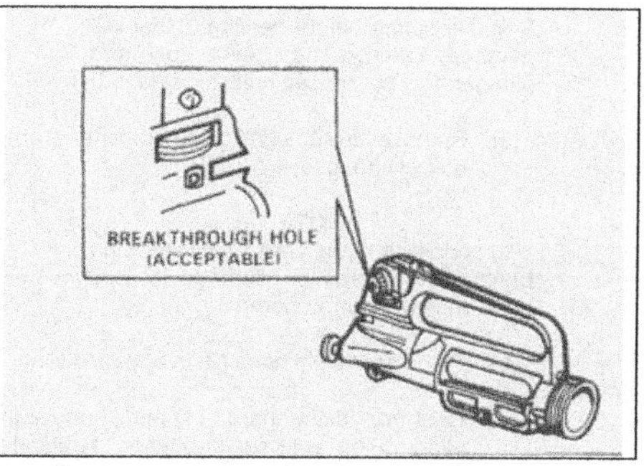

2. After curing, remove polyethylene sheet In accordance with Instructions by the manufacturer.

WARNING
When using solid film lubricant or dichloromethane, be sure the area is well ventilated.

CAUTION
Solid film lubricant is to be used only as an exterior surface protective finish and touchup If solid film lubricant comes in contact with recoiling parts or functional surfaces of the rifle, remove immediately by washing with technical dichloromethane.

3. Wash area with technical dichloromethane (methylenechloride) ; to remove dirt and grease, and foreign material.

4. Roughen area to be refinished with abrasive cloth and clean surface again Do not touch the area with fingers.

5. Repair shiny surfaces by spraying a coat of solid film lubricant in accordance with Instructions supplied by the manufacturer Dry 24 hours before handling.

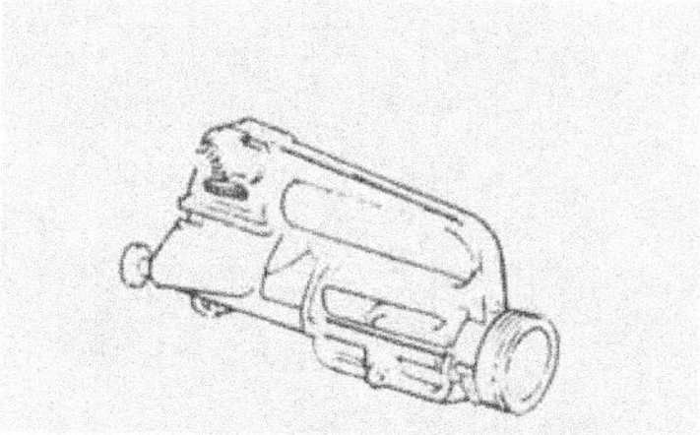

SHINY SURFACES (REPARABLE).

6. Straighten bent front sight guards (1) as follows:

NOTE

Remove spring before heating. (Heat will damage spring.) The sight post and plunger may be reused unless damaged.

(a) Remove front sight post, detent, and helical spring (see p 2-43).

NOTE

Use copper or brass caps (jaw Inserts on bench vise to prevent damage to front sight base (2) during clamping.

(b) Place front sight base (2) in a bench vise.

(c) Heat front sight guards (1) and bend with pliers. The front sight guards (1) should be put back as nearly as possible to the original position. Allow front sight housing to air cool.

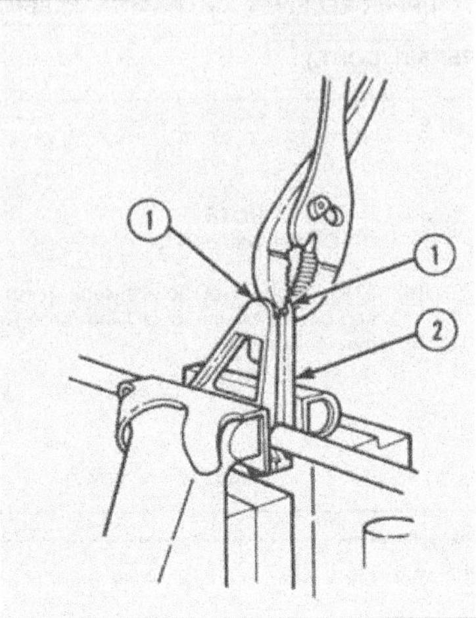

WARNING

Dry cleaning solvent is flammable and toxic and should be used in a well ventilated area. The use of rubber gloves is necessary to protect skin when washing rifle parts.

(d) Roughen any damaged surface of front sight guards with abrasive cloth and clean with dry cleaning solvent. Always wear rubber gloves when using dry cleaning solvent.

3-10. UPPER RECEIVER AND BARREL ASSEMBLY (CONT.).

c. REPAIR (CONT.)

CAUTION

Do not allow solid film lubricant to flow into front sight post threaded well.

(e) Apply solid film lubricant to cover the damaged finish.

(f) If front sight guards cannot be straightened utilizing the above procedures, reduce the rifle barrel assembly.

7. Slightly bent barrels may be straightened as follows'.

(a) Check straightness using straightness gage 8448202 (p 3 45) If the barrel falls the straightness test, and the gage remains in the barrel In the area of the front sight assembly, perform step (b) to determine If it may be straightened.

(b) With the gage remaining in the bore, hold the rifle in a vertical position with the end of the barrel into which the gage was inserted pointing up Ensure that if when the gage passes through the barrel it will not be damaged Using hand pressure' ONLY, flex the portion of the barrel between the front sight assembly and the compensator in all four directions (left, right, forward, and back) If the barrel is only slightly bent, the gage will drop through when the barrel is flexed in one of these directions Note the direction which allowed the gage to drop through the' barrel.

CAUTION

Remove the gage from the barrel before continuing

NOTE

If the gage does not pass through the barrel when it is flexed, replace the rifle barrel assembly.

(c) Place the barrel in a vise using appropriate protective laws . Clamp the barrel between the front sight assembly and the compensator approximately 1 inch (2.54 cm) from the front sight assembly The rifle barrel assembly should be in a horizontal position with the side noted in step (b) toward you.

CAUTION
Do not apply pressure to the receiver.

(d) Grasp the BARREL near the receiver so that when force is applied the barrel will flex In the same direction as noted In step (b).

(e) Give the barrel a sharp jerk of approximately 20 to 40 pounds of force (f) Remove the barrel from the vise and recheck straightness (step (al).

(g) If gage still will not pass through the barrel, perform step (b) to determine direction of bend If the barrel is still bent in the same direction as before, perform steps (c) through (f) using slightly more force. If the barrel Is now bent In the opposite direction, replace the rifle barrel assembly.

(h) If the gage passes freely through the barrel, the barrel shall be considered straight and continue In service.

(i) If the barrel has been straightened, the rifle must be targeted (p 3-45).

d. REASSEMBLY

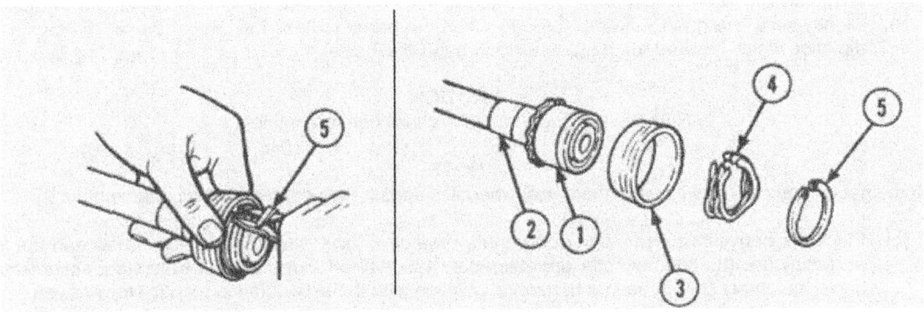

1. Position barrel nut (1) by sliding it to the rear of rifle barrel assembly (2) as far as possible
2. Slide handguard slip ring (3) over barrel nut (1).
3. Press slip ring spring (4) from both sides and insert It Into handguard slip ring (3)
4. Install retaining ring (5) against slip ring spring (4) using retaining ring pliers Snap retaining ring (5) to barrel nut (1).

NOTE
After cleaning, apply molybdenum disulfide grease to threads of barrel nut assembly before Installation.

3-10. UPPER RECEIVER AND BARREL ASSEMBLY (CONT).

d. REASSEMBLY (CONT)

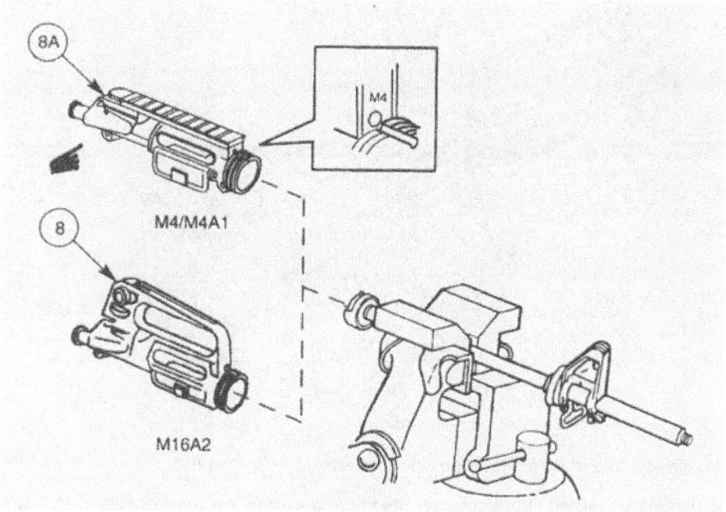

NOTE
The alignment pin (6) must not show any signs of looseness.

5. Position rifle barrel assembly (2) with alignment pin (6) up. Using barrel remover fixture (7), clamp rifle barrel assembly (2) in vise.

NOTE
The slot should fit the alignment pin perfectly with very little or no rotational play present. Note the play in a new rifle barrel assembly and new upper receiver assembly and use this as a guide.

6. Align upper receiver assembly (8) or (8A) using alignment pin (6) and the slot in upper receiver assembly (8) or (8A). Install over end of rifle barrel assembly (2).

7. Wipe upper receiver threads clean and ensure there are no burrs. Apply molybdenum disulfide grease to the threads before installation.

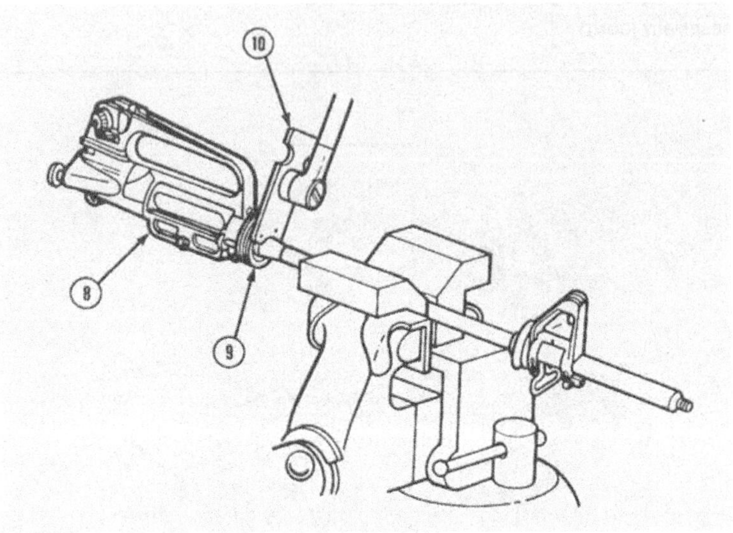

8. Engage threads of barrel nut assembly (9) with upper receiver assembly (8).

9. Using combination wrench (10) and torque wrench, torque barrel nut assembly (9) to 30 ft-lb (40.5 N-m). Torque is measured when both wrenches are used together.

NOTE

Three times torquing procedure provides for a better thread fit and prevents barrel nuts from becoming loose. Do not use the torque wrench for loosening.

10. Make certain all three drive pins on combination wrench are engaged with barrel nut assembly (91 Loosen and repeat torque operation. Then loosen the barrel nut again.

3-10. UPPER RECEIVER AND BARREL ASSEMBLY (CONT).

d. REASSEMBLY (CONT)

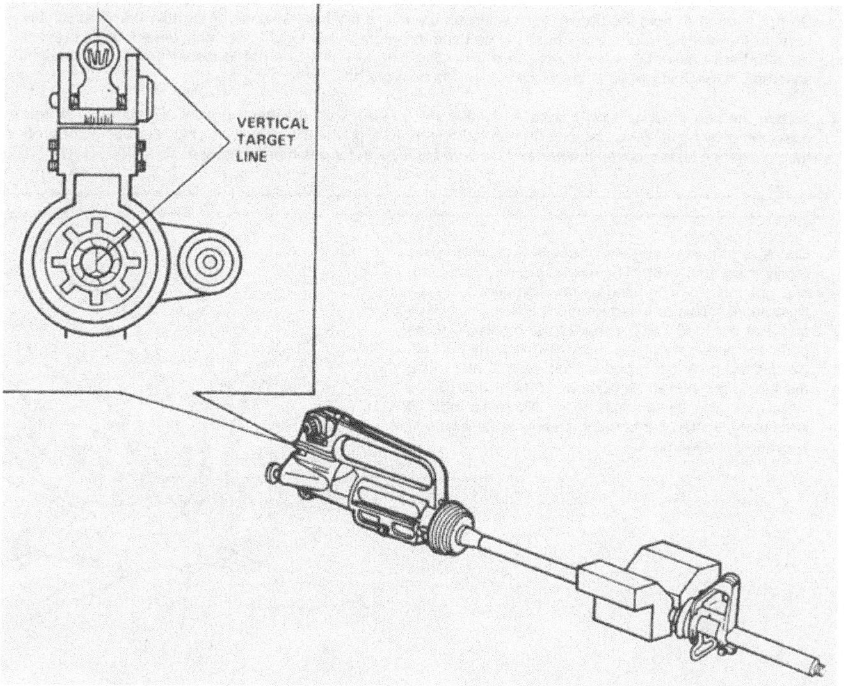

11. Front sight post must be Installed. Loosen the vise and align the bore on a distant vertical target Center the target In the bore from 12 o'clock through 6 o'clock The, front sight post should be on line and vertical with the target. Tighten vise Adjust the rear sight windage until a proper sight picture is obtained on the vertical target The rear sight aperture will be approximately In the center of the rear sight base If the rifle barrel assembly is properly aligned in the upper receiver assembly

NOTE

If rifle barrel assembly (usually new) is not properly aligned In the upper receiver assembly (usually an old part), excessive windage will be present and the upper receiver assembly will require replacement to obtain the proper fit between the alignment pin and slot.

CAUTION

Do not torque over 80 ft-lb (108 N-m) while tightening the barrel nut assembly to the next hole, to allow for proper alignment of gas tube.

NOTE

Do not attempt to hold the upper receiver assembly with a pry bar, however, If the rifle barrel assembly turns In the holding fixture, a pry bar may be used through the front sight assembly base to help prevent the rifle barrel assembly from turning In the holding fixture. Use care not to distort or bend front sight assembly or retaining pins. Use "buddy system" to hold pry bar.

12. Torque the barrel nut assembly again to 30 ft-lb (40.5 N-m) while maintaining sight alignment. The barrel nut assembly may be tightened beyond 30 ft-lb (40.5 N-m) to align the barrel nut assembly serrations for proper gas tube clearance Never loosen the barrel nut assembly to align for gas tube clearance.

13. Check alignment of barrel nut assembly (9) with upper receiver assembly (8). The front 8 inches (20.32 cm) of a gas tube may be used as an alignment tool (see Illustration). This Is Inserted Into the bolt carrier key and then Inserted into the rear of the receiver. If the parts of the barrel nut assembly are properly aligned, the tool will pass freely and lay top dead center along the top of the barrel. A number 15 twist drill (0.180 inch) may also be needed as an alignment tool. If necessary, tighten barrel nut assembly to next hole to allow proper alignment.

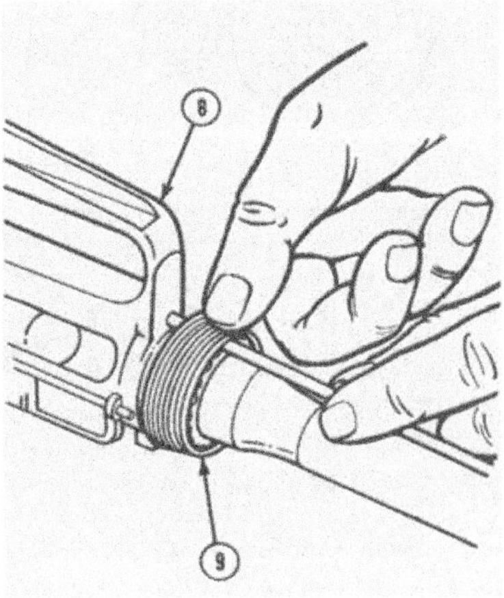

3-10. UPPER RECEIVER AND BARREL ASSEMBLY (CONT).

d. REASSEMBLY (CONT)

NOTE
The peel washer may be heated to remove thin sections. Always place thin section to rear.

14. Install peel washer (11) and compensator (12) on rifle barrel assembly (2).

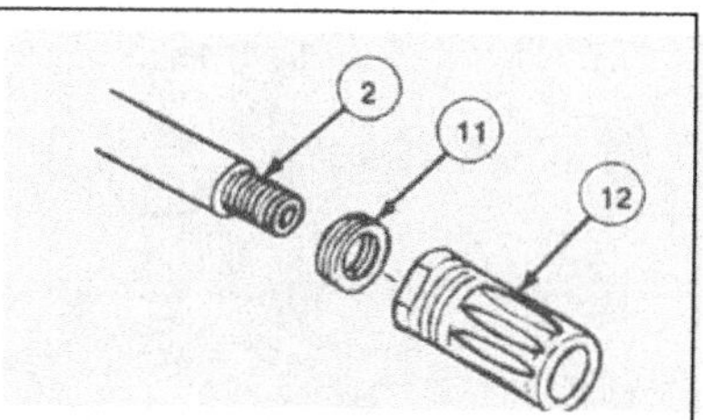

15. Torque compensator (12) to 15-20 ft-lb (20-27 N-m) using combination wrench (10) and torque wrench. Torque is measured when both wrenches are used together. Some portion of the openIng of the third or middle slot must be straight up (align with the front sight post) (12.1) at proper torque level. Thin sections of the peel washer may be removed or added as required. Save unused sections.

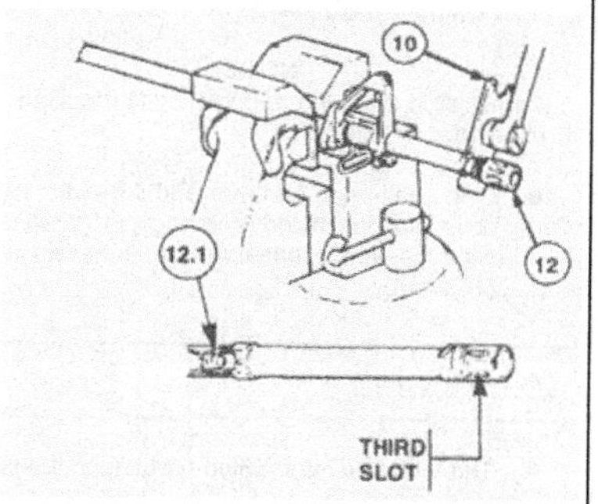

16. Slide gas tube (13) through the barrel nut assembly (9) and then slide forward, Inserting gas tube into hole In the front sight assembly base (14).

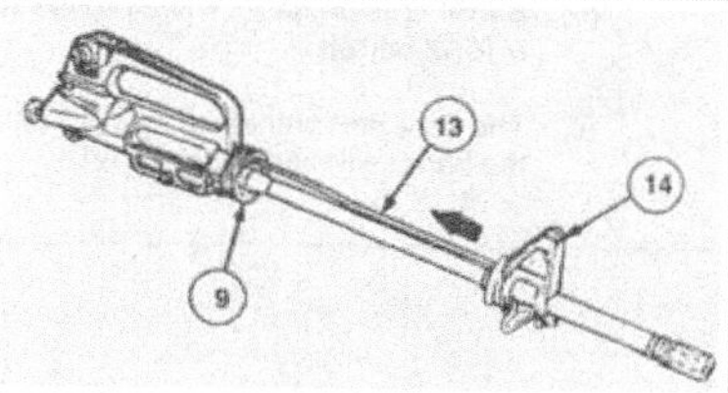

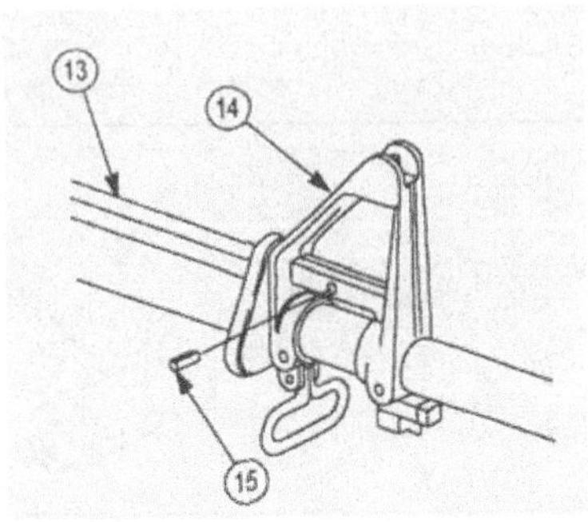

17. Align the holes in gas tube (13) and front sight assembly base (14).

NOTE

To assist in installing spring pin (1) modified nose pliers may be used.

18. Using ball-peen hammer and 5/64-inch diameter drive pin punch, drive spring pin (15) into front sight assembly base (14) to secure gas tube (13).

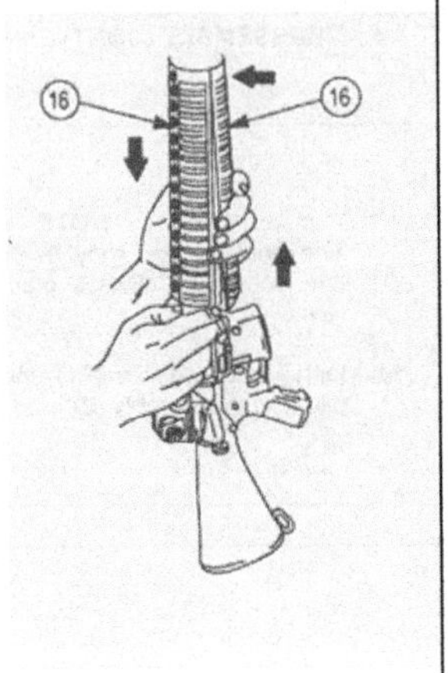

19. Install upper and lower handguard assemblies (16).

e. TEST

1. The following information pertains to the use of breech, bore, and other gages:

(a) All M1 6A2 barrels and chambers are chromed.

(b) Barrel erosion gage, PN 8448496 (normally used on M16A1 fully-chromed barrels), can be used to gage M16A2 barrels.

(c) The bore straightness gage, PN 8448202, is required for use on all barrels. The gage must pass through the barrel without being forced.

3-10. UPPER RECEIVER AND BARREL ASSEMBLY (CONT).

e. TEST (CONT)

2. Use barrel erosion gage (1) PN 8448496. Install key and bolt carrier assembly with bolt assembly and firing pin removed. Hold rifle vertical with receiver up. Insert gage into rear of key and bolt carrier assembly. The reject line must be read at the rear edge of the key and bolt carrier assembly.

3. If the reject mark passes beyond the rear surface of the key and bolt carrier assembly, the barrel is unserviceable and shall be replaced.

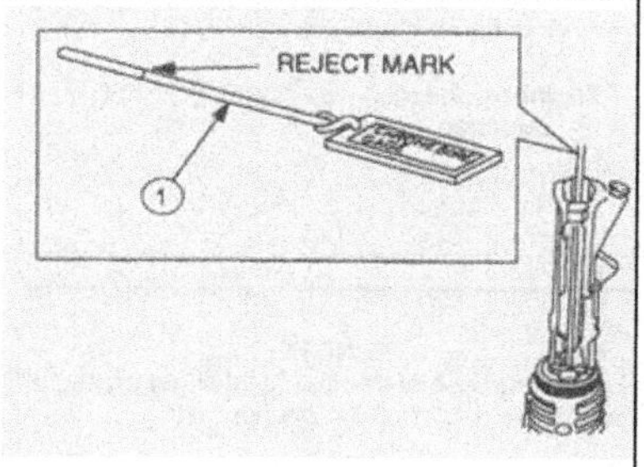

NOTE
Ensure barrel is clean prior to performing the following test.

4. Check straightness of bore using straightness gage (2) PN 8448202. Put gage in barrel. Tilt barrel and allow gage to fall through. Catch gage.

5. Gage must pass freely through barrel. If gage does not pass through barrel, recheck as follows. Hold upper receiver and barrel assembly (3) in vertical position with muzzle pointed down; insert gage into chamber end of barrel. Release gage and catch it as it exits muzzle end. If gage passes freely through the barrel, barrel is acceptable. If it does not, the barrel must be straightened or replaced. (See page 3-38 for straightening instructions/ procedures.)

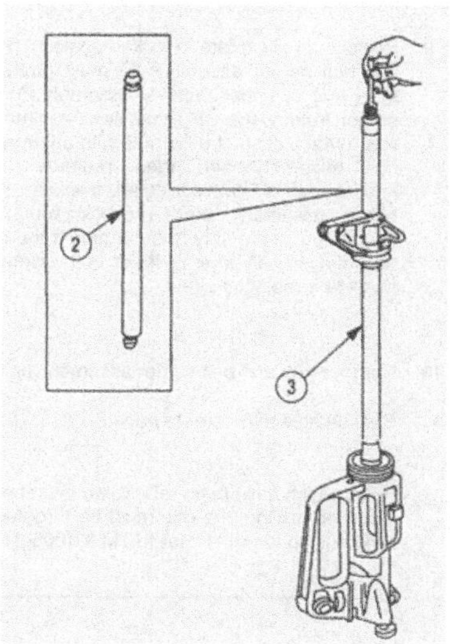

6. Assemble charging handle assembly (4), bolt assembly, and key and bolt carrier assembly (5) into upper receiver assembly (6).

7. Insert headspace gage (7) PN 7799734 in chamber.

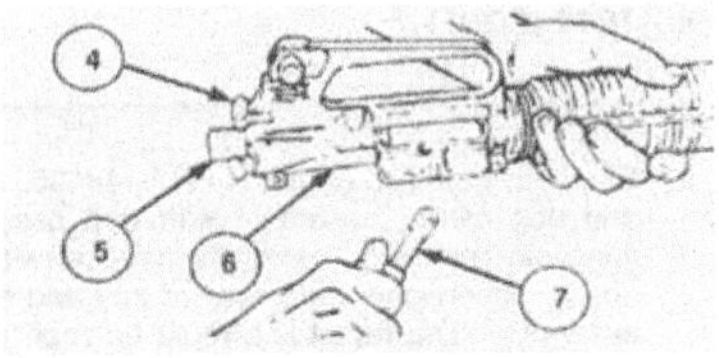

NOTE

For the purpose of this test "light finger pressure" is defined as 8 1/2 to 8 3/4 pounds.

8. Check headspace by pressing key and bolt carrier assembly (5) and charging handle assembly (4) forward using light finger pressure.

9. Bolt should not rotate to locked position. Key and bolt carrier assembly (5) must protrude from rear of upper receiver assembly (6) for proper headspace. If excessive headspace, first replace old bolt assembly with an unused bolt assembly and then recheck. If headspace is not corrected, replace rifle barrel assembly; then recheck with the original bolt assembly to determine if the bolt assembly is still good or if the bolt assembly should be replaced also.

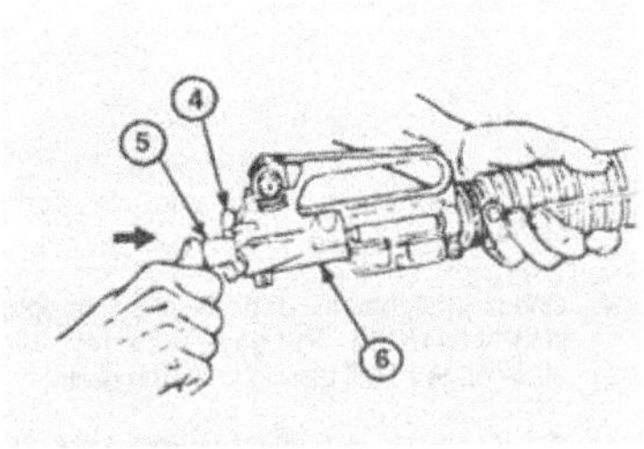

10. Remove key and bolt carrier assembly, bolt assembly, charging handle assembly, and headspace gage.

11. Reassemble rifle, refer to page 3-77.

NOTE

Rifles which have been rebarreled must be function-fired with seven rounds of 5.56mm ball ammunition. After rebarreling, the rifle must be targeted with three rounds of 5.56mm ball ammunition at 25 meter range using target. Refer to TM 9-1005-319-10 and FM 23-9

3-10. UPPER RECEIVER AND BARREL ASSEMBLY (CONT).

e. TEST (CONT)

12. (a) If the test fails using finger pressure, remove the gage and perform the test again as follows: With the muzzle down, stack 8 1/2 to 8 3/4 pounds of trigger weights (8) on a locally fabricated spacer/weight (9) on the bolt carrier assembly (10) Insert headspace gage (11) and test per above instructions.

(b) Remove trigger weights (8), spacer/weight (9), bolt carrier assembly (10), charging handle (12), and headspace gage (11).

(c) If excessive headspace, first replace bolt assembly and then recheck If headspace is not corrected, replace barrel assembly; then recheck with the original bolt to determine if the bolt is still good or If the bolt should be replaced also.

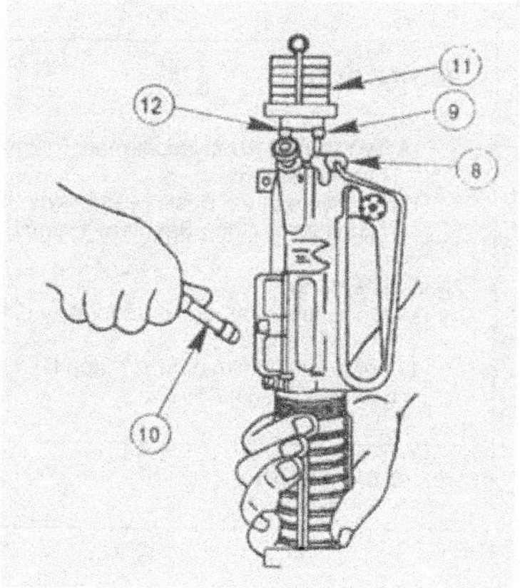

3-11. UPPER RECEIVER ASSEMBLY AND REAR SIGHT ASSEMBLY.

This task covers:

a. Disassembly
b. Inspection
c. Repair
d. Lubrication
e. Reassembly
f. Mechanical Zero Procedures (A.F. Only)

INITIAL SETUP:

Tools
(ARMY) Small Arms Repairman Tool Kit (item 3, app B)
Field Maintenance Basic Less Power Small Arms Shop Set (item 1, app B)

Materials/Parts
Cleaner, lubricant, and preservative (CLP) (item 9, app D)
Lubricant, solid film (item 21, app D)
Index screw (9349065)

Equipment Conditions
3-29 Upper receiver assembly removed.

General Safety Instructions
To avoid injury to your eyes, use care when removing and installing spring-loaded parts.
When using solid film lubricant or dichloromethane, be sure the area is well ventilated.

a. DISASSEMBLY

WARNING
To avoid injury to your eyes, use care when removing and installing springloaded parts.

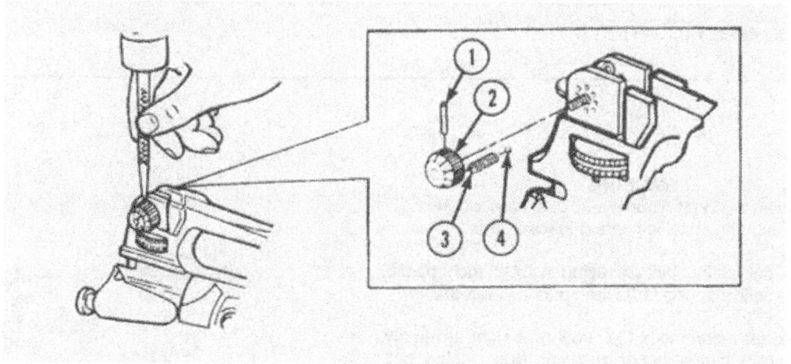

CAUTION

Be sure to catch small parts (1, 2, 3, and 4). AIR FORCE ONLY: Use magnet to keep from losing small parts (1, 3, and 4).

1. Drive out spring pin (1) using a hammer and 1/16 inch punch.

2. Catch rear sight assembly windage knob (2), helical spring (3), and ball bearing (4).

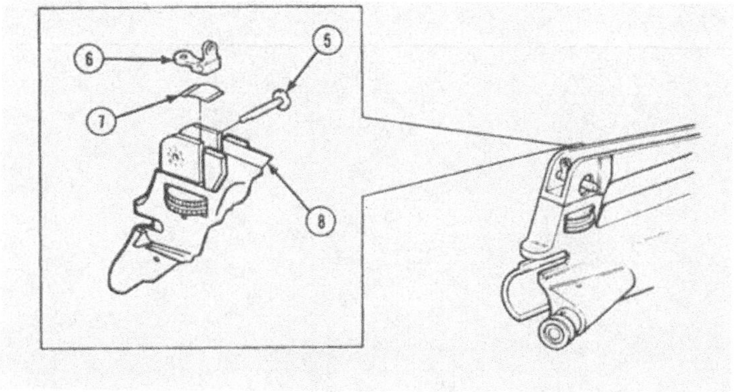

3. Using a flat-bladed screwdriver, remove rear sight assembly windage screw (5)

4. Remove sight aperture (6) and flat spring (7) from upper receiver (8).

3-11. UPPER RECEIVER ASSEMBLY AND REAR SIGHT ASSEMBLY (CONT).

a. DISASSEMBLY (CONT)

WARNING

To avoid injury to your eyes, use care when removing and installing spring-loaded parts.

5. Drive out spring pin (9) using a 3/32 Inch punch Catch helical spring (10) when punch is withdrawn.

6. Rotate elevation index (11) until rear sight assembly base (12) clears upper receiver (8). Catch ball bearing (13) and helical spring (14) as rear sight base clears.

7. Push elevation index (11) out with thumb using slight rotation motion. Catch ball bearing (15) and helical spring (16).

8. Use 1 16 Inch allen wrench to remove index screw (17). Discard Index screw (1 7) Separate elevation index (11) from elevation knob (18) by hand.

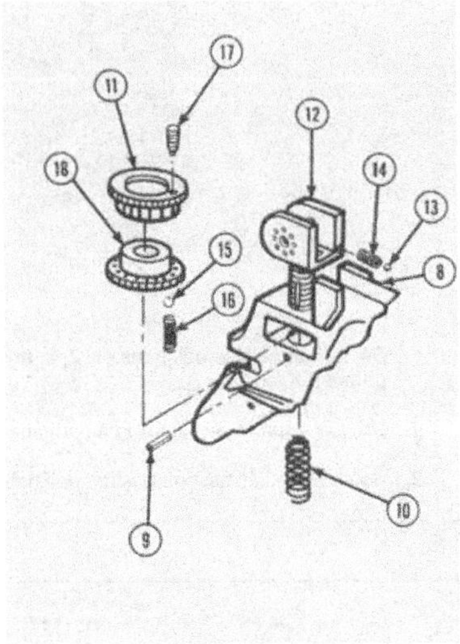

9. Remove pin (1 9) using 3,32 Inch drive pin punch and hand hammer .

WARNING

To avoid Injury to your eyes, use care when removing and installing spring-loaded parts.

10. Remove forward assist assembly (20) and helical spring (21) from upper receiver (8). For further disassembly of forward assist assembly see page 3-59.

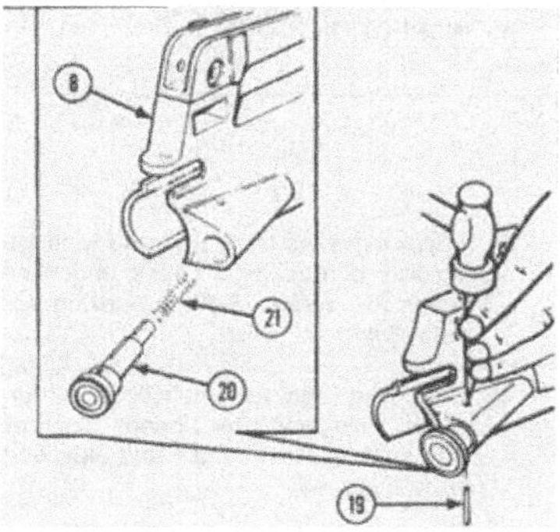

b. INSPECTION

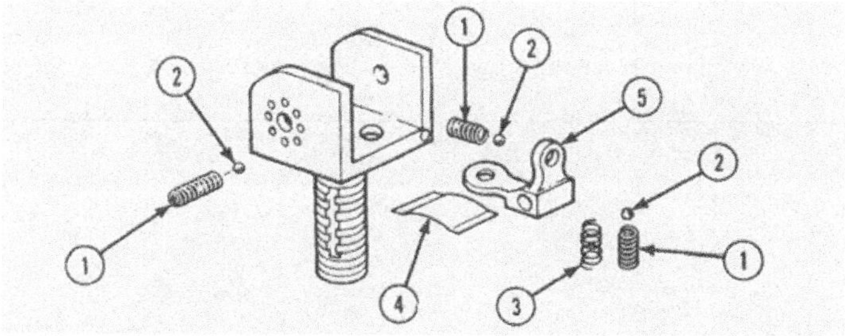

1. Check rear sight parts for serviceability. Inside of apertures should be round and distinct. Replace If defective

2. Visually inspect rear sight assembly helical springs (1) ball bearings (2), and helical spring (3) for breaks, bends, and missing parts Ball bearings should be smooth and round. Replace If defective

3. Check upper receiver for cracks, corrosion, and damage Clear drain hole with a piece of wire. Repair (p 3-29) or replace If defective.

4. Check that flat spring (4) retains sight aperture (5) firmly In either position. Replace flat spring (4) If sight aperture is not firm.

3-11. UPPER RECEIVER ASSEMBLY AND REAR SIGHT ASSEMBLY (CONT).

b. INSPECTION (CONT)

5. Check elevation Index 161 and windage knob (71 for legibility of markings Check underside of windage knobs for cracks. Detent indexing surfaces should be well formed.

6. Check rear sight assembly base (81 for serviceability Clear drain holes for springs Threaded portion of rear sight assembly base and elevation knob should be well formed.

7. Inspect rear sight guards for bends, if bent refer to the following page for repair procedures.

8. Inspect all parts for damage and wear Replace all defective parts.

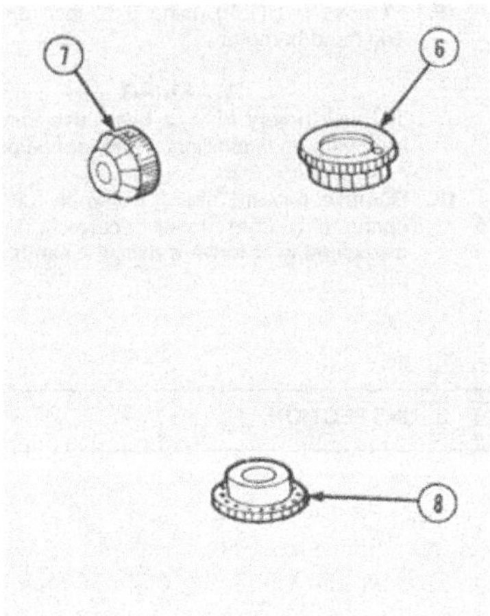

c. REPAIR

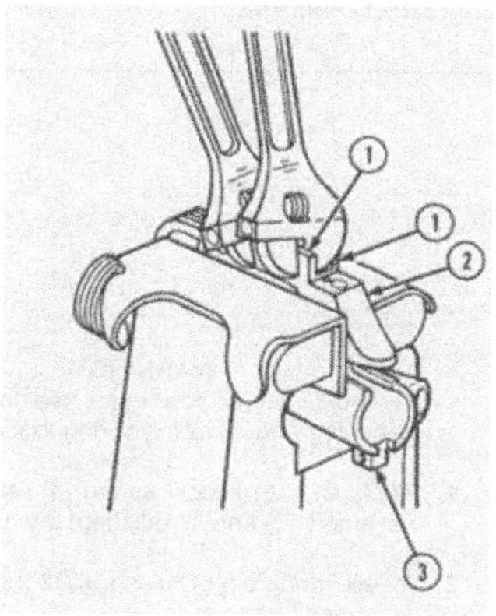

1. To straighten bent rear sight assembly guards (1), remove rear sight assembly components. Place carrying handle (2) In a vise using jaw clamps. Tighten vise to firmly hold upper receiver (3).

2. Using two adjustable wrenches, gradually bend guards (1) to straighten. When bending the guards (1), gradually bend beyond the straight point as the guard will partially return when bending pressure is stopped.

3. After straightening, use a flat file to remove any nicks, kinks, or burrs that remain on the inside of guards (1).

CAUTION
Do not use wire brush on aluminum surfaces

4. Apply solid film lubricant to brightened area for final protective coating.

5. Replace rear sight assembly components and check that rear sight assembly functions properly. If rear sight assembly functions check out, return upper receiver assembly to service.

6. If rear sight guards cannot be straightened utilizing the above procedures, replace the upper receiver

d. LUBRICATION

Lubricate upper receiver assembly and rear sight assembly. Apply CLP to helical springs and ball bearings (three each) and threaded portion of screws before installation. Lubricate helical springs and ball bearings through their respective drain holes.

3-11. UPPER RECEIVER ASSEMBLY AND REAR SIGHT ASSEMBLY (CONT).

e. REASSEMBLY

WARNING
To avoid injury to your eyes, use care when removing and Installing spring-loaded parts.

1. Apply CLP to helical spring (1) and forward assist assembly (2) and Install them into upper receiver (3).

2. Install spring pin (4) using 3, 32 Inch drive pin punch and hand hammer

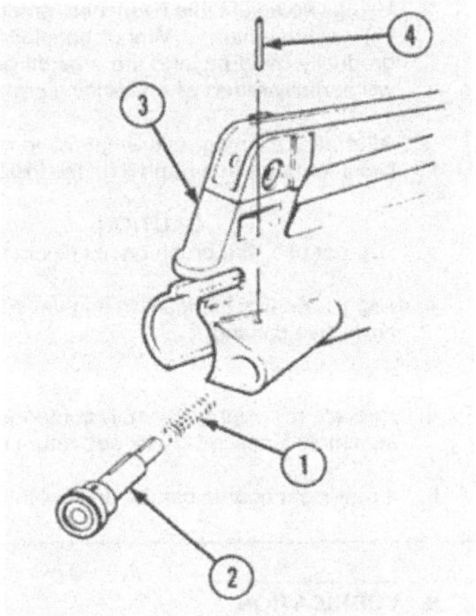

3. Assemble elevation knob (5), elevation index (6), and new index screw (7) using 1 16 Inch alien wrench. Do not overtighten Index screw as scale will require adjustment

4. Install ball bearing (8) and helical spring (9) using needle nose pliers or tweezers

5. Depress ball bearing (8) with a punch Inserted through access hole, and slide elevation knob assembly In upper receiver (3) from the side Center elevation knob assembly

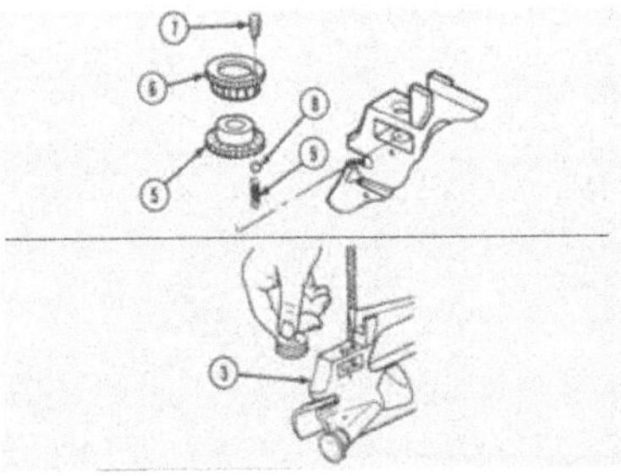

WARNING

To avoid injury to your eyes, use care when removing and installing spring-loaded parts.

NOTE

All springs are identical when new. Once disassembled from the rifle, their free length may vary due to different amounts of compression when installed. If the length of springs varies, use longest as #22, and the shortest as #9

6. Insert threaded portion of rear sight assembly base (10) into upper receiver (3) and rotate elevation knob assembly until threads engage.

7. Insert helical spring (11) and ball bearing (12) in their hole as rear sight assembly base is lowered into upper receiver as elevation knob assembly is further rotated. Rotate elevation knob assembly until rear sight assembly base is all the way down. Then come up 22 clicks before installing spring pin. Check spring action of helical spring (11) on upper receiver.

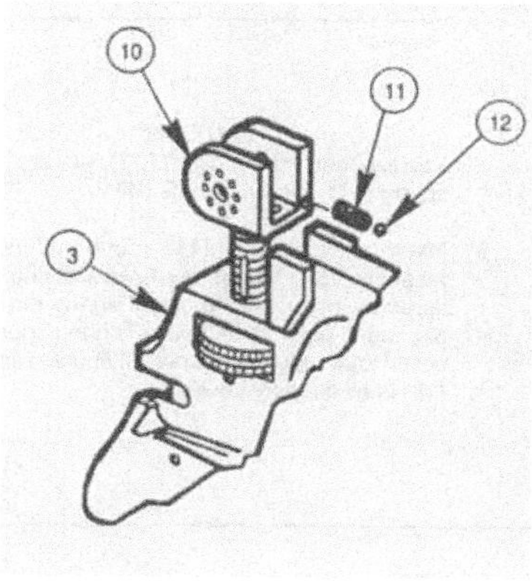

3-11. UPPER RECEIVER ASSEMBLY AND REAR SIGHT ASSEMBLY (CONT).

e. REASSEMBLY (CONT)

CAUTION

Ensure that spring pin (1 3) passes over helical spring (14), not through its coils

8. Insert helical spring (14) through underside of upper receiver (15). Compress helical spring with a small tip screwdriver (16) to install spring pin (1 3) Spring pin must pass over helical spring, not through its coils Rotate elevation knob (17) until rear sight base (18) Is all the way down.

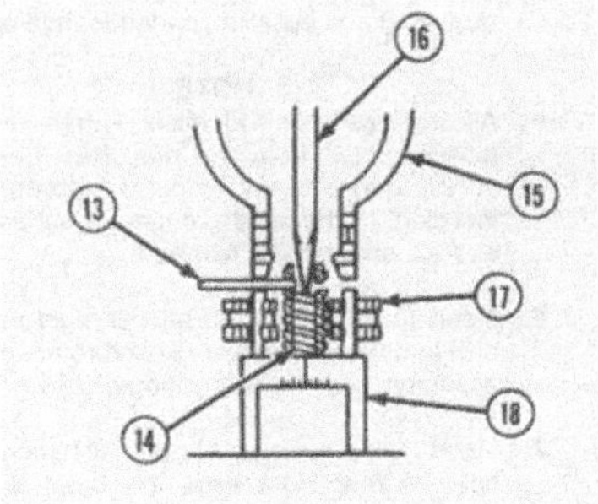

9. Install flat spring (19) and sight aperture (20) In rear sight assembly base (10). Install windage screw (21) with screwdriver.

10. Insert helical spring (22) and ball bearing 123) In windage knob (24)

NOTE

Tilt upper receiver toward wind age knob during positioning to prevent loss of ball bearing.

11. Position windage knob on shaft of windage screw (211 Align holes In windage knob with hole of shaft In windage screw and hold In alignment with a small screwdriver or punch Install spring pin (25).

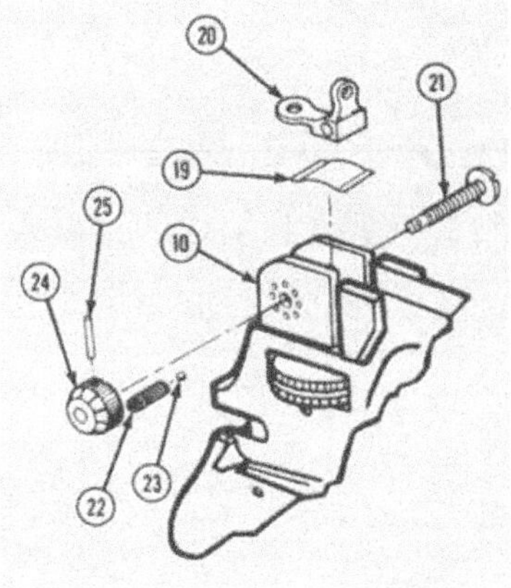

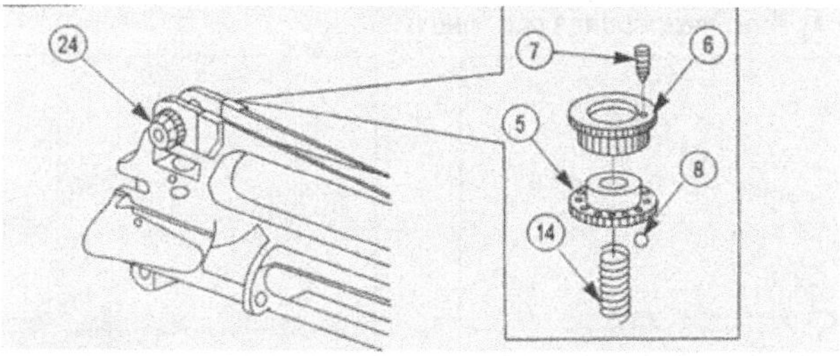

12. Rotate and test elevation index (6) and windage knob (24) for ease of functioning.

13. Inspect elevation knob zero as follows:

(a) Rotate elevation knob (5) counterclockwise until the rear sight assembly is all the way down. If a whole click is not felt as the rear sight assembly stops, the rear sight assembly has bottomed out and will not pivot freely.

(b) Position elevation knob back slightly to its last whole click so the rear sight assembly base is under tension of the ball bearing (8) and helical spring (14). The 300 meter mark should align with the mark on to receiver.

(c) If the 300 meter mark is not aligned with the mark on receiver, slip the range scale in the following manner:

(1) Position the 300 meter mark with the mark on the receiver.

(2) Insert a 1/16 inch allen wrench through the access hole of the rear sight assembly base and into the index screw (7).

(3) Loosen the index screw three turns and leave the wrench in place.

(4) Rotate lower portion of elevation knob

(5) counterclockwise until it stops (range scale should not have moved). Elevation knob should be positioned on its last whole click. (5) Tighten index screw (7) and remove wrench.

(6) Check for proper setting.

NOTE

After the rifle is assembled, the rear sight is centered, and placed at the 300 meter mark, perform the owing check: While looking at a light background, obtain good sight alignment. If the hole in the rear sight aperture appears oval Instead of round, the rear sight base or upper receiver should be replaced. To determine which part requires replacement, replace the rear sight base first. If this does not resolve the problem, replace the upper receiver.

14. Assemble rifle, refer to page 3-77

3-11. UPPER RECEIVER ASSEMBLY AND REAR SIGHT ASSEMBLY (CONT).

f. MEACHANICAL ZERO PROCEDURES (A.F. ONLY)

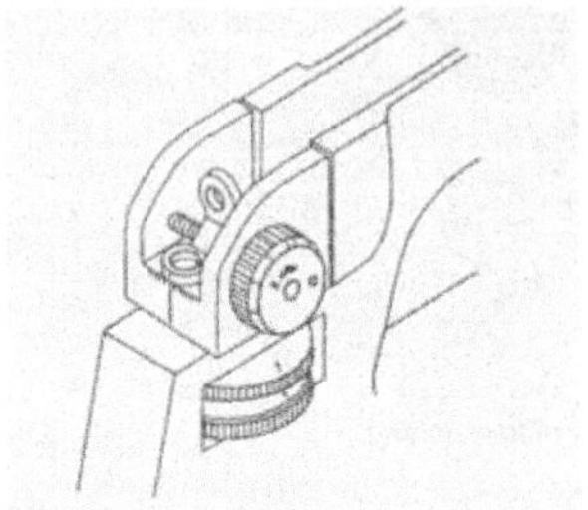

1. Center rear sight by moving windage knob in the appropriate direction.

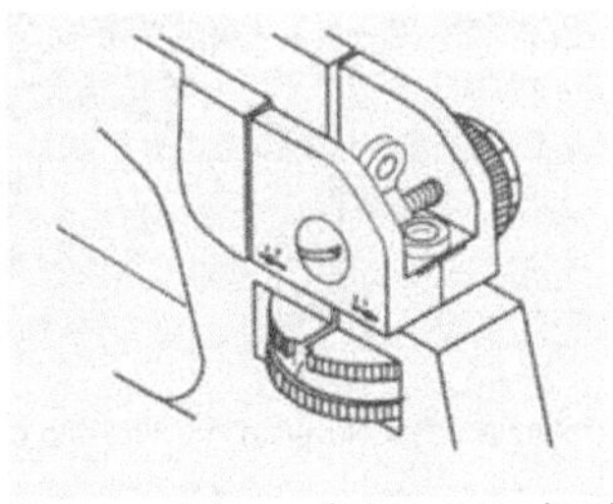

2. Always push in on windage screw head after making rear sight adjustments.

3. Visually check rear sight to ensure it is centered after making adjustments. Also, ensure the rear sight is set in the shortrange (unmarked aperture) position.

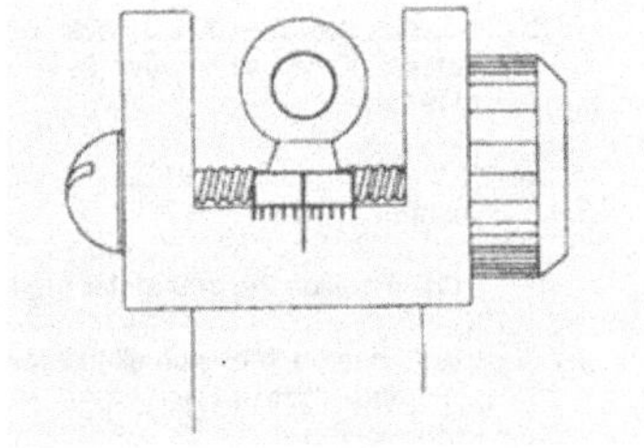

NOTE

This procedure, when used in conjunction with front sight mechanical zero adjustment (p 2-54), will give an approximate battle site zero to most M16A2 rifles. The above steps can also be used before firing a new or newly assigned rifle. Use the procedure to check rifles stored in preferred packaging during routine inspections. This will help ensure people armed with the rifles will stand a better chance of hitting an enemy if the rifles must be used before a live fire zero can be made. Whenever possible, zeroing of the rifle should be accomplished using ball ammunition on a 25 meter zeroing target using the "L" aperture.

After the rifle is assembled, the rear sight is centered, and placed at the 300 meter mark, perform the following check: While looking at a light background sight through the rear sight and obtain good sight alignment with the front sight placed in the center of the rear sight aperture. If while performing this check the hole in the rear sight aperture appears oval instead of round, the rear sight base or upper receiver should be replaced. To determine which part requires replacement, replace the rear sight base first. If this does not resolve the problem, replace the upper receiver.

3-12. FORWARD ASSIST ASSEMBLY.

This task covers:

a. Disassembly
b. Inspection
c. Repair
d. Lubrication
e. Reassembly

INITIAL SETUP:

Tools
(ARMY) Small Arms Repairman Tool Kit (item 3, app B)
Field Maintenance Basic Less Power Small Arms Shop Set (item 1, app B)

Materials/Parts
Cleaner, lubricant, and preservative (CLP) (item 9, app D) | a

Equipment Conditions
3-48 Forward assist assembly removed

General Safety Instructions
To avoid injury to your eyes, use care when removing and installing spring-loaded parts

a. DISASSEMBLY

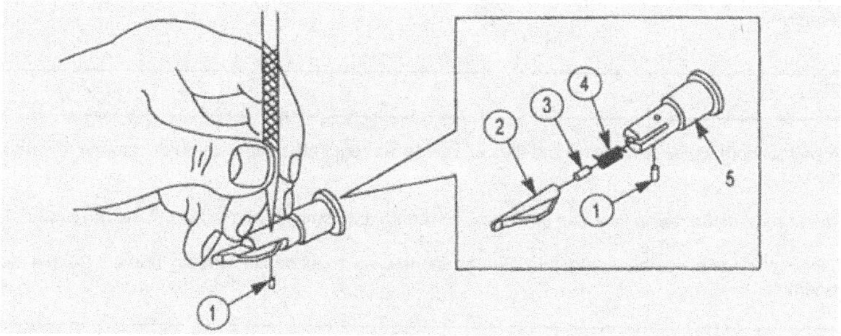

WARNING
To avoid injury to your eyes, use care when removing and installing spring-loaded parts.

1. Drive out spring pin (1) using 1/16 inch drive pin punch and hand hammer

2. Remove forward assist pawl (2), pawl detent (3), and helical spring (4) from plunger assembly (5)

3-12. FORWARD ASSIST ASSEMBLY (CONT)

b. INSPECTION

1. Inspect forward assist pawl (1) for burrs, chips, and cracks. Minor burrs may be removed using fine files or stones, as required. Replace forward assist pawl if defective.
2. Inspect pawl detent (2) for burrs and cracks. Minor burrs may be removed using fine files or stones, as required. Replace pawl detent if defective.
3. Inspect helical spring (3) for kinks, breaks, and wear. Replace helical spring if defective.
4. Inspect plunger assembly (4) for wear, burrs, chips, and breaks. Minor burrs may be removed using fine files or stones, as required. Replace forward assist assembly if defective.
5. Inspect spring pin (5) for wear. Replace if defective.

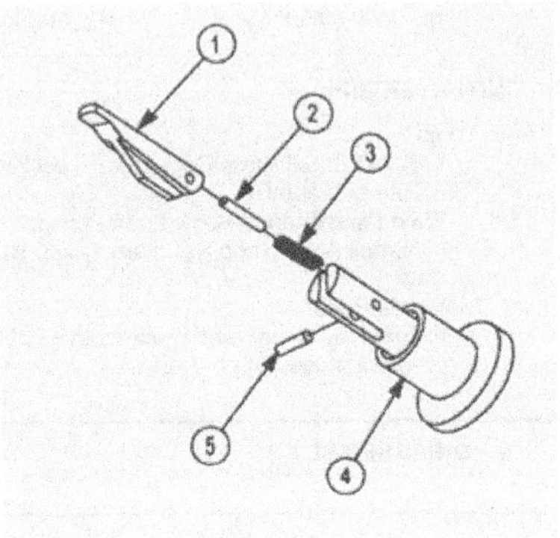

c. REPAIR

1. Repair forward assist pawl using fine files or stones, as required, to smooth burrs. Do not deform forward assist pawl.
2. Repair pawl detent using fine files or stones, as required, to smooth burrs. Do not deform pawl detent.
3. Repair plunger assembly using fine files or stones, as required, to smooth burrs. Do not deform plunger assembly.

d. LUBRICATION

Lubricate helical spring, pawl detent, and forward assist pawl with CLP (p 2-33) before installation.

e. REASSEMBLY

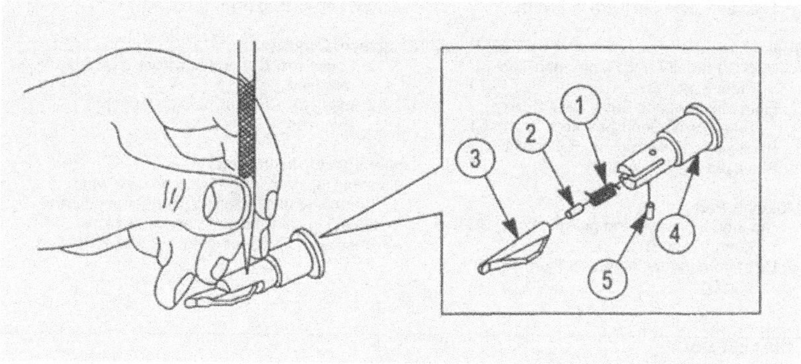

WARNING
To avoid injury to your eyes, use care when removing and installing spring-loaded parts.

1. Install helical spring (1), pawl detent (2), and forward assist pawl (3) into plunger assembly (4).
2. Align holes and install spring pin (5) using 1/16 inch drive pin punch and hand hammer. Spring pin must be flush or slightly below flush after reassembly.
3. Assemble rifle, refer to page 3-77.

3-13. LOWER RECEIVER AND BUTTSTOCK ASSEMBLY.

This task covers:

a. Disassembly
b. Inspection
c. Repair
d. Test
e. Reassembly

INITIAL SETUP

Test Equipment
Tool and Gage Set (item 2, app B)

Tools
(ARMY) Small Arms Repairman Tool Kit (item 3, app B)
Field Maintenance Basic Less Power Small Arms Shop Set (item 1, app B)
Pivot pin removal tool (fig E-3, app E)
Slave pin (fig E-5, app E)

Materials/Parts
Cleaner, lubricant, and preservative (CLP) (item 9, app D)
Dichloromehtane, technical (item 15, app D)
Lubricant, solid film (item 21, app D)
Screw, self-locking (item 6, p C-11)

Equipment Conditions
3-15 Lower receiver and buttstock assembly removed
2-57 Buttstock assembly and pistol grip removed

General Safety Instructions
To avoid injury to your eyes, use care when removing and installing spring-loaded parts.
When using solid film lubricant or dichloro methane, be sure the area is well ventilated.

a. DISASSEMBLY

WARNING

To avoid injury to your eyes, use care when removing and installing spring-loaded parts.

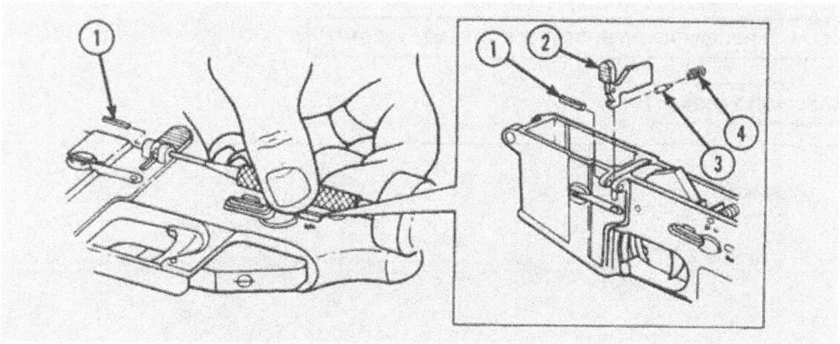

1. Remove spring pin (1) using 3/ 32 inch drive pin punch and hand hammer.

2. Remove bolt catch (2), bolt catch plunger (3), and bolt catch spring (4).

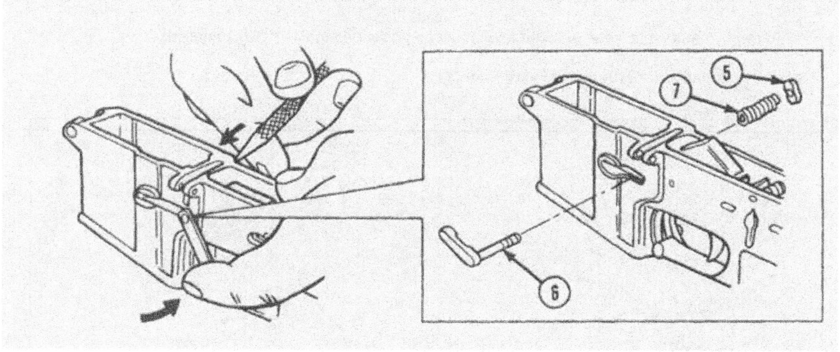

WARNING

To avoid injury to your eyes, use care when removing and installing spring-loaded parts.

3. Using drive pin punch, press in on magazine button (5) and turn magazine catch 16) counterclockwise to unscrew and remove.

4. Remove magazine button (5) and magazine catch spring (7).

3-13. LOWER RECEIVER AND BUTTSTOCK ASSEMBLY (CONT).

a. DISASSEMBLY (CONT)

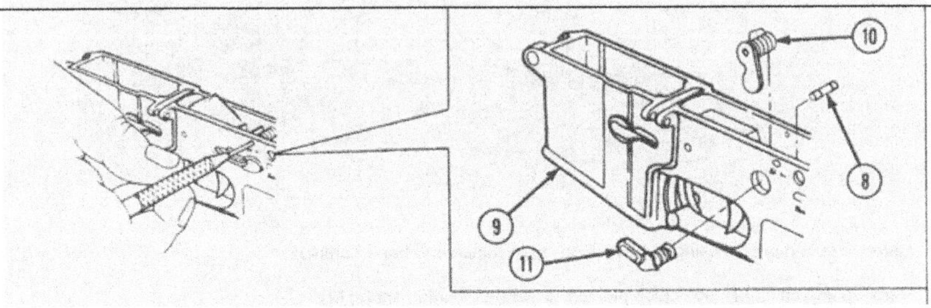

5. Use a rounded tip drive pin punch to push automatic sear pin (8) out of lower receiver (9).

NOTE
To remove automatic sear, selector lever must be positioned to BURST (if installed).

6. Remove automatic sear (10) and selector lever (11).

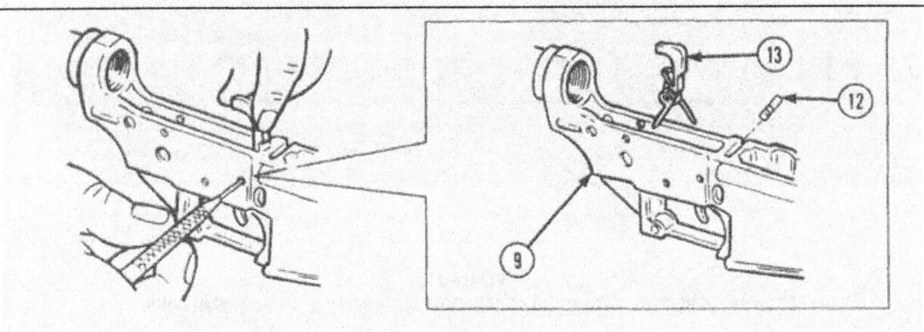

NOTE
To remove (hammer assembly should be forward), place selector lever (if installed) to SEMI position.

7. Use drive pin punch to push hammer pin (12) from lower receiver (9).

8. Remove hammer assembly (13). If further disassembly is required, see page 3-73.

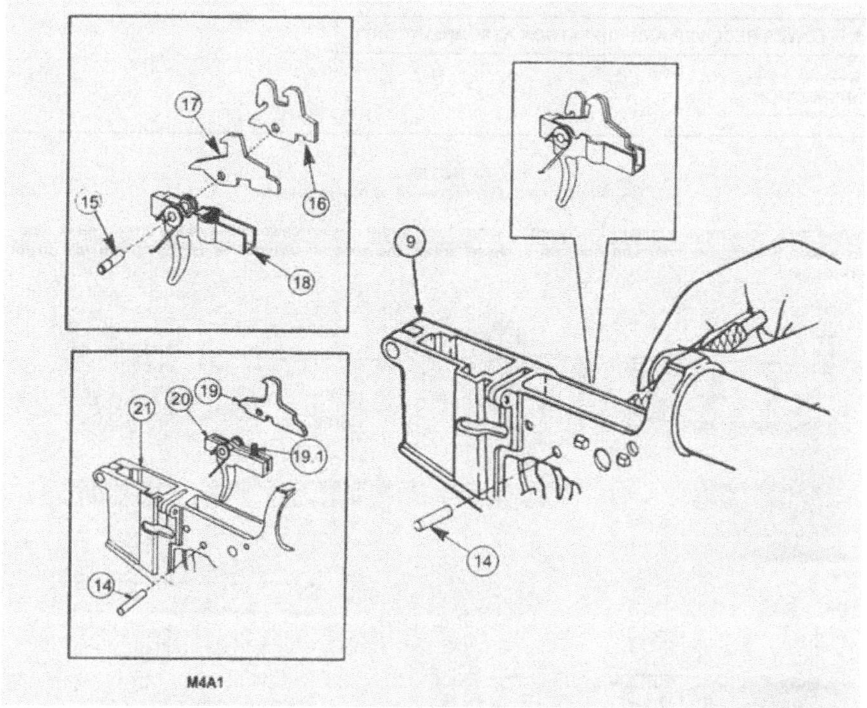

NOTE

Use of fabricated slave pin will allow removal of the following parts as a unit.

9. Remove trigger pin (14) by pushing from the left side of lower receiver (9) with fabricated slave pin (15) and a drive pin punch.

10. Remove semiautomatic disconnector (16), burst disconnector (17), and trigger assembly (18). If further disassembly of trigger assembly is required, see page 3-75.

M4A1 ONLY

11. Remove disconnector (19), disconnector spring (19.1), and trigger assembly (20) from lower receiver and extension subassembly (21).

3-13. LOWER RECEIVER AND BUTTSTOCK ASSEMBLY (CONT).

b. INSPECTION

NOTE

Refer to page 3-35 for repair of corroded surfaces.

Inspect lower receiver and buttstock assembly for corrosion in the lower receiver lobes of the pivot area or hinge pin area. If extensive corrosion appears in these areas, the receiver will not be repaired; turn rifle in for replacement.

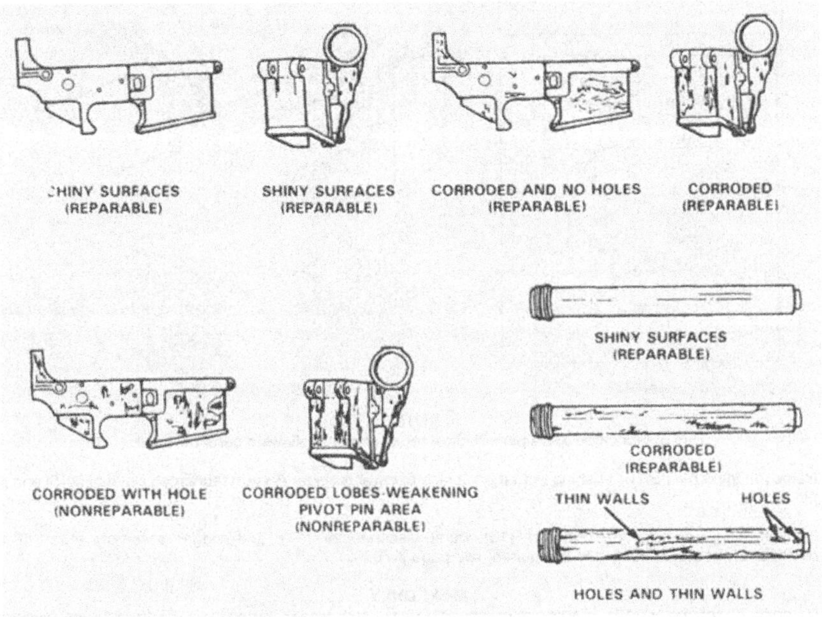

WARNING

When using solid film lubricant or dichloromethane, be sure the area is well ventilated.

CAUTION

Do not use a wire brush on aluminum surfaces.

NOTE

If a weapons lower receiver is missing one third or more of its exterior protective finish, resulting in an unprotected, light reflecting surface, it is candidate for overhaul. This missing finish will be considered a shortcoming. This shortcoming requires action to obtain a replacement weapon. Once a replacement has been received, evacuate the original weapon to depot for overhaul.

Solid Film Lubricant (SFL) is the only authorized touchup for the weapon and may be used on up to one third of the exterior finish of the weapon. **FOR ARMY CONUS USE ONLY AND AIR FORCE TRAINING WEAPONS ONLY:** SFL may be used as a touchup without limitation on the upper receiver and barrel assembly. This is to say that units which DO NOT fall under the category of Divisional Combat Units or rapid deployment type units may have up to 100 percent of the exterior surface of the upper receiver and barrel assembly protected with SFL if necessary.

c. REPAIR

Repair or replace all parts of lower receiver and buttstock assembly if defective.

3-13. LOWER RECEIVER AND BUTTSTOCK ASSEMBLY (CONT).

d. TEST

1. With the upper receiver attached to the lower receiver, and the pivot pin and takedown pins in place, perform the following test:

   (a) Apply hand pressure to push the upper receiver as far to one side as possible.

   (b) Attempt to insert a 0.020 inch thickness gage between the pivot pin lugs of the upper and lower receivers.

   (c) If the thickness gage penetrates to the pivot pin at all accessible locations, repair by replacement of the upper receiver (see (b) below) or replacement of rifle is required.

2. If the rifle fails the above test, remove the upper receiver and install a "NEW" upper receiver and perform the test again.

3. If the rifle now passes the above test, it shall be considered serviceable and continue in use.

4. If the rifle falls the test with a new upper receiver, this failure shall be considered a shortcoming. This shortcoming requires action to obtain a replacement rifle. Once a replacement has been received, evacuate the original rifle to depot for overhaul.

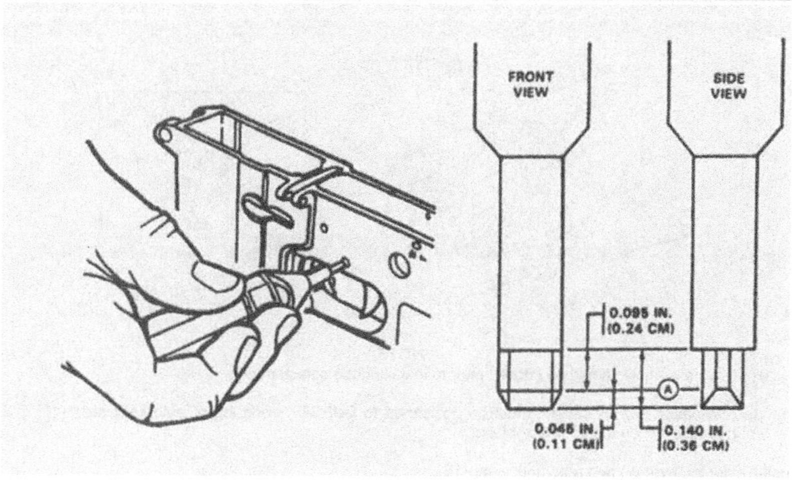

NOTE

If the lower receiver is not disassembled, visually inspect for broken or damaged parts, and to ensure that the hammer and trigger springs are correctly installed before beginning this test. It is not necessary to disassemble the lower receiver for the sole purpose of this visual Inspection. If broken or damaged parts are found, disassemble (p 3-62) and repair as authorized.

5. Test two hammer pin holes and two trigger pin holes using not-go plug gage PN 12006472. This test may be conducted by disassembly of the lower receiver (p 3-62) or by pushing the pin far enough to disengage the end of the pin from the side of the receiver which is being tested. If the lower receiver is not disassembled and the not-go plug gage enters any hole to first shoulder (A), the lower receiver must be disassembled and all four holes must be tested again.

6. Gently insert the not-go plug gage and rotate it 180 degrees. If the not-go plug gage passes through any one of the four pin holes, the rifle is unserviceable and will be turned in for replacement. The gage must extend through the wall thickness to be unserviceable.

7. After completion of gaging operation, visually inspect hammer and trigger springs to ensure proper location of spring legs.

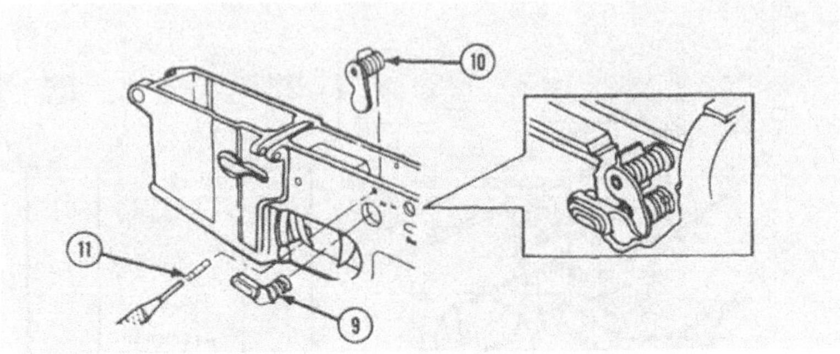

NOTE

Hammer assembly should be cocked prior to installing the selector lever.

Selector lever (9), if installed, must be positioned to BURST. Long leg of automatic sear (10) spring must rest on top of selector lever.

5. Install selector lever (9) and automatic sear (10).

6. Install automatic sear pin (11) (install on the right side) into receiver using drive pin punch. Push in until flush.

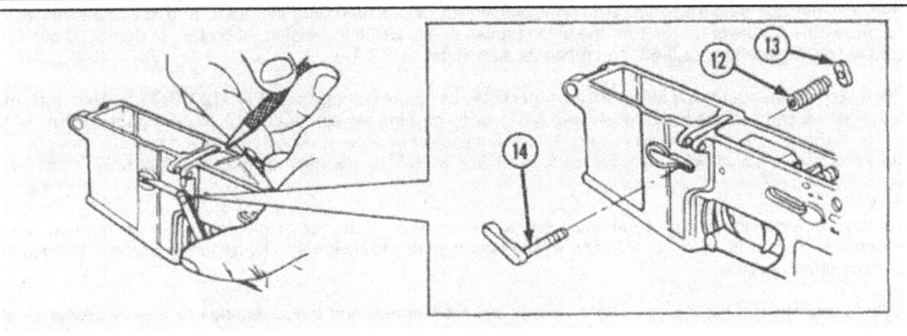

7. Install magazine catch spring (12) and magazine button (13).

NOTE

Drive pin punch should be larger than hole in magazine button.

8. Install magazine catch (14). Push in on magazine button (13) using a drive pin punch and turn magazine catch (14) clockwise until threaded end of magazine catch is flush with magazine button head.

3-13. LOWER RECEIVER AND BUTTSTOCK ASSEMBLY (CONT).

e. REASSEMBLY (CONT)

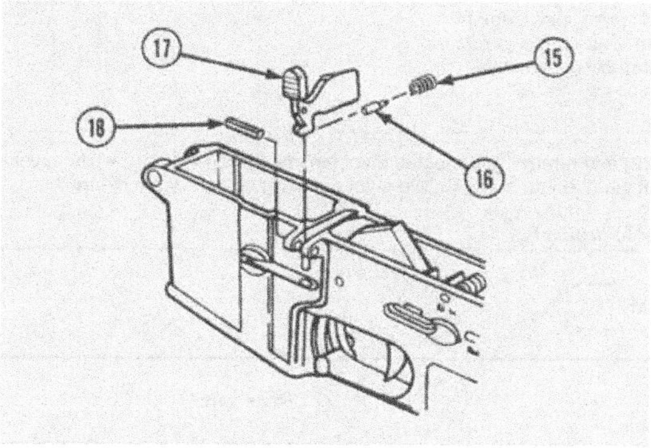

9. Install bolt catch spring (15), bolt catch plunger (16), and bolt catch (17).

10. Secure by installing spring pin (18) using 3/32 inch drive pin punch and hand hammer.

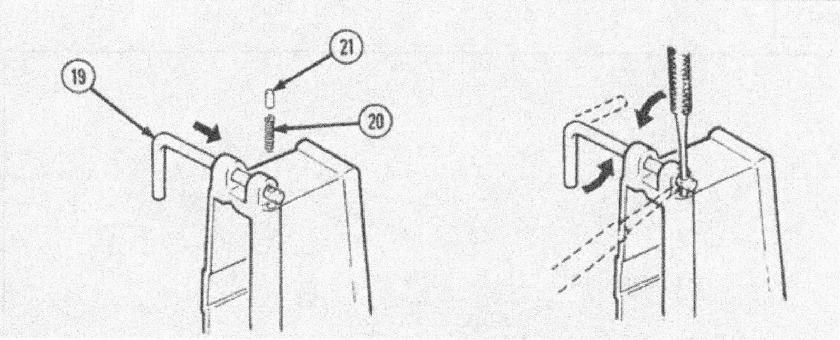

11. Install fabricated pivot pin installation tool (19). Insert helical spring (20) and pivot pin detent 21). Compress pivot pin detent in recess with punch and rotate tool. Remove punch.

WARNING

To avoid injury to your eyes, use care when removing or installing spring-loaded parts.

NOTE

Rounded end of pivot pin detent must be in the groove of the pivot pin (22) when assembly is complete.

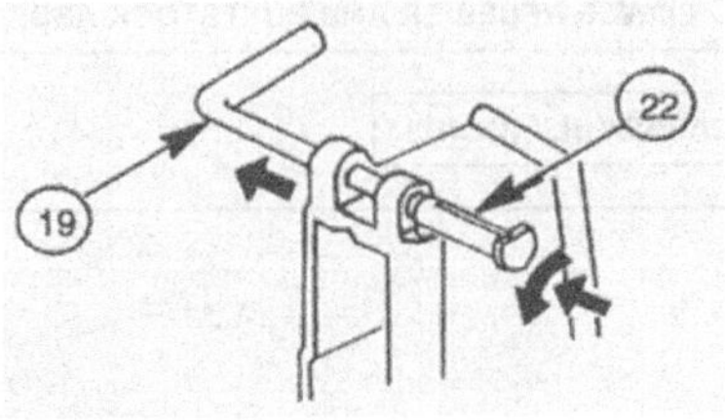

12. Position pivot pin (22) and removing fabricated pivot pin installation tool (19) while maintaining pressure, slide pivot pin (22) into hole. Rotate pivot pin to receive pivot pin detent.

13. Assemble rifle, refer to page 3-82.

3-14. HAMMER ASSEMBLY

This task covers:

a. Disassembly
b. Inspection/Repair
c. Reassembly

INITIAL SETUP

Equipment Conditions
3-62 Hammer assembly removed

a. DISASSEMBLY

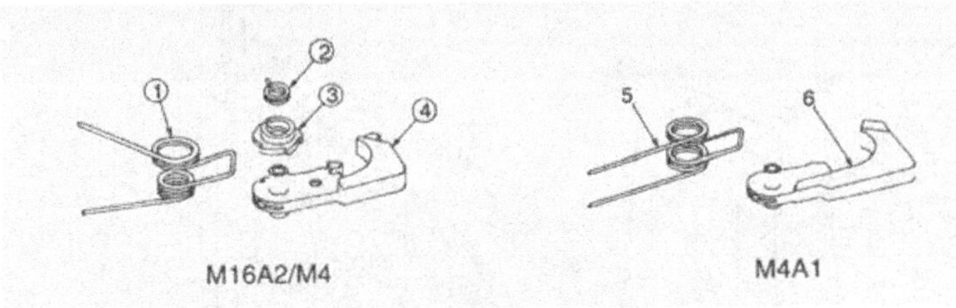

NOTE

M16A2 burst cam is black. M4 burst cam is nickel colored (shiny).

(M16A2/M4) remove hammer spring (1), cam clutch spring (2), and burst cam (3) from hammer and hammer pin retainer assembly (4). **(M4A1)** remove hammer spring (5) from hammer (6).

3-14. HAMMER ASSEMBLY (CONT).

b. INSPECTION/REPAIR

1. Inspect hammer spring for deformities, breaks, and bends. Pay special attention to the large coil. Replace hammer spring if defective.

2. Inspect cam clutch spring and burst cam for deformities, breaks, and bends; replace if defective.

3. Inspect hammer and hammer pin retainer assembly for chips and breaks. Hammer pin should click home under strong finger pressure. Install hammer pin into hole in hammer to check spring retention of the hammer pin. Replace hammer and hammer pin retainer assembly if defective.

c. REASSEMBLY

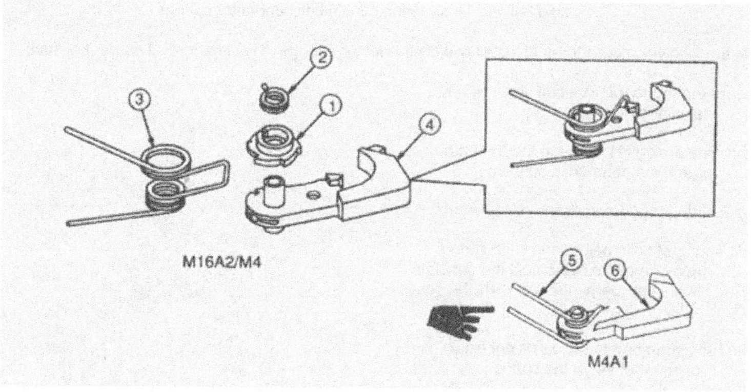

NOTE

M16A2 burst cam is black. M4 burst cam is nickel colored (shiny).

Cam clutch spring should be assembled with bend to the inside. The large loop of the hammer spring should be assembled over the burst cam.

(M16A2/M4) install burst cam (1), cam clutch spring (2), and hammer spring (3) on hammer and hammer pin retainer assembly (4). **(M4A1)** install hammer spring (5) on hammer (6).

3-15. TRIGGER ASSEMBLY AND TRIGGER SUBASSEMBLY (CONT).

b. INSPECTION/REPAIR

1. Inspect trigger spring for kinks, deformities, and weakness. Replace if defective.
2. Inspect disconnector springs for deformities, bends, breaks, and weakness. Replace if defective.
3. Inspect trigger for chips, wear, and cracks. Inspect for damaged searing surface on the trigger nose. Replace if defective.

c. REASSEMBLY

M16A2 AND M4 ONLY

NOTE

Use bolt carrier key tool to install disconnector springs.

1. Use the following procedures to install two disconnector springs (1) using the bolt carrier key tool:

(a) Secure the trigger (2) in a soft vise jaws or similar device.

(b) Place one spring (1) firmly on the tool with large diameter coils outward.

(c) Press spring (1) into recess to solid height.

(d) Hold spring at solid height and slide spring into one of the holes until the punch is flush and perpendicular with the recess wall.

(e) Turn the spring one to two turns opposite of coil winding of the spring.

(f) Discontinue winding when an audible click or snap is heard or felt. This indicates that the spring is seated.

(g) Hold the spring (1) in place when removing the tool to avoid unseating or damaging the spring.

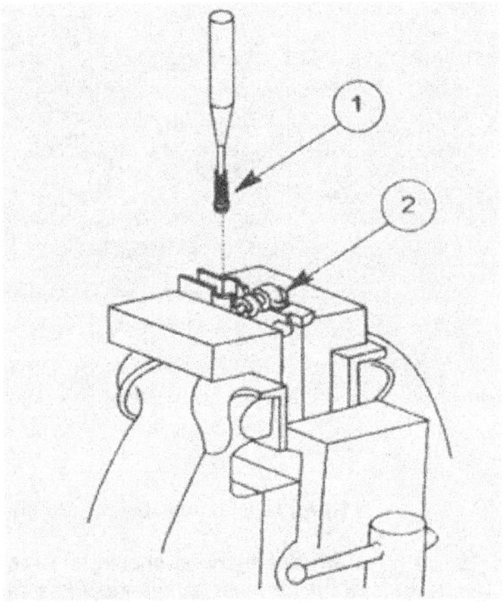

M4 CARBINE ONLY

NOTE

The semi and burst disconnector springs are not the same. The Semi disconnector spring (left side) (figure C-15, item 1B) is black while the burst disconnector spring (right side) (figure C-15, item 1A) is nickel (shiny). Ensure that the correct spring is installed on each side for proper functioning.

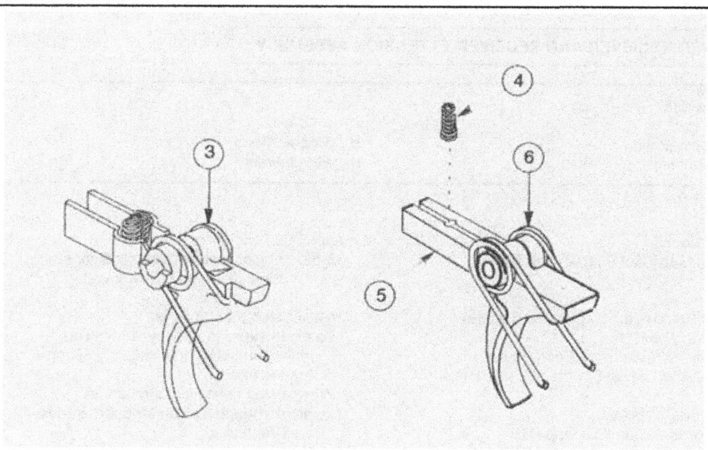

M16A2/M4 M4A1

M16A2 and M4

2. Install trigger spring (3).

M4A1 ONLY

3. Install disconnector spring (4) by inserting large end of spring into trigger (5).

4. Install trigger spring (6) on trigger (5).

3-16. LOWER RECEIVER AND RECEIVER EXTENSION ASSEMBLY.

This task covers:

a. Disassembly
b. Inspection
c. Repair/Modify
d. Reassembly

INITIAL SETUP

Tool Equipment
Tool and Gage Set (item 2, app B)

Tools
(ARMY) Small Arms Repairman Tool Kit (item 3, app B)
Field Maintenance Basic Less Power Small Arms Shop Set (item 1, app B)

Materials/Parts
Cloth, abrasive (item 13, app D)
Grease, molybdenum disulfide (item 19, app D)
Lubricant, solid film (item 21, app D)

Equipment Conditions
3-62 Lower receiver and receiver extension assembly removed

General Safety Instructions
To avoid injury to your eyes, use care when removing and installing spring-loaded parts.
When using solid film lubricant or dichloromethane, be sure the area is well ventilated.

a. DISASSEMBLY

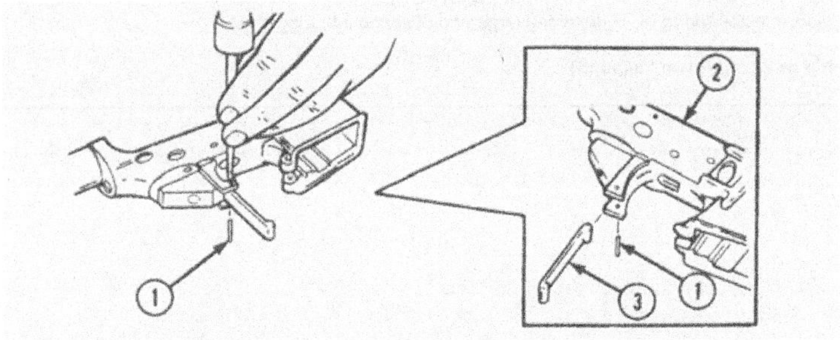

1. Remove spring pin (1) from lower receiver (2) using 1/8 inch drive pin punch and hand hammer.

2. Remove trigger guard (3).

3-16. LOWER RECEIVER AND RECEIVER EXTENSION ASSEMBLY (CONT).

a. DISASSEMBLY (CONT)

3. Use padding between lower receiver and brass vise laws. Grip the solid portion of the lower receiver with brass vise jaws which conform to the shape of the lower receiver in this area.

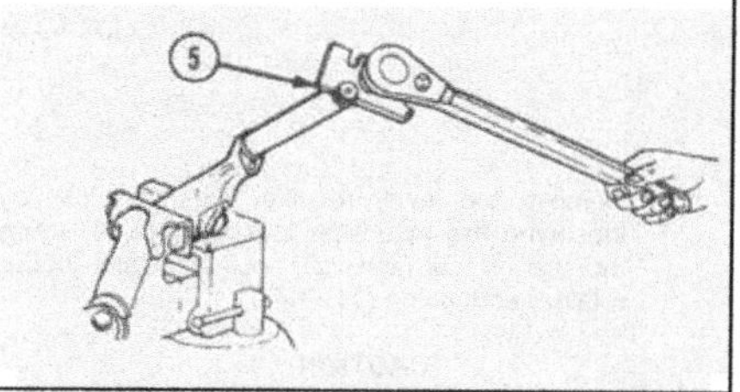

RIFLE ONLY

4. Clamp lower receiver (2) in a machinist's vise using vise jaw caps and tighten on solid portion just tight enough to hold.

WARNING

To avoid injury to your eyes, use care when removing and installing spring-loaded parts.

NOTE

As lower receiver extension is removed, catch buffer retainer and helical spring. Lower receiver is a serial number controlled item.

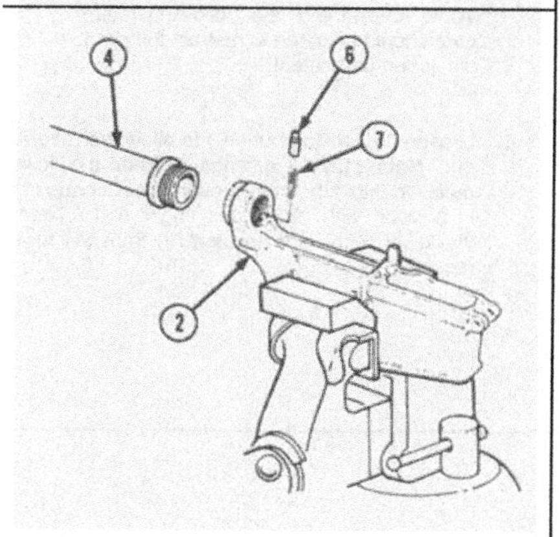

5. Remove lower receiver extension (4) from lower receiver (2) using combination wrench (5) and socket wrench handle. Catch buffer retainer (6) and helical spring (7).

CARBINE ONLY

6. Clamp lower receiver (8) in vise and tighten on solid portion just tight enough to hold.

NOTE

Use wooden vise jaws in place of brass vise jaw caps.

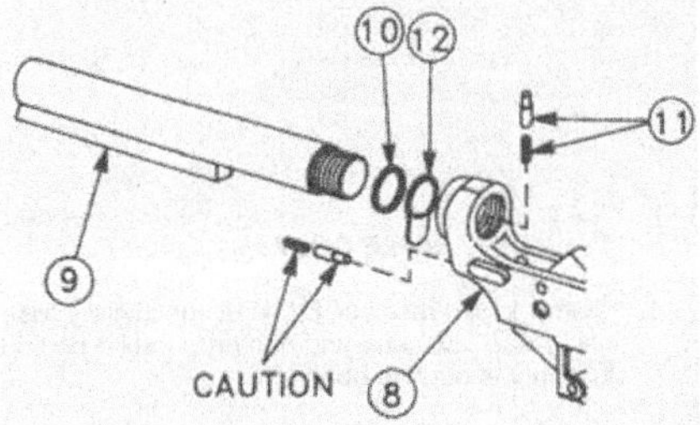

7. Remove the lower receiver extension (9), by loosening the extension locking nut (10) using the special tool (item 12, app C). Catch buffer retainer and spring (11).

CAUTION

While performing the following step, care should be taken to restrain the pivot pin spring and detent.

8. Loosen the locking nut (10) to allow the receiver end plate (12) to disengage from the lower receiver. Hold the buffer retainer and spring (11) in position with your index finger and unscrew the lower receiver extension (9) from the lower receiver (8).

b. INSPECTION

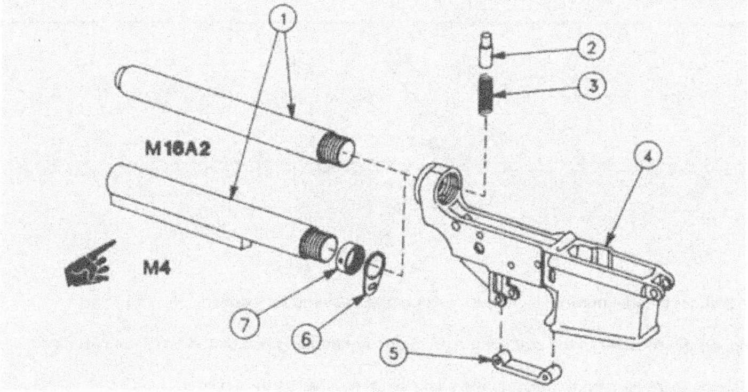

1. Inspect lower receiver extension (1) for corrosion, dents, and wear. Repair or replace if defective.
2. Inspect buffer retainer (2) for wear. Replace if defective.
3. Inspect helical springs (3) for deformities and breaks. Replace if defective.
4. Inspect lower receiver (4). See page 3-66, INSPECTION.
5. Inspect trigger guard (5) for deformities and check operation of plunger and spring. Replace trigger guard if defective.
6. **CARBINE ONLY:** Inspect receiver end plate (6) and locking nut (7) for damage. Replace if damaged.

c. REPAIR/MODIFY

1. Repair lower receiver extension by using abrasive cloth to remove light corrosion. Retouch using solid film lubricant.

NOTE

AIR FORCE ONLY: Only depot maintenance is authorized to restamp the serial number on rifle.

ARMY ONLY: Only direct support level is authorized to restamp serial number.

3-16. LOWER RECEIVER AND RECEIVER EXTENSION ASSEMBLY (CONT).

c. REPAIR/MODIFY (CONT)

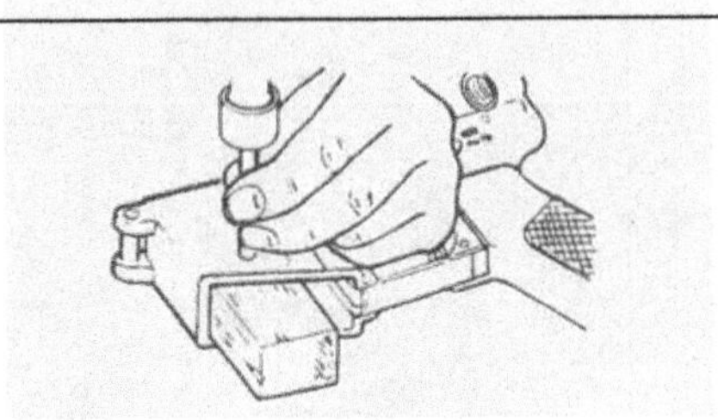

2. **ARMY ONLY:** If serial number is hard to read on rifle, restamp as follows:

   (a) Support the receiver In the stamping area to prevent bending and distortion of the receiver.

   (b) Exercise extreme care to restamp the same serial number as the original.

   (c) Restamp the serial number the same size as the original serial number.

NOTE

Most rifle serial numbers are 1/8 inch (0.31 cm) in height, or close enough that this size is acceptable for such restamping. In the event that a rifle has a serial number that cannot be reproduced by the use of the die sets contained In the Set D Field Maintenance Post, Camp, and Station Small Arms Shop Set, local purchase of an appropriate size die set is authorized.

d. REASSEMBLY

RIFLE ONLY

WARNING

To avoid injury to your eyes, use care when removing and installing spring-loaded parts.

1. Install helical spring (1) and buffer retainer (2) into lower receiver (3).

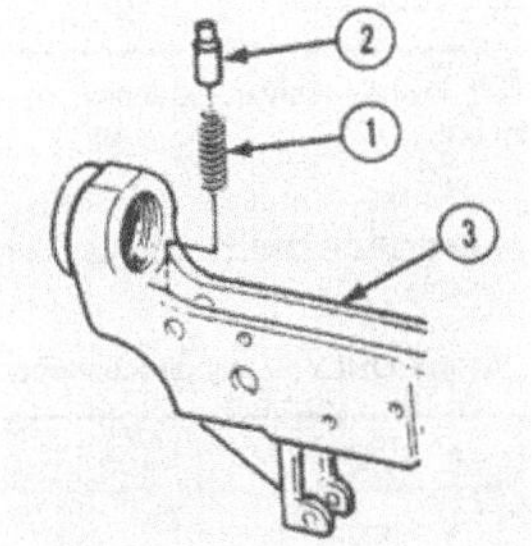

RIFLE ONLY

2. Lubricate threads of lower receiver (3) and lower receiver extension (4) with molybdenum disulfide grease (item 19, app D) before reassembly.

3. Install lower receiver extension (4) into lower receiver (3) while depressing buffer retainer.

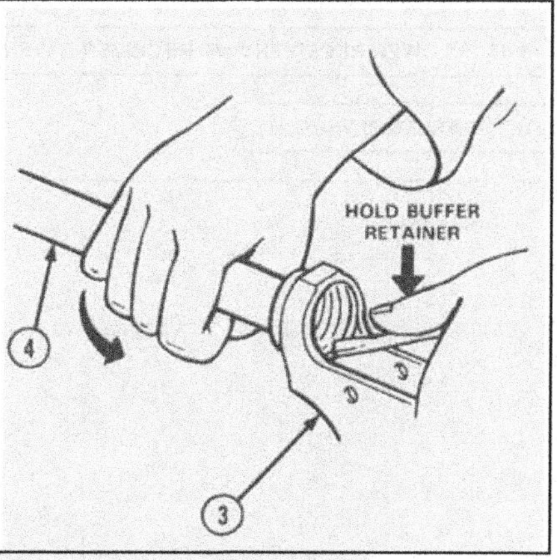

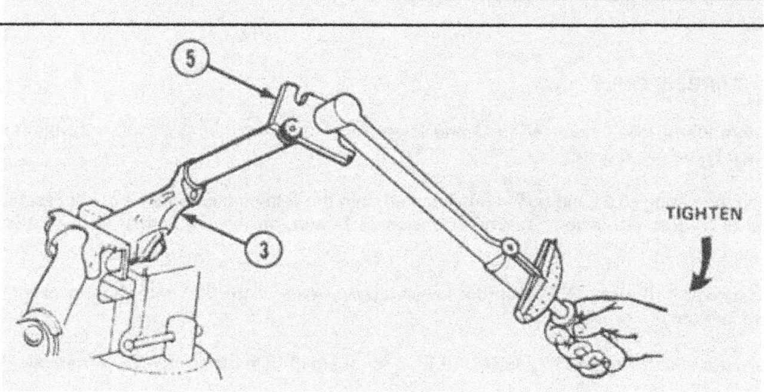

RIFLE ONLY

4. Use padding between lower receiver and brass vise jaws. Use vise jaws in vise and brass vise jaw caps, if available.

5. Clamp solid portion of lower receiver (3) in a machinist's vise using vise jaws. Grip the solid portion of the lower receiver with vise jaws which conform to the shape of the lower receiver in this area.

6. Using combination wrench (5) and torque wrench, torque lower receiver extension to 35-39 ft-lb (47 25 52 65 N•m). Torque is read when both wrenches are used to gether.

3-16. LOWER RECEIVER AND RECEIVER EXTENSION ASSEMBLY (CONT).

d. REASSEMBLY (CONT)

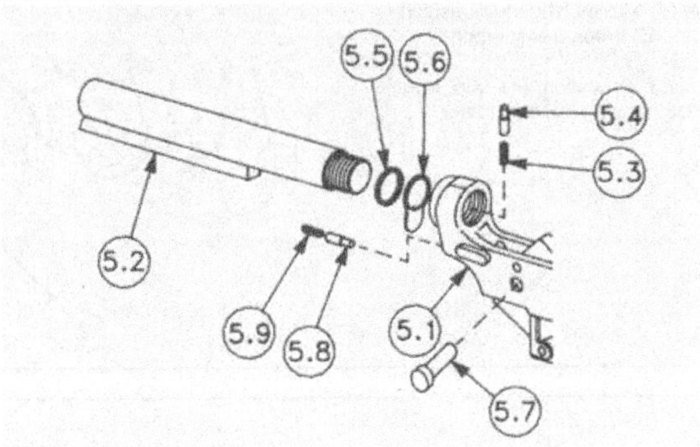

CARBINE ONLY

6A. Lubricate threads of lower receiver (5.1) and lower receiver extension (5.2) with molybdenum disulfide grease (item 19, app D) before reassembly.

6B. Pre-position the spring (5.3) and buffer retainer (5.4) into the retaining hole of the lower receiver (5.1). Screw the locking nut (5.5) onto the lower receiver extension (5.1) with the three notches on the locking ring (5.5) facing forward.

6C. Align the receiver end plate (5.6) onto the lower receiver extension (5.2) with the lug of the receiver end plate (5.6) facing forward.

6D. Pre-position the takedown pin (5.7), detent (5 8), and spring (5.9) in lower receiver assembly (5.1).

6E. Push down on the buffer retainer (5.4) and spring (5.3) and at the same time, screw the lower receiver extension (5.2) in until it retains the buffer retainer (5 4) in position.

6F. Align the lug of the receiver end plate (5.6) into the rear of the lower receiver (5.1). Screw the locking nut (5.5) forward until it contacts the receiver end plate (5.6).

6G. Using the special tool (Item 12, app C) tighten the locking nut (5.5) until snug.

6H. Using the special tool (item 12, app C), and torque wrench, torque locking nut (5.5) to 40 ±2 inch pounds.

6J. Stake the receiver end plate (5.6) in 2 places across from the notches in the locking nut (5.5).

3-16. LOWER RECEIVER AND RECEIVER EXTENSION ASSEMBLY (CONT).

d. REASSEMBLY (CONT)

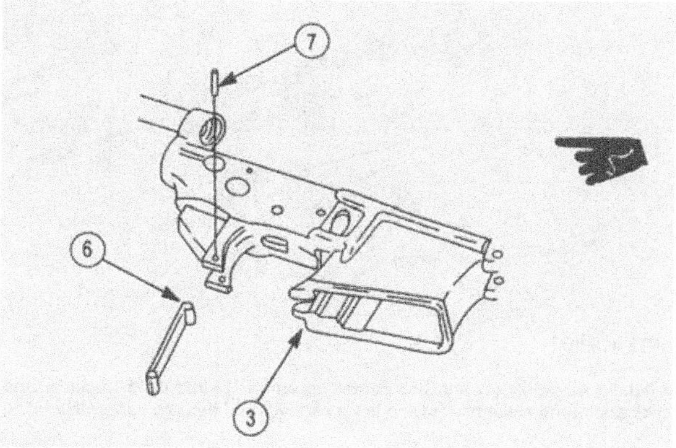

7. Install trigger guard (6) into lower receiver (3).

8. Install spring pin (7) using 1/8 inch drive pin punch and hand hammer.

9. **Deleted.**

3-17. MAJOR COMPONENTS OF M16A2 RIFLE

This task covers:

a. Reassembly   b. Test   c. Inspection

INITIAL SETUP

Test Equipment
Tool and Gage Set (item 2, app B)

Tools
(ARMY) Small Arms Repairman Tool Kit (item 3, app B)

Reference
TM 9-1005-319-10

General Safety Instructions
To avoid injury to your eyes, use care when removing and installing spring-loaded parts.
Live ammunition must not be near the work area.

a. REASSEMBLY

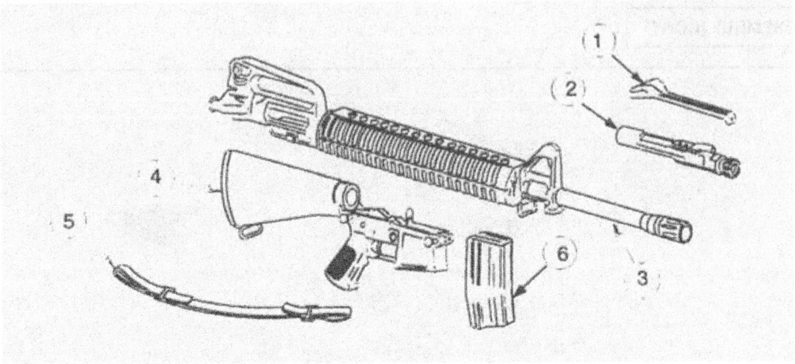

1. Refer to TM 9-1005-319-10.
2. Install charging handle assembly (1) and bolt carrier assembly (2) into upper receiver and barrel assembly (3). Join upper receiver and barrel assembly (3) and lower receiver and buttstock assembly (4).
3. Snap on sling (5) and install magazine (6).

b. TEST

Test trigger pull as follows:

M16A2 and M4 ONLY

(a) Clear the weapon. Place selector to BURST. Pull the trigger and hold it to the rear. Pull the charging handle to the rear and return to the bolt closed position three times. (This will place the BURST disconnector in the deep notch of the BURST cam.)

(b) Release the trigger. Place the selector to SEMI. Hold the rifle in the vertical position. Using trigger pull measuring fixture, 7274758, add weights until hammer trips. Determine weight applied.

(c) Hammer must not trip when 5.5 lbs (2.49 kg) have been applied, hammer must trip on applying 9.5 lbs (4.31 kg).

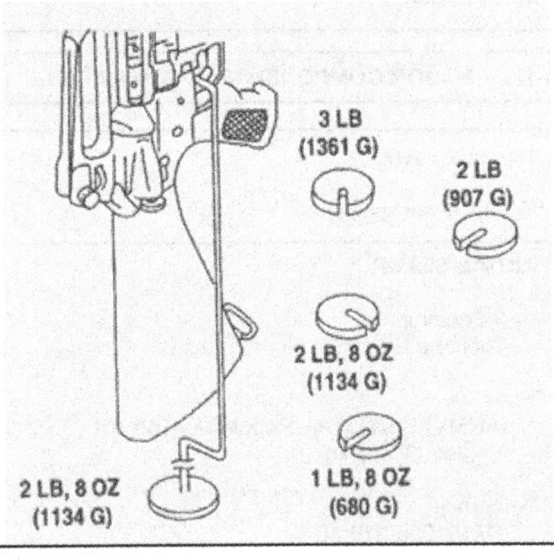

3-17. MAJOR COMPONENTS OF M16A2 RIFLE (CONT)

b. TEST (CONT)

M4A1 ONLY

(d) Clear the weapon. Cock weapon. Place weapon in SEMI position and hold weapon in vertical position.

(e) Using trigger pull measuring fixture 7274758 as shown, add weights until hammer trips. Determine weight applied.

(f) Hammer must not trip when 5.5 lbs (2.49 kg) have been applied. Hammer must trip on applying 8.5 lbs (3.86 kg).

ALL WEAPONS

(g) If weapon fails trigger pull test or excessive creep is present, replace trigger and/or hammer.

NOTE

Always gage trigger and hammer pin holes with not go plug gage 12006472 before replacing parts.

c. INSPECTION

Perform final inspection procedures below.

3-18. M16A2 RIFLE FINAL INSPECTION.

This task covers:

a. Final inspection
b. Test
c. Functional theory of three-round burst control

INITIAL SETUP

Test Equipment
Tool and Gage Set (item 2, app B)

References
TM 9-1005-319-10

General Safety Instructions
Live ammunition must not be near the work area.

Air Force rifles will be inspected in accordance with AFM 36-2227, Vol. 1.

a. FINAL INSPECTION I

1 Visually inspect general appearance of rifle should look almost new. All metal surfaces are to have a dull, rustor corrosion-resistant finish with no burrs or deep scratches.

2 Visually inspect barrel for serviceability. Barrels must be straight, clean, free of rust, powder fouling, and free of bulges and rings. Fine pitting is allowable.

Using moderate hand pressure, check for rotational movement of the front sight in relation to the barrel. Also, using moderate hand pressure, check for rotational movement of the barrel in relation to the upper receiver. If movement between the front sight and the barrel exist, the barrel assembly must be replaced. If movement between the barrel and the upper receiver exist, the barrel assembly must be aligned and tightened (page 3-29).

3 Visually inspect rifle for missing parts. All parts must be attached and all modifications must be applies. Steel parts must be rust free. Spring pins must be secure and screws must be tight.

4 Functionally inspect key and bolt carrier assembly and gas tube alignment. Refer to TM 9-1005-319-10 and use the following procedures:

(a) Disengage the takedown pin and open the receiver.

(b) Remove bolt carrier assembly.

(c) Remove bolt assembly from bolt carrier assembly.

(d) Insert key and bolt carrier assembly into upper receiver and barrel assembly. The bolt assembly must not be installed while performing test.

(e) Slide key and bolt carrier assembly forward to detect binding between key and bolt carrier assembly and gas tube by feel. Badly bent gas tube could cause damage to both the key and bolt carrier assembly and the gas tube. A slightly bent gas tube will cause unnecessary wear of the key and bolt carrier assembly and gas tube.

(f) Correct slight binding by removing handguard assemblies and slightly bending gas tube in the handguard area while repeating step (e) above until no binding is detected. Badly bent gas tubes will be replaced and realigned.

(g) Remove key and bolt carrier assembly from upper receiver and barrel assembly.

(h) Reassemble bolt assembly into key and bolt carrier assembly.

(i) Reinstall key and bolt carrier assembly into upper receiver and barrel assembly.

5 Make a functional check on an assembled weapon with selector lever in SAFE, SEMI, and BURST/AUTO positions Any portion of this check may be used alone to determine the operational condition of any specific firing position selected

3-18 M16A2 RIFLE FINAL INSPECTION (CONT)

a. FINAL INSPECTION (CONT)I

6 Check rear sight assembly as follows:

(a) Rotate elevation knob counterclockwise until the rear sight assembly is all the way down. If a whole click Is not felt as the rear sight assembly stops, the rear sight assembly has bottomed out and will not pivot freely.

(b) Position elevation knob back slightly to its last whole click as the rear sight assembly base is under tension of the ball bearing and helical spring. The 300 meter mark should align with the mark on the receiver (c) If the 300 meter mark is not aligned with the mark on receiver, slip the range scale in the following manner:

(1) Position the 300 meter mark with the mark on the receiver.

(2) Insert a 1/16 inch alien wrench through the access hole of the rear sight assembly base and into the index screw.

(3) Loosen the index screw three turns and leave the wrench in place.

(4) Rotate lower portion of elevation knob counterclockwise until it stops (range scale should not have moved). Elevation knob should be positioned on its last whole click.

(5) Tighten index screw and remove wrench.

(6) Check for proper setting.

7 Pull charging handle assembly to rear. Check that chamber is clear. Leave hammer in cocked position.

8. Place selector lever is SAFE position and pull trigger. Hammer should not fall.

WARNING

If rifle falls any of the following inspections, continued use of the rifle could result In injury to, or death of, personnel.

NOTE

For the purpose of the following tests, "SLOW" is defined as 1/4 to 1/2 the normal rate of trigger release

9. Place selector lever in SEMI position.

10 Pull trigger. Hammer should fall.

11 Hold trigger to the rear, charge rifle and release the trigger with a slow, smooth motion without hesitations or stops, until the trigger is fully forward; an audible click should be heard. Hammer should not fall.

12 Repeat the above selector lever SEMI position test (steps 10 and 11) five times. The rifle must not malfunction during any of these five repetitions. If the rifle malfunctions during any of these five tests, refer to page 3-7.

M16A2 and M4 ONLY

13 Place selector lever in BURST position. Charge rifle and pull trigger. Hammer should fall.

14 While holding the trigger to the rear, pull charging handle assembly to the rear and release it three times. Hammer should not fall the third time the charging handle assembly is released. It may or may not fall the first and second time the charging handle assembly is released. When the burst disconnector reached the deep notch of the burst cam, the burst disconnector should have held the hammer to the rear while the trigger was in the pulled position. Release the trigger with a slow, smooth motion without hesitations or stops, until the trigger is fully forward. The hammer should not fall.

15 Pull trigger. Hammer should fall. This would be the first round of a three-round burst.

NOTE

A detailed explanation of the three-round burst control can be found in the following pages.

M4A1 ONLY

15.1 Place selector lever in AUTO position. Charge weapon and pull trigger. Hammer should fall.

15.2 Hold trigger to the rear, charge weapon, and release trigger. Pull trigger. Hammer should not fall. Automatic assembly sear should have released the hammer as the bolt closed.

ALL WEAPONS

16 With the hammer in the forward position, using moderate finger/thumb pressure, attempt to place the selector lever in the SAFE position. Selector lever should not go in the SAFE position.

17 Perform the following additional functional checks and adjustments on assembled weapon:

(a) Press magazine catch button. Make sure it functions properly.

(b) Press bolt catch. Make certain it operates smoothly and holds bolt in open position.

(c) Inspect front sight and rear sight assembly. Make certain they can be adjusted properly.

(d) Actuate forward assist assembly It must work freely.

3-18 M16A2 RIFLE FINAL INSPECTION (CONT)

a. FINAL INSPECTION (CONT)

(e) Inspect upper receiver and barrel assembly. Barrel assembly should not rotate within upper receiver assembly.

(f) Check that third or middle slot of compensator is straight up (TDC).

b. TEST

1 Check headspace using headspace gage PN 7799734. See page 3-45, TEST.

2 Check firing pin protrusion using firing pin protrusion gage PN 7799735. See page 3-45, TEST.

3 Check extent of barrel erosion using barrel erosion gage PN 8448496. See page 3-45, TEST.

4 Check barrel straightness using barrel straightness gage PN 8448202. See page 3-45.

c. FUNCTIONAL THEORY OF THREE-ROUND BURST CONTROL (M16A2 AND M4 ONLY).

NOTE

First become familiar with the functioning of the firing mechanism especially when in the SAFE and SEMI positions. You should also understand the role that the automatic sear plays when firing in the BURST position. Functioning of the mechanism is explained below In a step by step manner. This actually will seem to complicate something that is really very simple and happens in less than 1 second. The diagrams below and on the following pages do not show the associated springs for the sake of simplicity The positioning of the burst cam is shown in detail

Functional check of three-round burst is as follows:

NOTE
Assume the rifle is fully loaded with a live round in the chamber and the selector lever on BURST.

(a) Hammer is cocked

(b) Front hook of burst lever is in stop notch

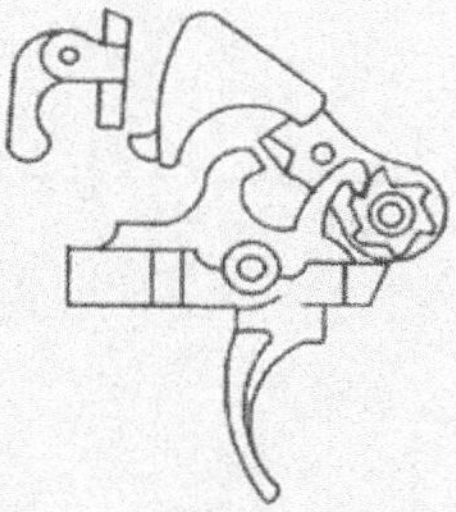

(c) Trigger Is pulled.

(d) Trigger nose drops and hammer falls firing the FIRST ROUND.

(e) Front hook of burst disconnector holds burst cam In place as hammer falls.

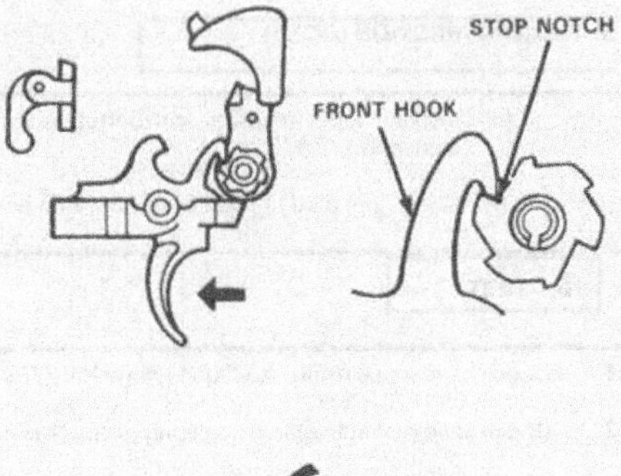

NOTE

Anytime the hammer falls forward, the clutch spring releases the burst cam and allows the front hook of the burst disconnector to keep It In place.

(f) As the key and bolt carrier assembly moves to the rear, the hammer is forced to the rear.

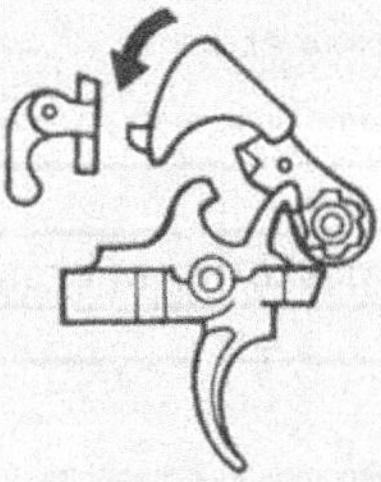

(g) The clutch spring of the burst cam clutches the burst cam and causes It to rotate one notch as the hammer is forced back.

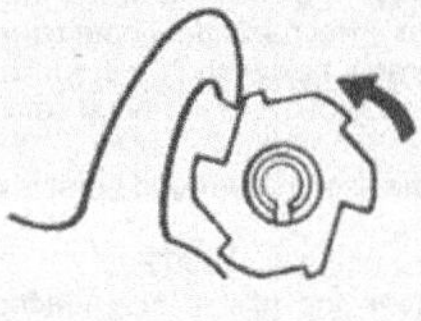

(h) When hammer is fully to the rear, the automatic sear catches It.

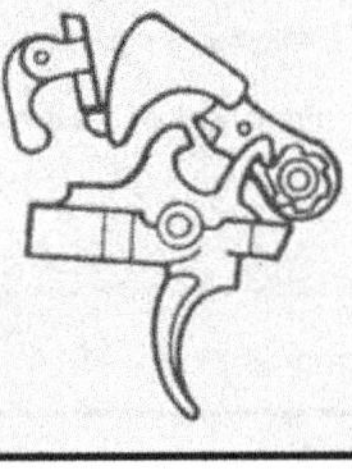

3-18 M16A2 RIFLE FINAL INSPECTION (CONT).

c. FUNCTIONAL THEORY OF THREE-ROUND BURST CONTROL (CONT)

(i) The front hook of the burst disconnector is now fully In the second notch.

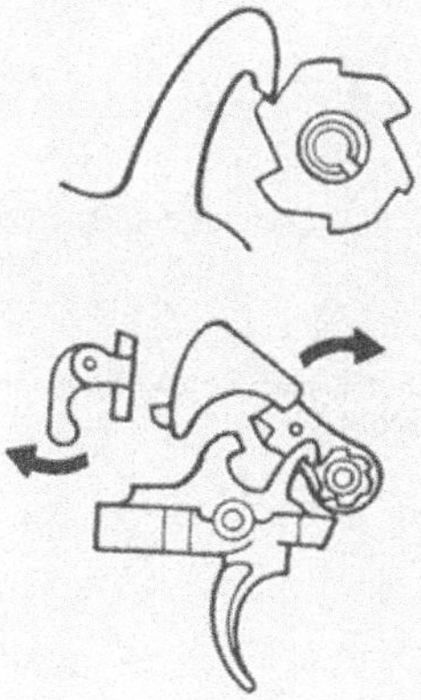

(j) As the key and bolt carrier assembly travels forward, the automatic sear releases the hammer and the hammer falls.

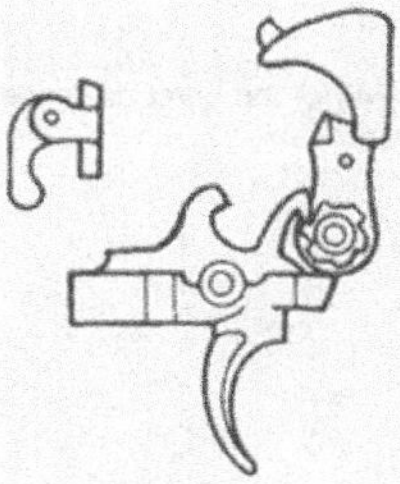

(k) When the hammer falls, the SECOND ROUND is fired.

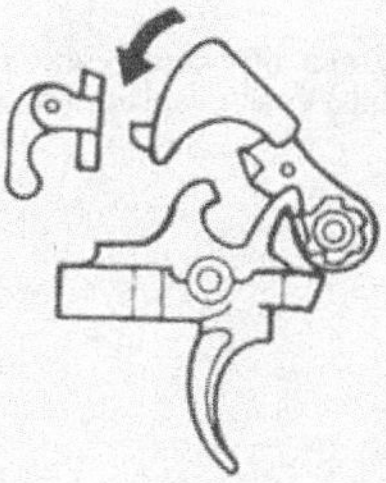

(l) As the key and bolt carrier assembly moves to the rear, the hammer is forced back to the rear.

(m) The clutch spring of the burst cam clutches against the burst cam and causes It to rotate one notch as the haiimmer is forced back

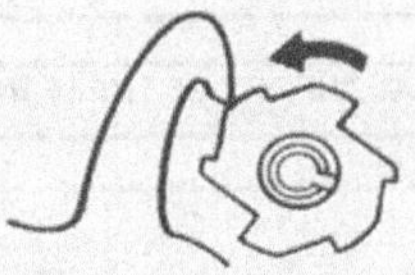

(n) When the hammer is fully to the rear, the automatic sear catches It

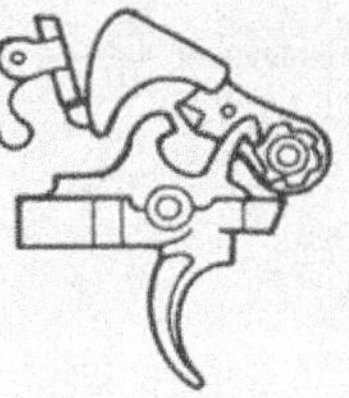

(o) The front hook of the burst disconnector IS now fully In the third notch

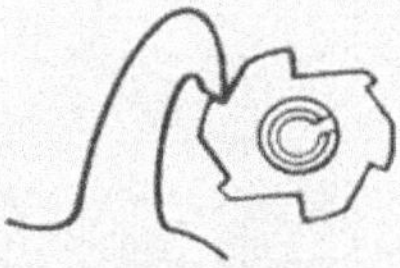

(p) As the key and bolt carrier assembly) travels forward, the automatic sear releases the hammer and the hammer falls

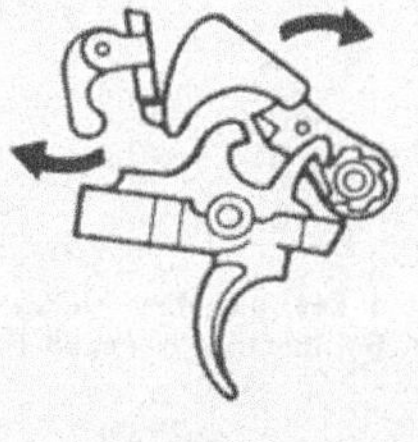

3-18. M16A2 RIFLE FINAL INSPECTION (CONT).

c. FUNCTIONAL THEORY OF THREE-ROUND BURST CONTROL (CONT)

(q) When the hammer falls, the THIRD ROUND Is fired

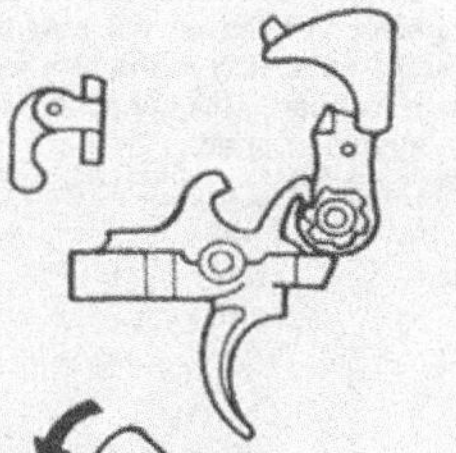

(r) As the key and bolt carrier assembly moves to the rear, the hammer is forced back to the rear

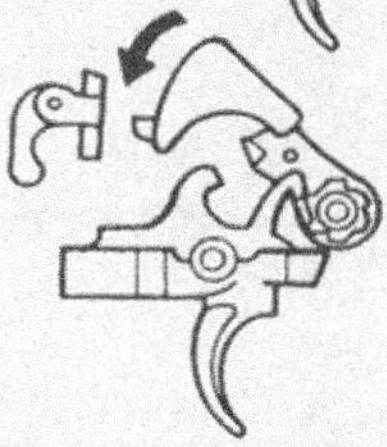

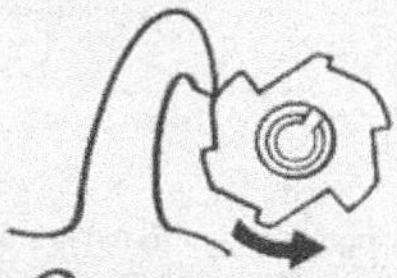

(s) The clutch spring of the burst cam clutches against the burst cam and causes it to rotate one notch as the hammer is forced back.

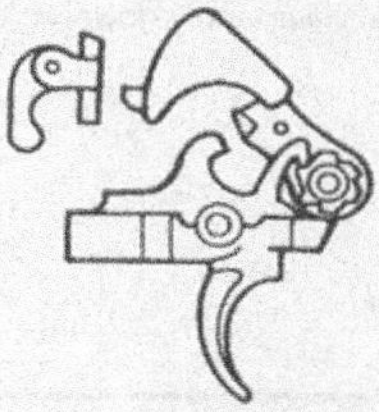

(t) When the hammer Is fully to the rear, It Is initially caught the, ;automatic sear However, the front hook of the first disconnector Is now fully in the next stop notch which is deeper than the others

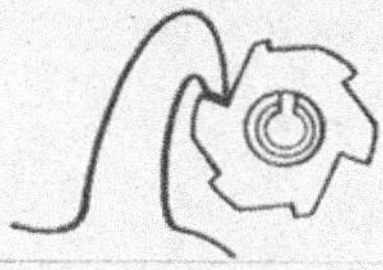

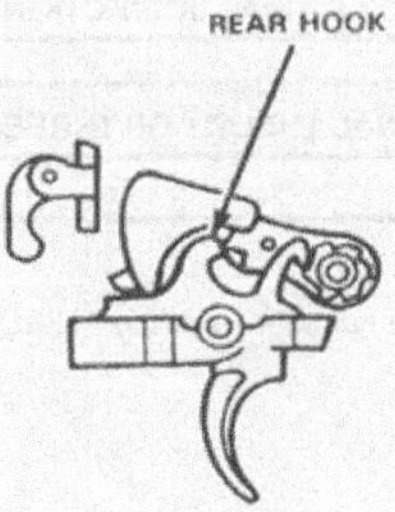

(u) Because a stop notch is deeper than the others, It allows the front hook of the burst disconnector further forward than before. This allows the rear hook of the burst disconnector to latch on the rear hammer notch. This holds the hammer fully to the rear even though the trigger is still to the rear. This happens when the burst Is over and the firing Is stopped.

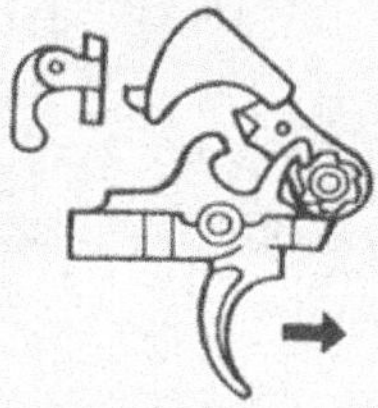

(v) Once the trigger is released, the trigger nose comes up and holds the hammer back.

NOTE

Pulling the trigger to the rear and holding it back again will fire another three round burst This will continue until the magazine is empty However the trigger must be released between each burst

3-19 MI 6A2 RIFLE AND M4/M4A1 CARBINE ANNUAL GAGING REQUIREMENTS.

This task covers.

a Inspection b Gaging

INITIAL SETUP

Test Equipment
Tool and Gage Set (item 2, app B)

Tools
(ARMY) Small Arms Repairman Tool Kit (item 3, app B)
Field Maintenance Basic Less Power Small Arms Shop Equipment (item 1, app B)

References
DA PAM 738-750
TM 9-1005-319-10
AFR 50-36
AFTO Form 105

General Safety Instructions
- To avoid injury to your eyes, use care when removing and installing spring-loaded parts
- Initial gaging is required 1 year from receipt of the weapons.
- All rifles and carbines must be gaged at least once annually for safety.
- All Army Reserve and Army National Guard M16A2 rifles must be inspected and gaged at least once every 2 years, after the initial inspection/gaging procedures have been accomplished. This 2 year interval may be maintained unless preventive maintenance checks and services (PMCS), or other physical evidence, indicates that an individual unit's rifles require inspection/gaging at a more frequent interval. If it is determined that a yearly inspection is necessary for an individual unit, only that unit will be affected. This will not affect other units in regard to the interval of inspection.
- It is recommended that training units inspect/gage all rifles at the end of each training cycle. Training units will inspect/gage all rifles at least once annually.
- Air Force M16 rifles will be inspected in accordance with AFR 50-36, Vol 1, Chapter 5.

a. INSPECTION

1 Visually inspect general appearance of weapon Overall appearance will be approximately that of a new weapon. For inspection criteria refer to final inspection, page 3-84 All visual and functional Inspection requirements must be met.

NOTE

To perform the following tests, disassemble weapon only as far as allowed in chapter 2 (Unit Maintenance Instructions). unless a deficiency is uncovered

2 Perform a general inspection of weapon per section III of this chapter Repair as required and authorized

b. GAGING

1	Gage bolt carrier assembly for firing pin protrusion using firing pin protrusion gage PN 7799735. See page 3-20, TEST.
2.	Gage bolt carrier assembly for firing pin hole wear using not-go plug gage PN 12620101. See page 3-24, TEST.
3	Inspect chamber in upper receiver and barrel assembly using chamber reflector tool PN 8448201 See page 3-33, INSPECTION/CLEANING.
4	Gage barrel in upper receiver and barrel assembly using barrel erosion gage PN 8448496 and bore straightness gage PN 8448202. See page 3-45, TEST.
5	Check headspace in upper receiver and barrel assembly by inserting headspace gage PN 7799734 in chamber. See page 3-45, TEST.
6	Gage pivot pin lug area clearance in lower receiver assembly using 0.020 thickness gage. See page 3-68.
7	Gage hammer and trigger pin holes in lower receiver assembly using taper plug gage PN 12006472. See page 3-68, TEST.
8	Gage trigger pull using trigger pull measuring fixture PN 7274758. See page 3-83, TEST.
9	Document inspection with DA PAM 738-750, AFTO Form 105, or NAVMC 11003 when completed.

Section IV. PREEMBARKATION INSPECTION OF MATERIEL IN UNITS ALERTED FOR OVERSEAS MOVEMENT

NOTE

This section refers to all the weapons (MI 16A2 rifle, M4, and M4A1) unless otherwise stated.

3-20 PURPOSE: This section establishes standards for overseas shipment (pre-embarkation in section criteria) for all weapons. These standards are provided to ensure that the user is furnished equipment which will perform its mission without early failure or major maintenance problems.

3-21 SCOPE.

a. The standards prescribed provide for a high percentage of remaining life in affected rifles; therefore, rifles designated for overseas shipment must qualify under the standards contained in the following paragraph, table, and in referenced DA publications, before they can be approved for shipping action.

b. Provisions of this standard apply to all US Army agencies/activities selecting or preparing rifles for shipment to US troops overseas. It also applies to CONUS troops preparing rifles for shipment overseas. Provisions do not apply either to rifles being prepared for shipment to MAP/MAS recipients unless specifically prescribed by MAP/MAS transaction for the materiel or to rifle being returned to CONUS from overseas. The maintenance instructions and standards contained herein do not apply to rifles once the material has arrived at the overseas destination. At that time, maintenance instructions contained in the applicable TM's will be used.

c. This applies to rifles which are the logistic responsibility of the US Army Armament, Munitions and Chemical Command.

3-22. GENERAL.

a. Only rifles which have been classified as serviceable condition code A, B, or C under AR 725-50 will be considered for overseas shipment. All items of equipment for which equipment serviceability criteria have been published must, as a minimum, be rated green under the ESC as a prerequisite to overseas shipment. In addition to the condition code standard, as enumerated above, and the required ESC rating prescribed herein, the rifle being considered for overseas shipment must meet the requirements of this section. The ESC will be discontinued as new operator manuals are revised which will be used to determine serviceability condition of rifle.

b. Waivers to provisions can only be granted by the gaining command of any particular end Item being considered for issue, deployment, or shipment. The issuing services may recommend issue or shipment of rifles not meeting the provisions when all the following conditions exist:

(1) Repair parts in required quantities cannot be obtained from the supply system prior to delivery of the end item.

(2) The gaining command concurs in the receipt of the end item for storage until required repair parts become available. The gaining command must also state that capability, facilities, and funds are available to perform the necessary work when parts become available.

(3) Department of the Army approval is obtained on a case-by-case basis.

(4) Required repair parts are requisitioned by the issuing command for delivery to the Framing command.

c. All Department of the Army MWO's applicable to the specific rifle being considered for shipment overseas must have been applied.

d. Refer to SB 746-1 for pertinent publications relating to equipment processing and marking information

e. Refer to AMC-P 310-9 for publications containing applicable overhaul standards.

3-23. SHIPMENT OR ISSUE.

a. Organizational Repair Parts, Tools, and Equipment. Rifles must be complete with all Items required by applicable Department of the Army publications, including those In the basic Issue items list of the appropriate operator's manual.

b. Publications. Operator publications applicable to the equipment log book must accompany the equipment. All log book entries must be complete and up-to-date including those covering any repairs, replacements, or adjustments made to the rifle In complying with this section.

c. Documentation. Prepare DA Form 2408-9 (Equipment Control Record) at time of overseas shipment or Issue to another stock record or property book account, in accordance with the provisions of DA PAM 738-750.

d. Preparation. Process rifles for shipment as required by shipping documents and pertinent regulations.

3-24. DISPOSITION. Disqualified rifles which do not qualify for shipment will either be redistributed within the camp, post, or station, be repaired, or become candidates for overhaul, cannibalization, or other disposition as required by existing regulations.

3-25. GENERAL INSPECTION CRITERIA.

WARNING

Before starting an Inspection, be sure to clear the rifle Do not actuate the trigger until the rifle has been cleared. Inspect the bore and chamber to ensure that It is empty and free from obstructions, and check to see that no ammunition Is In position to be Introduced.

a. Before Inspection, the materiel must be thoroughly cleaned of all grease, dirt, or other foreign matter that might Interfere with Its proper function or the use of gages and tools during inspection.

b. Materiel must be free of burrs, rust, or corrosion on functional surfaces.

3-25. GENERAL INSPECTION CRITERIA (CONT).

c. Parts must not be cracked, bent, distorted, or damaged and must be free of detrimental wear or looseness

d. Minor defects In metal components do not normally affect their acceptability For example, scratches and tool marks are ordinarily of no Importance

e. Inspect finish of metal surface

(1) General. Satisfactory metal surfaces for rifles range from black to light gray. A worn shiny metal surface is objectionable only when It is capable of reflecting light No rifle will be reelected unless exterior parts have a shine All rear sights must have a dull gray or black finish of all surfaces that would cause a glare

(2) M16A2 Rifle. Minor loss of finish (shiny spots, nicks, scratches) on exterior surfaces of the barrel and flash suppressor shall not be cause for rejection of M16A2 rifles located In hands of troops at training centers Large shiny surfaces, nicks, scratches, etc., can be restored by the use of solid film lubricant (Item 21, app D) Rifles (small arms) missing in excess of one-third or more of the exterior finish resulting In an unprotected, light reflecting surface, are considered candidates for overhaul The only authorized level of maintenance to phosphate finish small arms Is depot.

f. Plastic components must not be cracked or damaged In such a way as to Interfere with their structural strength Surface cracks, bruises, or dents that do not affect their strength will not be cause for rejection. Cracks will be cause for rejection. Criteria for determining which cracks are repairable are on page 2-1 4.

g. Barrels must be clean and free of corrosion such as that caused by moisture and powder fouling Standards of serviceability are indicated In (1) through (9).

(1) Pits In the chamber are allowable If they do not cause extraction difficulties

(2) Pits as wide as a land and 3'8 Inch (O 95 cm) or less in length are allowable for 5 56-mm barrels Pits not greater than the width of a land and less than 3 8-inch (O 95 cm) long are permissible

(3) Scattered or uniformly fine pits, or fine pits In a densely pitted area are allowable

(4) Tool marks are acceptable regardless of length They will appear as lines running laterally In the grooves, or may run spirally across the top of lands

(5) Ringed bores or bores ringed sufficiently to bulge the outside surface of the barrel are cause for rejection. However, faint rings or shadowy depressions do not indicate an unserviceable barrel and will not be cause for rejection. Gap in lined barrels will not be classified as a ringed bore.

(6) Lands that appear dark due to coating of gilding metal from projectiles will not be cause for rejection.

(7) Breech bore diameter will be checked on unlined barrels using the appropriate breech bore gage.

(8) Barrel erosion gages are provided for lined barrels. Bore wear will be checked using barrel erosion gage for the M16A2. For detailed instructions in the use of the above gage and for serviceability limits, refer to page 3-46.

(9) Flaking or checking (fine cracks) of chromium plate in barrels or chambers will not be cause for rejection, unless accompanied by pitting to the degree that extraction difficulties are encountered or accuracy is unacceptable.

h. Springs must be free of distortion and broken coils. Springs must have sufficient tension to perform their intended function.

i. Screw heads must be in serviceable condition and threads must not be stripped. Internal threads must not be stripped.

j. The sear, hammer, and/or cocking notches must be in good condition. Chipped engaging corners will be cause for rejection. Slight wear on functional surfaces, including engaging corners, shall be acceptable, providing the minimum trigger pull requirements and selector lever checks are met in accordance with instructions on page 3-83.

k. Chips, flat spots, or bent striker points on firing pins will be cause for rejection.

l. The cartridge engaging surfaces on extractors must not be chipped or deformed.

m. Evidence of any damage to sights will be cause for a sight alignment check. Rear sight bases should have no movement.

n. Rear sight elevating and windage mechanisms must operate with distinct clicks, without binding. Sights must have sufficient tension to retain their setting during firing. Graduations and numerals must be legible. Graduation filler Is not required.

o. Safeties must positively position in both the ON and OFF position. When in the ON or safe position, the rifle must not fire when the trigger is squeezed; when in the OFF or fire position, the rifle must fire when the trigger is squeezed.

3-25. GENERAL INSPECTION CRITERIA (CONT).

p. All locking devices such as latches, magazine latches, or detents must be positive in action and must not become disengaged due to normal handling and firing. Retaining pins and similar devices must not be subject to accidental loss during use or transportation.

q. Each rifle must be hand functioned to check for unusual binding, positive cocking action, and general operation. Dummy ammunition must be used to assure positive feeding, chambering, extraction, and ejection action.

Table 3-1. 5.56MM Rifle M16A2

Item	Standard
RIFLE:	
General	Clear rifle of any ammunition and Inspect In accordance with paragraph 3-25.
Barrel and barrel extension	Check barrel erosion. Use barrel erosion gage 8448496 for chrome lined barrels. Stripping of lands and grooves shall not be cause for rejection unless so determined by barrel erosion gage.
	Visually inspect, using chamber reflector tool 8448201.
	Pits 1/8 inch (0.31 cm) In length and those pits large enough to extend from the body of the chamber into the shoulder stop area and forcing cone area are cause for rejection. Large pits are defined as those 1/8 inch (0.31 cm) or more in diameter as determined by visual Inspection. Only closed flash suppressors are acceptable.
	Check barrel for straightness using bore straightness gage 8448202. Gage must pass freely through the bore to be acceptable, either dropped from the muzzle or chamber end.
Front sight and gas tube	Inspect gas tube for proper alignment with carrier key. Gas tube must not bind when mating with the key.
	Evidence of gas leaks around the front sight connection of the gas tube shall be cause for rejection until rifle has been function fired to determine if the loss of gas is sufficient to cause malfunction.
	If function firing malfunctions occur, repairs are necessary.
	Inspect front sight for damage.

Table 3-1. 5.56MM Rifle M16A2, M4/M4A1 Carbine - Cont

Item	Standard
Bolt carrier group	Inspect bolt for elongated or oversized firing pin hole using plain cylinder gage 12620101. Firing pinholes which permit the plain cylinder plug gage to fully penetrate at any position on the circumference will be rejected. Bolt face with a cluster of pits which are touching or tightly grouped, covering an area measuring approximately 1/8 inch (0.31 cm) across will be rejected. Bolts which contain pits extending into the firing pinhole will not be rejected unless firing pin hole gaging check determines rejection. Bolts which contain individual pits or scattered pits will not be cause for rejection. Only phosphated bolt carriers are acceptable. Both phosphated and chrome plated bolts are acceptable.
Bolt locking lugs and bolt cam pin hole	Inspect for cracks in the locking lugs and cam pin hole area. Use a black light, if available; otherwise, use a glass of no more than 3X magnification or use inspection penetrant (item 25, app D). Use instructions contained in kit for application. If cracks are detected, the bolts will be replaced. **NOTE** Particular attention must be given to the area where the lugs meet the bolt body and around the side walls of the cam pin hole. Bolt rings must not be broken. Ring gaps must be properly spaced approximately 1/3 turn apart and not in line. Firing pin protrusion must be not less than 0.028 inch (0.071 cm) or more than 0.036 inch (0.091 cm). (Use firing pin protrusion gage 7799735.) Socket head capscrews must be staked. Carrier key must not be dented where end mates with gas tube. Repair or replace damaged carrier keys.
Headspace	Inspect headspace using headspace gage 7799734. Excessive headspace will be cause for rejection.
Trigger pull	Inspect trigger pull using trigger measuring fixture 7274758. Trigger pull must be minimum 5.5 pounds (2.49 kg), maximum for M16A2 and M4 is 9.5 pounds (4.31 kg) and maximum for M4A1 is 8.5 pounds (3.86 kg). Test trigger pull, refer to page 3-83, steps a, b, c, and d.

3-25. GENERAL INSPECTION CRITERIA (CONT).

Table 3-1. 5.56MM Rifle M16A2, M4/M4A1 Carbine - Cont

Item	Standard
Lower receiver group	Inspect hammer and trigger pin holes using plain cylinder plug gage 12006472 Penetration of the gage in any one or more of the four holes will be cause for rejection. Inspect for cracks, corrosion, or mutilation which would affect functioning. Small dents or gouges will not be cause for rejection. Inspect receiver for corrosion in the lobes of the pivot or hinge pin area. Width between lobes shall not exceed 0 515 inch (1.30 cm). Inspect receiver for break through of metal. Inspect receiver and receiver extension for initial loss of protective coating.
Action spring	Free length of spring shall be between 11-3/4 and 13-1/2 inches (29.84 and 34.29 cm) **M16A2 ONLY** and 10-1/16 and 11-1/4 inches (25.56 and 28.58 cm) **M4/M4A1 ONLY**.
Handguard	Inspect handguard assembly for breaks, separations, and cracks. Breaks and separations of material which prevent proper retention or interfere with functioning of the weapon will be cause for handguard rejection and replacement. Cracks up to 1 inch (2.54 cm) in length are acceptable provided they do not extend into the handguard retaining flange (1) (critical area). **M16A2 ONLY** Each handguard assembly may have up to two of the three front retaining tabs (2) missing. If all three front tabs are missing, the handguard assembly must be replaced. **M4/M4A1 ONLY** No tabs can be missing. **ALL WEAPONS** Replace severely cracked handguards. Handguards that have a heat shield loose enough to rattle when installed on the weapon must be replaced.
Stock assembly	Inspect buttstock assembly for dents, cracks, and chips. Check for breaks and separation of material which could prevent proper functioning of weapon. M4/M4A1 inspect for proper functioning. Repair as required

Table 3-1. 5.56MM Rifle M16A2, M4/M4A1 Carbine - Cont

Item	Standard
	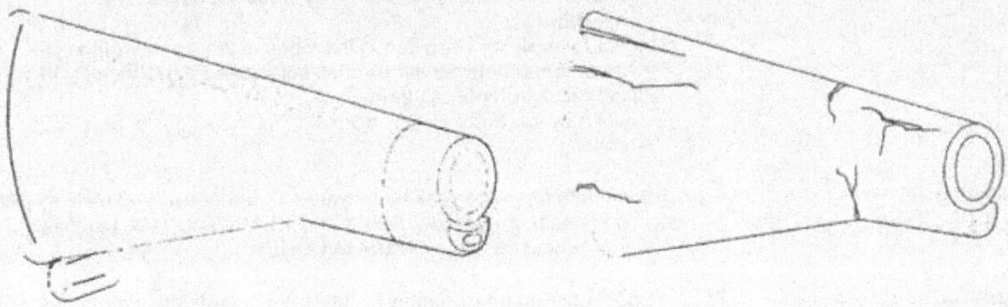
	M16A2 ONLY Under the following conditions, hairline cracks originating from buttplate end of buttstock are acceptable. No chipped away material is allowed. a. One hairline crack, not to exceed 1 inch (2.54 cm) in length, per side of buttstock. b. Two additional hairline cracks up to 0.22 inch (0.55 cm) in length, per side of buttstock. Buttstocks with unauthorized markings stamped into their surfaces will be replaced. Unauthorized markings, scratched, etched, carved, etc., are acceptable if they do not extend into the fiber of the buttstock which may weaken it. These marks may lie at any location on the buttstock. Cracks in the critical area at the front end of the buttstock are not acceptable and these buttstocks must be replaced.

CHAPTER 4

MAINTENANCE OF AUXILIARY EQUIPMENT

CHAPTER OVERVIEW

This chapter contains information and instructions to keep auxiliary equipment used with your weapon in good repair.

Section I. AUXILIARY EQUIPMENT REPAIR

4-1 GENERAL.

a The following items of auxiliary equipment are used in conjunction with the Weapons:

(1) 40mm Grenade Launcher M203, NSN 1010-00-179-6447, (M16A2 ONLY)

(2) Lock Plate, NSN 1005-00-233-9031.

(3) Top Sling Adapter, NSN 1005-00-406-1570.

(4) Blank Firing Attachment M15A2, NSN 1005-00-118-6192, (M16A2 ONLY), Blank Firing Attachment M23, NSN 1005-01-361-8208 (M4 ONLY)

Change 4**(5)** ARMY ONLY: Conversion Kit, M261 (caliber .22 rimfire adapter), NSN 1005-01-010-1561.

(6) Bayonet-Knife M7, NSN 1005-00-073-9238.

(7) Bayonet-Knife Scabbard M10, NSN 1095-00-223-7164.

(8) M9 Multi-Purpose Bayonet System, NSN 1005-01-227-1739.

(9) Night Vision Sight, Individual Served Weapon, AN/PVS-4, NSN 5855-00-629-5334 (M16A2 ONLY)

(10) M2 Practice Bolt, NSN 1005-01-184-4041 (M16A2 ONLY)

b Refer to TM 9-1010-221-23&P for unit maintenance for the Grenade Launcher M203.

c ARMY ONLY: Refer to TM 9-6920-363-12&P for unit maintenance for the M261 Conversion Kit (caliber rimfire adapter) M16, M16A1, and M16A2 Rifles.

d Refer to TM 9-1005-237-23&P for repair instructions and repair parts for Bayonet-Knife M7 and Bayonet Knife Scabbard M10 and M9 Multi-Purpose Bayonet System.

e Refer to TM 11-5855-213-23&P for maintenance for the Night Vision Sight, Individual Served Weapon, AN/PVS-4.

f Refer to TM 9-6920-746-12&P for unit maintenance for the M2 Practice Bolt.

4-2. LOCK PLATE

This task covers:

a Installation
b Removal
c Inspection

INITIAL SETUP

Tools
(ARMY) Small Arms Repairman Tool Kit
(item 3, app B)

WARNING

The lock plate prevents the selector lever from being placed in BURST and will be installed at the discretion of the unit commander. It is mandatory for use in civil disturbance (riot control). Do not keep live ammunition near work area.

a. INSTALLATION

1 Using a screwdriver, reach inside pistol grip (1) and remove screw (2) and lockwasher (3).

WARNING

To avoid injury to your eyes, use care when removing and installing spring-loaded parts.

2 Carefully remove pistol grip (1). Hold detent helical spring (4) in place.

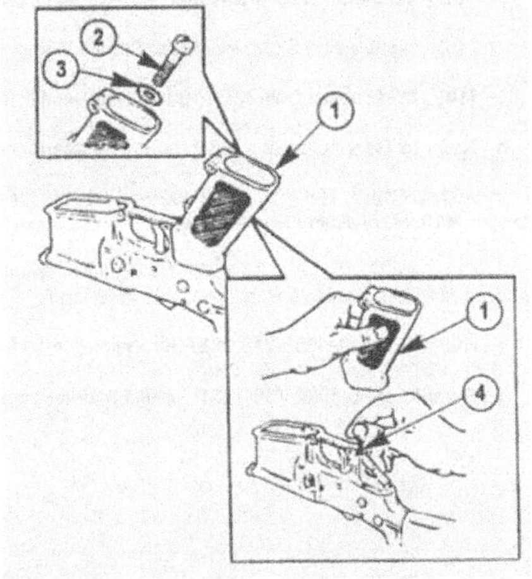

3 Install lock plate (5) with the detent helical spring (4) passing through the hole In the right side of the lock plate and the arm on the outside of the receiver The selector lever just point to the SAFE position

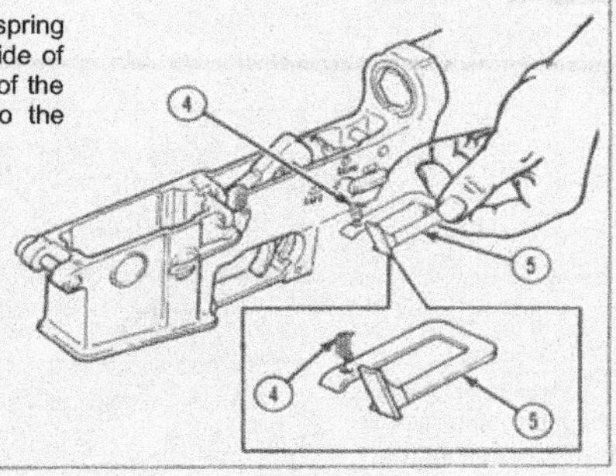

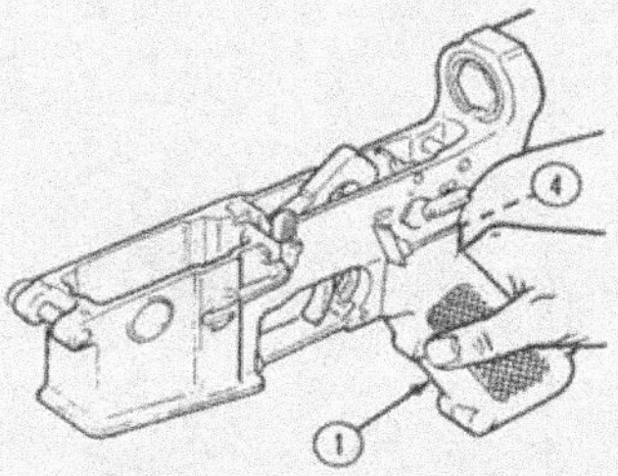

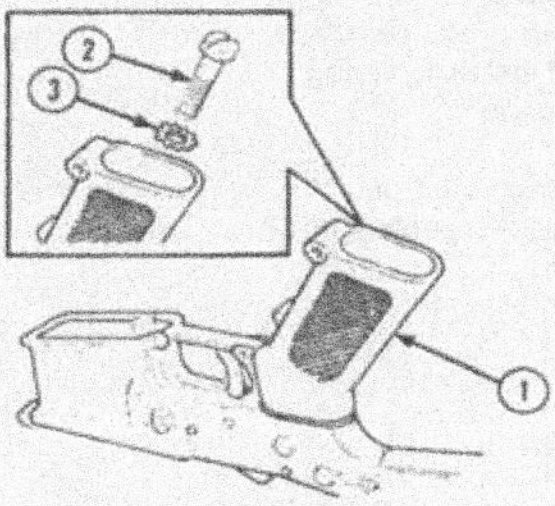

4 Carefully compress detent helical grip (1)

5. Using a screwdriver secure pistol by Installing lockwasher 131 spring (4) and position pistol grip (1) and screw 12)

4-2. LOCK PLATE (CONT).

b REMOVAL

1. Using a screwdriver, reach inside pistol grip (l) and remove screw (2) and lockwasher (3)

WARNING

To avoid Injury to your eyes, use care when removing and installing spring loaded parts

2 Carefully remove pistol grip (1) Hold detent helical spring (4) In place

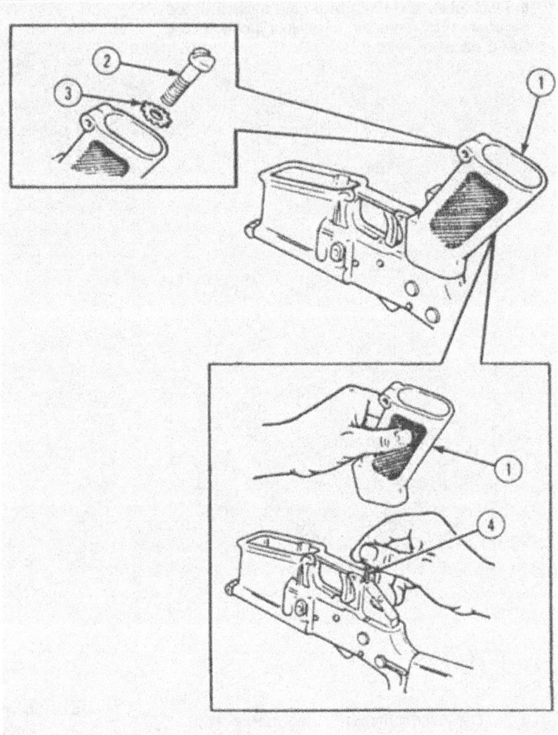

3 Remove lock plate (51 while continuing to hold detent helical spring (4) In place The selector lever must point to the SAFE position

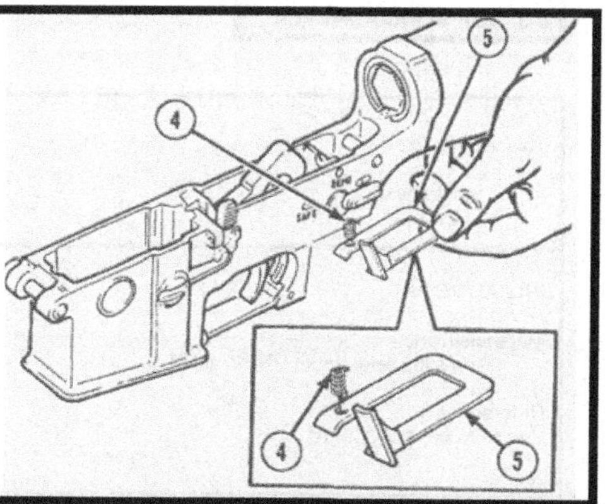

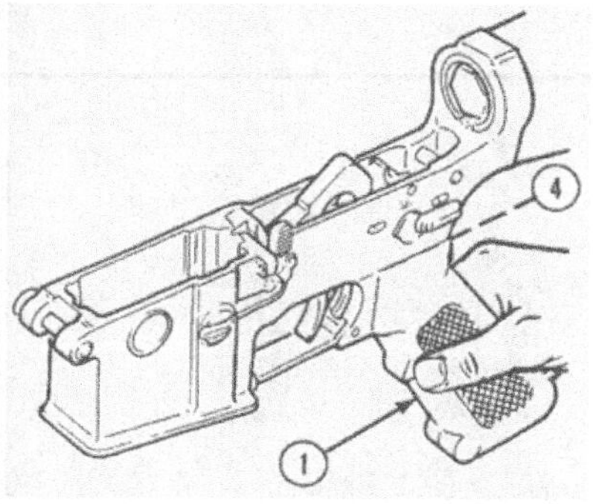

4 Carefully compress detent helical grip (I)

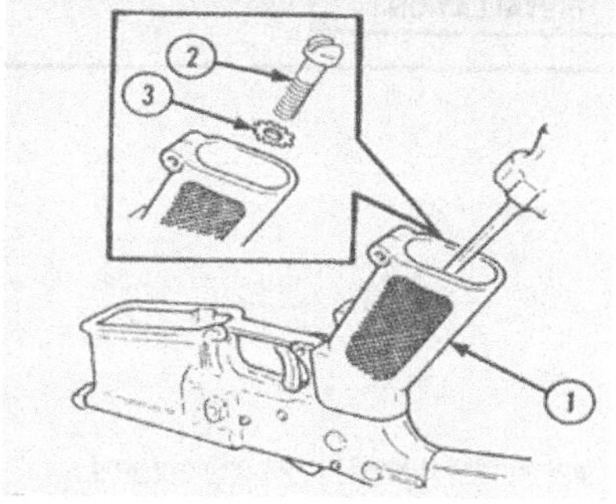

5. Using a screwdriver, secure pistol by installing lockwasher (3) spring (4) and position pistol grip (1) and screw (2)

c. INSPECTION

Inspect lock plate for serviceability and broken arm Replace If unserviceable or if arm is broken off

4-3. TOP SLING ADAPTER

This task covers:

a Installation
b Removal
c Inspection

INITIAL SETUP

Materials/Parts
Top sling adapter kit PN 8448471

References
TM 9-1005-319-10

a. INSTALLATION I

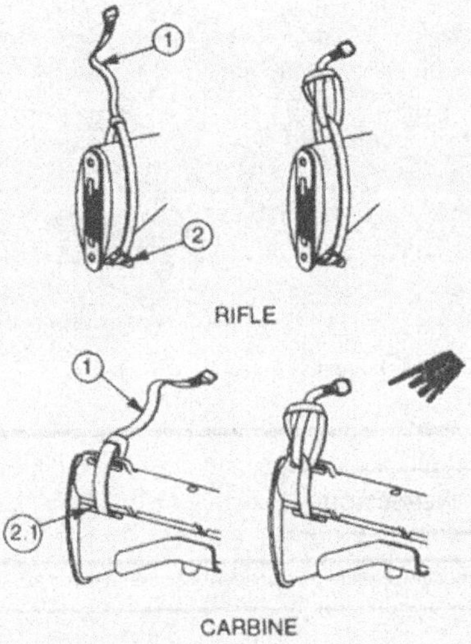

1 Refer to TM 9-1005-319-10 to remove sling.

RIFLE ONLY

2 Install top sling adapter strap (1) through sling swivel (2) and tie.

CARBINE ONLY

2.1. Install top sling adapter strap (1) through sling opening (2.1) and tie.

3 Remove upper handguard assembly.

4 Use pliers to install clamp (3) on front sight base (4).

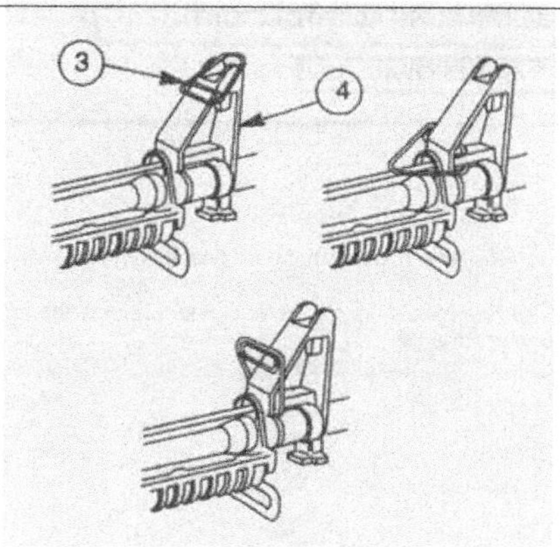

5 Attach sling (5) to top sling adapter strap (1) and to clamp (3).

6 Reinstall upper handguard assembly (6).

7 Adjust sling (5).

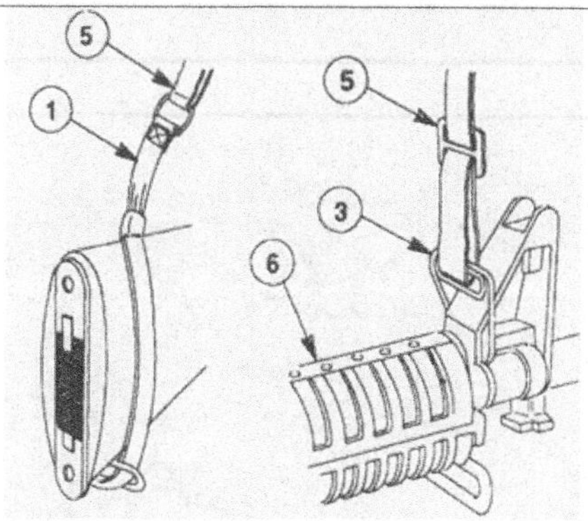

4-3. TOP SLING ADAPTER (CONT).

b. REMOVAL

1 Remove upper handguard assembly.

2 Remove sling (1) from clamp (2) and top sling adapter strap (3).

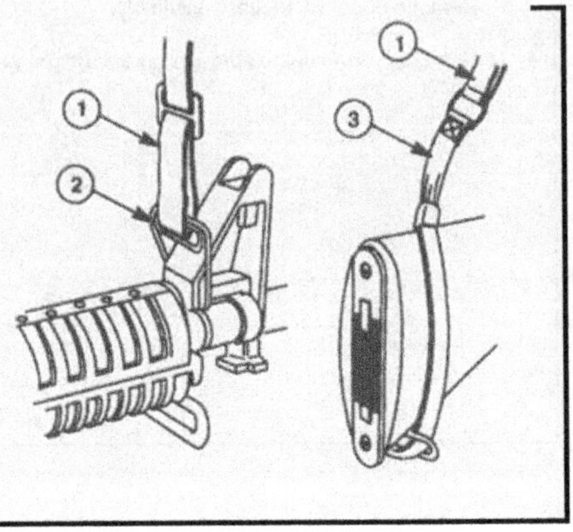

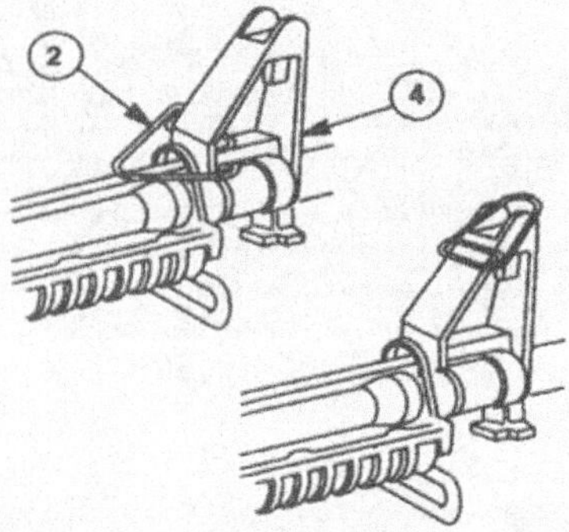

3 Use pliers to remove clamp (2) from front sight base (4). (5).

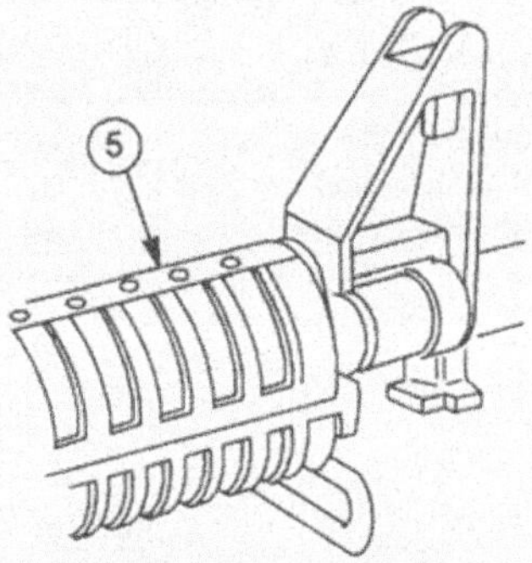

4 Reinstall upper handguard assembly

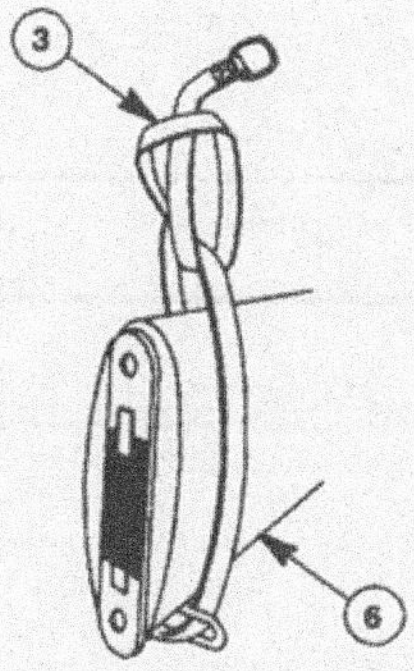

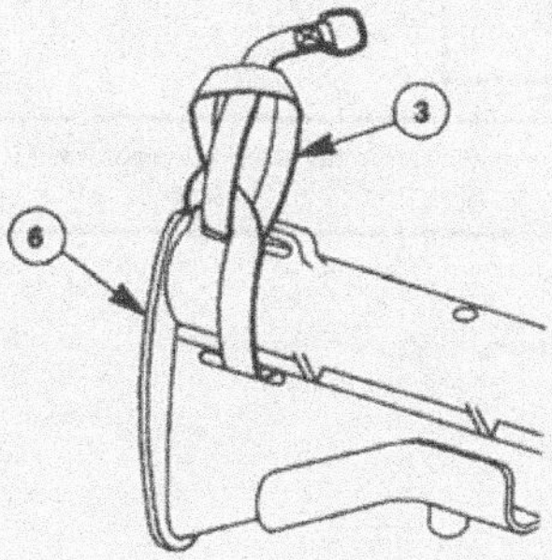

RIFLE

CARBINE

5 Untie top sling adapter strap (3) and remove from buttstock (6).

6 Refer to TM 9-1005-319-10 and install sling.

4-3. TOP SLING ADAPTER (CONT)

c. INSPECTION

Visually Inspect top sling adapter strap and replace If it is badly worn or damaged

4-4. BLANK FIRING ATTACHMENTS M15A2 (M16A2) AND M23 (M4/M4A1).

This task covers:

a Installation
b Removal
c Cleaning
d Inspection
e Repainting
f Replacement

INITIAL SETUP

Materials/Parts
Cleaner, lubricant, and preservative (CLP) (item 9, app D)
Coating compound, enamel (Red - M16A2) (item 23.2, app D)
Coating compound, enamel (Yellow - M4/M4A1) (item 23.3, app D)

General Safety Instructions
Do not keep live ammunition near the work area.

Only blank cartridge M200 is to be used when the blank firing attachment is attached to the weapon.
Do not fire blank ammunition at a representative enemy at distances of less than 20 feet (6.10m).
The unburned propellant grains can cause injury within this distance.

a. INSTALLATION

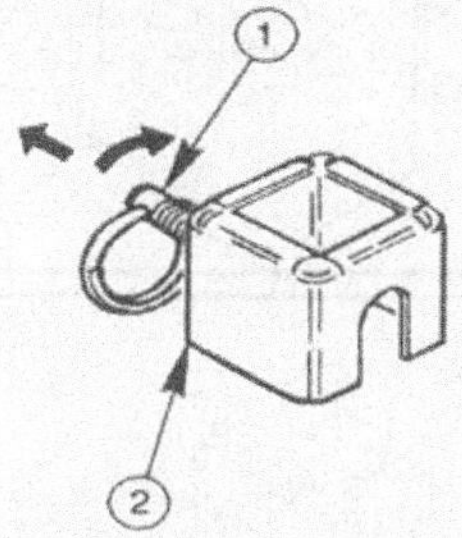

NOTE
M23 blank firing attachment is stamped "M4 Carbine Only" painted yellow and may be used on the M4 and M4A1 carbines. M15A2 BFA is painted red and is used on the M16A2 rifle.

1 Unscrew and pull slide (1) all the way out on blank firing attachment (2)

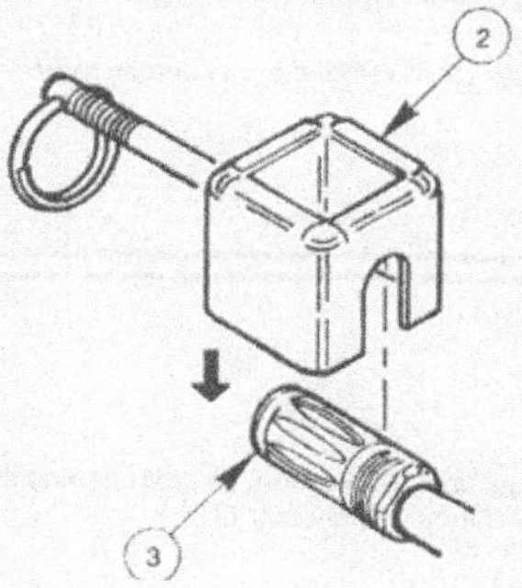

2 Hook blank firing attachment (2) behind the first groove of the compensator (3).

4-4 BLANK FIRING ATTACHMENTS M15A2 (M16A2) AND M23 (M4/M4A1)(CONT).

a. INSTALLATION (CONT)

CAUTION

Do not use tools to tighten the blank firing attachment. **USE HANDS ONLY.**

3 Push slide (2) into compensator (3) and hand tighten.

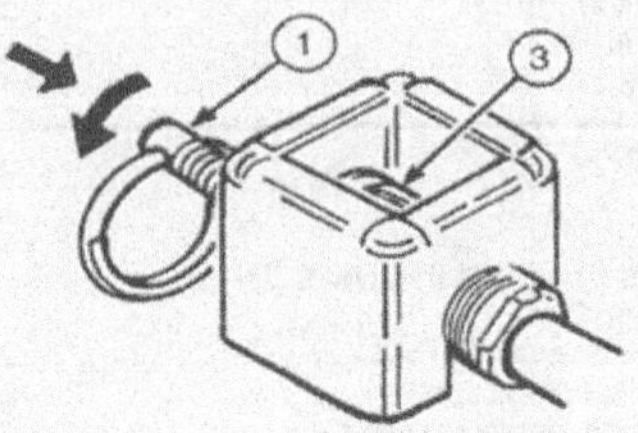

NOTE

Check for tightness after firing approximately 50 blank rounds.

b. REMOVAL

CAUTION

Do not use tools to tighten the blank firing attachment. USE HANDS ONLY.

1 Unscrew slide (1) to remove from compensator (2).

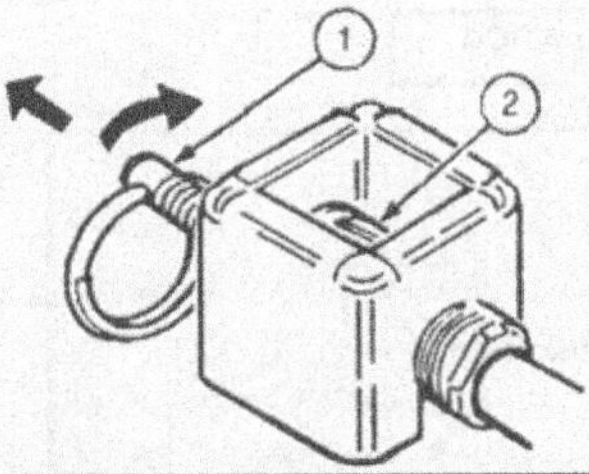

2. Unhook blank firing attachment (3) from behind the first groove of the compensator (2)

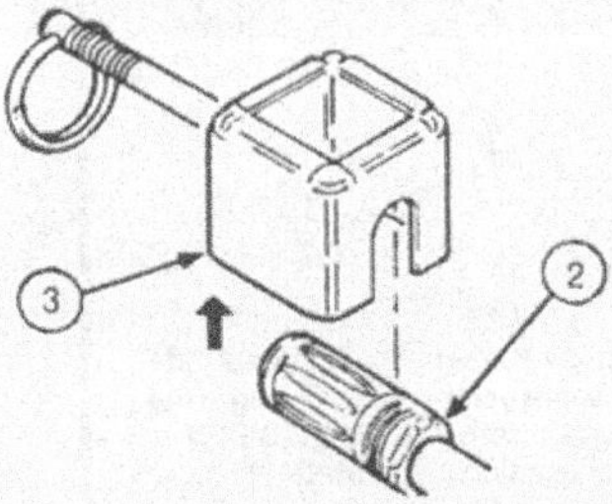

3 Screw slide (1) all the way in on blank firing attachment (3).

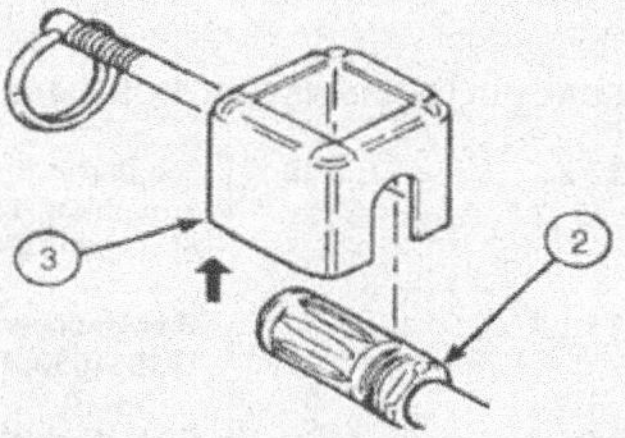

c. CLEANING

Clean blank firing attachment with CLP, wipe dry, and coat with CLP.

d. INSPECTION

Inspect blank firing attachment for cracks or distortion. Be sure the parts in the slide are clear and clean. If blank firing attachment is cracked or distorted, it is unserviceable.

e. REPAINTING

Repaint blank firing attachment using enamel coating compound (Red for M16A2 rifle or Yellow for M4/M4A1 carbine). Painting is the only repair authorized.

f. REPLACEMENT

Replace blank firing attachment if unserviceable.

APPENDIX A
REFERENCES

A-1. TECHNICAL BULLETINSINSTRUCTIONS/MANUALS/ORDERS.

TB 9-1000-247-34	Standards for Overseas Shipment or Domestic Issue of Small Arms, Aircraft Armament, Toward Howitzers, Mortars, Recoilless Rifles, Rocker Launchers and Associated Fire Control Equipment
TB 43-0002-73	Maintenance Expenditures Limits for FSC Group 10; FSC Classes 1005, 1010, 1015, 1030, 1055, 1090, and 1095
TI 05538A-15/2A	Early Warning Indications of Malfunctions
TI 05538A-35/12A	Checking Muzzle and Breech Erosion in Barrel
TI 8005-24/20	Prefire Inspection, Small Arms Weapons, Ordnance Material
TM 9-1005-237-23&P	Organizational and Direct Support Maintenance Manual (Including Repair Parts and Special Tools List) for Bayonet-Knife M6 and M7, with Bayonet-Knife Scabbard M10 and M9 Multipurpose Bayonet System
TM 9-1005-319-10	Operator's Manual for 5.56mm, M16A2 Rifle and M4/M4A1 Carbine
TM 9-1010-221-23&P	Organizational, Direct Support and General Support Maintenance Manual (Including Repair Parts and Special Tools List) for Launcher, Grenade: 40mm M203 W/E (NSN 1010-00-179-6447)
TM 9-6920-363-12&P	M261 Conversion Kit
TM 9-6920-746-12&P	M2 Practice Bolt
TM 11-5855-213-23&P	Night Vision Sight, Individual Served Weapon, AN/PVS-4
TM 750-244-7	Procedures for Destruction of Equipment in Federal Supply Classifications 1000, 1005, 1015, 1020, 1025, 1030, 1055, 1090, and 1095 to Prevent Enemy Use
TM 4700-15/1	Equipment Record Procedures
TO 00-35D-54	Materiel Deficiency Reporting and Investigating System
TO 11 A-13-10-7	Storage and Maintenance Procedures for Small Arms Ammunition
TO 11W-1-10	Historical Data Recording of Inspection, Maintenance, and Firing Data for Ground Weapons
TO 33K-1-100	(TMDE) Interval Calibration and Repair Reference Guide and Work Unit Code Manual

A-2. FIELD MANUALS.

FM 21-11 ....First Aid for Soldiers

FM 23-9 ....M16A1 and M16A2 Rifle and Rifle Marksmanship

A-3 PAMPHLETS.

DA PAM 25-30...Consolidated Index of Army Publications and Blank Forms

DA PAM 738-750....................................The Army Maintenance Management System (TAMMS)

A-4 RELATED PUBLICATIONS.

AFM 36-2227, Vol 1................................Combat Arms Training Management and Range Operation

AMC-P 310-9..Equipment Publications Listing

AR 725-50 ....Requisitioning, Receipt, and Issue System

AR 750-1 ....Army Materiel Policies

CTA 8-100 ....Army Medical Department Expendable/Durable Items

CTA 50-970 ....Expendable/Durable Items (Except: Medical, Class V, Repair Parts and Heraldic Items)

DOD 4160.21-M-1 ....Defense Demilitarization Manual

FMFM 1-3A...Field Firing Techniques

SB 746-1...Publications for Packing Army General Supplies

SPI 00-856-6885 ....Special Packaging Instructions for M16 Rifle

A-5 FORMS.

AFTO Form 22 ....Technical Order System Publications Improvement Report

AFTO Form 105 ....Inspection, Maintenance, and Firing Data for Ground Weapons

DA Form 2028 ....Recommended Changes to Publications and Blank Forms

DA Form 2028-2 ....Recommended Changes to Equipment Technical Manuals

DA Form 2404 ....Equipment Inspection and Maintenance Worksheet

DA Form 2408-9 ....Equipment Control Record

DD Form 314 ....Preventive Maintenance Schedule and Record

NAVMC 10772 .... Recommended Changes to Technical Publications

NAVMC 11003 .... Ordnance Serialized Items Subsidiary Records

SF 364 .... Report of Discrepancy (ROD)

SF 368 .... Product Quality Deficiency Report

A-6 SUPPLY CATALOGS/SUPPLY LISTS.

SC 4933-95-CL-A11 .... Shop Set, Small Arms Field Maintenance

SC 5180-95-CL-A07 .... Tool Kit, Small Arms Repair

APPENDIX B

MAINTENANCE ALLOCATION CHART

Section I. INTRODUCTION

B-1. GENERAL.

a. This section provides a general explanation of all maintenance and repair functions authorized at various maintenance levels

b. The Maintenance Allocation Chart (MAC) In section II designates overall authority and responsibility for the performance of maintenance functions on the identified end Item or component. The application of the maintenance functions to the end Item or component will be consistent with the capacities and capabilities of the designated maintenance levels

c. Section III lists the tools and test equipment (both special tools and common tool sets) required for each maintenance function as referenced from section II.

d. Section IV contains supplemental instructions and explanatory notes for a particular maintenance function.

B-2. MAINTENANCE FUNCTIONS.

Maintenance functions will be limited to and defined as follows: (except for ammunition MAC')

a. Inspect. To determine the serviceability of an Item by comparing Its physical, mechanical, and or electrical characteristics with established standards through examination leg , by sight, sound, or feel).

b. Test. To verify serviceability by measuring the mechanical, pneumatic, hydraulic, or electrical characteristics of an Item and comparing those characteristics with prescribed standards

c. Service. Operations required periodically to keep an Item In proper operating condition, I e to clean (includes decontaminate, when required), to preserve, to drain, to paint, or to replenish fuel, lubricants, chemical fluids. or gases.

d. Adjust. To maintain or regulate, within prescribed limits, by bringing Into proper or exact position, or by setting the operating characteristics to specified parameters

---

'Exception is authorized for ammunition MAC to permit the redesignation/redefinition of maintenance function headings to more adequately identify ammunition maintenance functions The heading designations and definitions will be Included In the appropriate technical manual for each category of ammunition

B-2. MAINTENANCE FUNCTIONS (CONT).

e. Align. To adjust specified variable elements of an item to bring about optimum or desired performance

NOTE

Air Force Precision Measurement Equipment Laboratory (PMEL) person will calibrate small arms Inspection gages In accordance with TO 33K-1-100

f. Calibrate. To determine and cause corrections to be made or to be adjusted on Instruments or test, measuring, and diagnostic equipments used In precision measurement Consists of comparisons of two Instruments, one of which Is a certified standard of known accuracy, to detect and adjust any discrepancy in the accuracy of the Instrument being compared

g. Remove/Install. To remove and Install the same item when required to perform service or other maintenance functions. Install may be the act of emplacing, seating, or fixing Into position a spare, repair part, or module (component or assembly) In a manner to allow the proper functioning of an equipment or system.

h. Replace. To remove an unserviceable Item and Install a serviceable counterpart In Its place "Replace"' is authorized by the MAC and is shown as the 3d position code of the SMR code.

i. Repair. The application of maintenance services2, including fault location troubleshooting3, removal 'installation, and disassembly'assembly4 procedures, and maintenance actions" to Identify troubles and restore serviceability to an Item by correcting specific damage, fault, malfunction, or failure In a part, subassembly, module (component or assembly), end Item, or system.

j. Overhaul. That maintenance effort (service/action) prescribed to restore an item to a completely serviceable operational condition as required by maintenance standards In appropriate technical publications (i.e , DMWR). Overhaul is normally the highest degree of maintenance performed by the Army Overhaul does not normally return an Item to like new condition

k. Rebuild. Consists of those services/actions necessary for the restoration of unserviceable equipment to a like new condition In accordance with original manufacturing standards Rebuild is the highest degree of materiel maintenance applied to Army equipment The rebuild operation Includes the act of returning to zero those age measurements (hours miles, etc) considered In classifying Army equipment/components.

---

[2]Servlces--Inspect, test, service, adjust, align, calibrate, and or replace
[3]Fault locate troubleshoot The process of investigating and detecting the cause of equipment malfunctioning; the act of isolating a fault within a system or unit under test (UUT).
[4]Disassemble assemble encompasses the step-by-step taking apart (or breakdown) of a spare/functional group coded Item to the level of Its least competency identified as maintenance significant (.e , assigned an SMR code) for the level of maintenance under consideration '
[5]Actionswelding, grinding, riveting, straightening, facing, remachinery, and or resurfacing

B-3. EXPLANATION OF COLUMNS IN THE MAC, SECTION II.

a. Column 1, Group Number. Column 1 lists functional group code numbers, the purpose of which is to identify maintenance significant components, assemblies, subassemblies, and modules with the next higher a assembly. End item group number shall be "OO."

b. Column 2, Component Assembly. Column 2 contains the names of components. Assemblies, subassemblies, and modules forwhich maintenance is authorized.

c. Column 3, Maintenance Function. Column 3 lists the functions to be performed on the it em listed in Column 21. (for detailes explanation of these functions, see paragraph B-2.)

d. Column 4, Maintenance Level. Column 4 specifies, by the list ing of a work time figure in the appropriate subcolumn(s), the level of maintenance. If the number of complexity of the t asks within the listed maintenance function vary at different maintenance levels, appropriate work time figures will be shown for each level. The work time figure represents the average time required to restore an item (assembly, subassembly, component, module, end it4m, or system) or a serviceable condition under typicalfield operating conditions. This time includes preparation time (including any necessary disassembly assembly time), troubleshooting fault location time, and quality assurance quality control time in addition to the time required to perform the specific tasks identified for the maintenance functions authorized i the maintenance allocation chart. The symbol designations for the varies maintenance levels are as follows:

C..Operator or Crew
O..Unit Maintenance
F..Direct Support Maintenance
H..General Support Maintenance
L..Specialized Repair Activity (SRA)[6]
D..Depot Maintenance

e. Column 5, Tools and Equipment. Column 5 specifies, by code, those common tool sets (not individual tools) and special tools., TMDE, and support equipment required to perform the designated function.

f. Column 6, Remarks. This column shall, when applicable, contain a lett4er code, in alphabetic order, which shall be keyed to the remarks contained in Section IV.

[6]This maintenance level is not included in Section II, column (4) of the Maintenance Allocation Chart. To identify functions to this level of maintenance, enter a work time figure in the "H" column of section II, column (4), and use an associated reference code in the Remarks column (6). Key the c ode to Section IV . Remarks, and explain the SRA complete repair application there. The explanatory remarks)s) shall reference the specific Repair Parts and Special Tools List (RPSTL) TM which contains additional SRA criteria and the authorized spare/repair parts.

B-4. EXPLANATION OF COLUMNS IN TOOL AND TEST EQUIPMENT REQUIREMENTS, SECTION III.

a. Column 1, Reference Code. The tool and test equipment reference code correlates with a code used In the MAC, Section II, Column 5.

b. Column 2, Maintenance Level. The lowest level of maintenance authorized to use the tool or test equipment.

c. Column 3, Nomenclature. Name or identification of the tool or test equipment.

d. Column 4, National Stock Number. The National stock number of the tool or test equipment.

e. Column 5, Tool Number. The manufacturer's part number.

B-5. EXPLANATION OF COLUMNS IN REMARKS, SECTION IV.

a. Column 1, Reference Code. The code recorded In column 6, Section II.

b. Column 2, Remarks. This column lists information pertinent to the maintenance function being performed as indicated In the MAC, Section II.

Section II. MAINTENANCE ALLOCATION CHART FOR M16A2 RIFLE AND M4/M4A1 CARBINES

(1) GROUP NUMBER	(2) COMPONENT ASSEMBLY	(3) MAINTENANCE FUNCTION	(4) MAINTENANCE LEVEL UNIT C	UNIT O	DS F	DS H	DEPOT D	(5) TOOLS AND EQPT	(6) REMARKS CODE
00	M16A2 5.56MM RIFLE, AND M4/ M4A1 5.56MM CARBINES	Inspect	0.1	0.3	0.7			2	
		Test			0.2			2,3	
		Service	0.2	0.3	0.3				
		Replace		0.1					
		Overhaul							
00.1	CARRYING HANDLE ASSEMBLY M4/M4A1	Inspect	0.1	0.1	0.1				
		Service	0.1						
		Remove/ Install	0.1	0.1	0.1				
		Replace			0.3			1,3	
		Repair		0.1	0.2			1,3	
00.101	Rear Sight Assembly M4/ M4A1	Inspect			0.1				
		Replace			0.3			1,3	
		Repair			0.2			1,3	
01	BOLT CARRIER ASSEMBLY	Inspect	0.1	0.1	0.1				
		Service	0.1	0.1					
		Remove/ Install	0.1						
		Replace			0.1			1,2	
		Repair	01	0.1				3	
0101	Bolt Assembly	Inspect	0.1	0.1	0.1				
		Test			0.1			2	
		Service	0.1						
		Remove/ Install	0.1						
		Replace			0.1			2	
		Repair		0.1	0.1			3	

*'Worktimes are included in DMWR 9-1005-249

MAINTENANCE ALLOCATION CHART (CONT)

(1) GROUP NUMBER	(2) COMPONENT ASSEMBLY	(3) MAINTENANCE FUNCTION	(4) MAINTENANCE LEVEL UNIT C	UNIT O	DS F	DS H	DEPOT D	(5) TOOLS AND EQPT	(6) REMARKS CODE
0102	Key and Bolt Carrier Assembly	Inspect	0.1		0.1				
		Service	0.1						
		Remove/ Install	0.1						
		Replace			0.1			1	
		Repair		0.1	0.1			1,3	
02	HANDLE ASSEMBLY	Inspect	0.1						
		Service	0.1						
		Remove/ Install	0.1						
		Replace			0.1		**		
		Repair			0.1			3	
03	UPPER RECEIVER AND BARREL ASSEMBLY	Inspect	0.1	0.2	0.1				
		Test			0.2			2	
		Service	0.2		0.1			1	
		Remove/ Install	0.1						
		Replace			0.5			1,2	
		Repair		0.2	0.5			1,2,3	A,B
0301	M16A2 Barrel Assembly and M4/ M4A1 Replacement Barrel	Inspect		0.1	0.1				
		Replace			0.3			1,2	
		Repair	0.1					3	
0302	Upper Receiver Assembly	Inspect			0.1				
		Replace			0 5			1.2	
		Repair			0.3			1.3	
030201	Forward Assist Assembly	Inspect			0.1				
		Replace			0.1				
		Repair			0 1			1.3	
030202	Rear Sight Assembly M16A2	Inspect			0 1				
		Replace			0.3			1,3	
		Repair			0 2			1,3	

MAINTENANCE ALLOCATION CHART (CONT)

(1) GROUP NUMBER	(2) COMPONENT ASSEMBLY	(3) MAINTENANCE FUNCTION	(4) MAINTENANCE LEVEL UNIT C	UNIT O	DS F	DS H	DEPOT D	(5) TOOLS AND EQPT	(6) REMARKS CODE
04	LOWER RECEIVER AND BUTTSTOCK ASSEMBLY	Inspect	0.1	0.1	0.1				
		Test			0.1			2	
		Service	0.2	0.2					
		Repair		0.2	0.3			1,2,3	C,D,E
0401and 0401A	Buttstock Assembly	Inspect		0.1					
		Remove/ Install		0.1				3	
		Replace		0.1				3	
		Repair		0.1				3	
0402	Hammer Assembly	Inspect			0.1				
		Remove/ Install			0.1				
		Replace			0.1				
		Repair			0.1				
0403	Trigger Assembly	Inspect			0.1				
		Remove/ Install			0.1			1,3	
		Replace			0.1			1,3	
		Repair			0.1			1,3	
040301	Trigger Subassembly M16A2 and M4	Inspect			0.1				
		Replace			0.1			1,3	
		Repair			0.1			1,3	
0404 and 0404A	Lower Receiver and Receiver Extension Assembly	Inspect			0.1				
		Test			0.1			2	
		Repair			0.2			1,3	

Section III. TOOL AND TEST EQUIPMENT REQUIREMENTS FOR M16A2 RIFLE AND M4/M4A1 CARBINES

TOOL OR TEST EQUIPMENT REF CODE	(2) MAINTENANCE CATEGORY	(3) NOMENCLATURE	(4) NATIONAL/NATO STOCK NUMBER	(5) TOOL NUMBER
1 (ARMY ONLY)	F	Shop Set, Small	4933-00-754-0664	SC 4933-95-CL-A11
2 (ARMY ONLY)	F	Tool and gage Set, DS/GS Maintenance for 5.56mm Rifle, M16 Series	4933-00-056-7106	8426685
3 (ARMY ONLY)	O	Tool Kit, Small Arms Repairman	5180-00-357-7770	SC 5180-95-CL-A07
4 (A.F ONLY)	F	Torque Wrench, ft-lb	5120-00-640-6365	A-A-411
5 (A.F ONLY)	F in.-lb	Torque Wrench,	5120-00-230-6380	T-E-12A
6 A.F ONLY)	F	Trigger Weights	4933-00-647-3696	7274758

Section IV. REMARKS

Reference Code	Remarks
A	Tool, Front Sight Post Removal and Installation
B	Depressor, Front Sight Detent
C	Tool, Pivot Pin Removal
D	Tool, Pivot Pin Installation
E	Pin, Slave

APPENDIX C

REPAIR PARTS AND SPECIAL TOOLS LIST

Section I. INTRODUCTION

C-1. SCOPE.

This RPSTL lists and authorizes spares and repair parts; special tools; special test, measurement, and diagnostic equipment (TMDE); and other special support equipment required for performance of unit and direct support maintenance of the M16A2 rifle and M4/M4A1 carbines. It authorizes the requisitioning, issue, and disposition of spares, repair parts, and special tools as indicated by the source, maintenance, and recoverability (SMR) codes.

C-2. GENERAL.

In addition to Section I, Introduction, this Repair Parts and Special Tools List is divided into the following sections:

a. Section II Repair Parts List. A list of spares and repair parts authorized by this RPSTL for use in the performance of maintenance. The list also includes parts which must be removed for replacement of the authorized parts. Parts lists are composed of functional groups in ascending alphanumeric sequence, with the parts in each group listed in ascending figure and item number sequence. Bulk materials are listed by item name in FIG BULK at the end of the section. Repair parts kits or sets are listed separately in their own functional group within Section II. Repair parts for repairable special tools are also listed in the section.

b. Section III Special Tools List. A list of special tools, special TMDE, and other special support equipment authorized by this RPSTL (as indicated by Basis of Issue (BOI) information in DESCRIPTION AND USABLE ON CODE (UOC) column) for the performance of maintenance.

c. Section IV Cross-Reference Indexes. A list, in National item identification number (NIIN) sequence, of all National stock numbered items appearing in the listings, followed by a list in alphanumeric sequence of all part numbers appearing in the listing. National stock numbers and part numbers are cross-referenced to each illustration figure and item number in alphanumeric sequence and cross-references NSN, CAGEC, and part number.

C-3. EXPLANATION OF COLUMNS (SECTIONS II AND III).

a. ITEM NO. (Column (1)). Indicates the number used to identify items called out in the illustration.

b. SMR CODE (Column (2)). The Source, Maintenance, and Recoverability (SMR) code is a 5-position code containing supply/requisitioning information, maintenance level authorization criteria, and disposition instruction, as shown In the following breakout:

C-3. EXPLANATION OF COLUMNS (SECTION II AND III) (CONT).

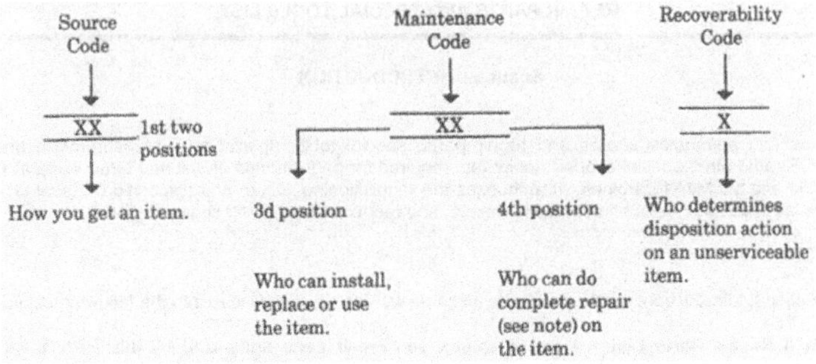

*Complete Repair: Maintenance capacity, capability, and authority to perform all corrective maintenance tasks of the "Repair" function in a use/user environment In order to restore serviceability to a failed item.

(1) Source Code. The source code tells you how to get an item needed for maintenance, repair, or overhaul of an end item/equipment. Explanations of source codes follow:

Code	Explanation
PA PB PC* * PD PE PF PG	Stocked items; use the applicable NSN to request/requisition Items with these source codes They are authorized to the level indicated by the code entered In the 3d position of the SMR code. * *NOTE: Items coded PC are subject to deterioration.
KD KF KB	Items with these codes are not to be requested/requisitoned Individually They are part of a kit which is author->zed to the maintenance level indicated In the 3d positon of the SMR code The complete kit must be requisitioned and applied.
MO-(Made at Unit/ AVUM Level) MF- (Made at DS/ AVIM Level) MH-(Made at GS Level) ML- (Made at Specialized Repair Act (SRA)) MD-(Made at Depot)	Items with these codes are not to be requested/requisitioned Individually They must be made from bulk material which Is identified by the part number In the DESCRIPTION AND USABLE ON CODE (UOC) column and listed In the Bulk Material group of the repair parts list in this RPSTL If the Item Is authorized to you by the 3d position code of the SMR code, but the source code Indicates It is made at a higher level, order the Item from the higher level of maintenance.

Code	Explanation
AO- (Assembled by unit/AVUM Level) AF- (Assembled by DS/AVIM Level) AH- (Assembled by GS Level) AL- (Assembled by SRA) AD- (Assembled by Depot)	Items with these codes are not to be requested/requisitioned Individually. The parts that make up the assembled Item must be requisitioned or fabricated and assembled at the level of maintenance indicated by the source code. If the 3d position code of the SMR code authorizes you to replace the Item, but the source code Indicates the Item Is assembled at a higher level, order the Item from the higher level of maintenance.

XA-Do not requisition an "XA"-coded item. Order Its next higher assembly. (Also, refer to the NOTE below.)

XB--If an "XB" item is not available from salvage, order It using the CAGEC and part number given.

XC-installation drawing, diagram, instruction sheet, field service drawing, that is identified by manufacturer's part number.

XD-Item is not stocked. Order an "XD"-coded Item through normal supply channels using the CAGEC and part number given, If no NSN Is available.

NOTE

Cannibalization or controlled exchange, when authorized, may be used as a source of supply for items with the above source codes, except for those source coded "XA" or those aircraft support Items restricted by requirements of AR 750-1.

(2) Maintenance Code. Maintenance codes tell you the level(s) of maintenance authorized to USE and REPAIR support items. The maintenance codes are entered In the third and fourth positions of the SMR code as follows:

(a) The maintenance code entered in the third position tells you the lowest maintenance level authorized to remove, replace, and use an Item. The maintenance code entered In the third position will indicate authorization to one of the following levels of maintenance

Code	Application/Explanation
C	-Crew or operator maintenance done within unit or aviation unit maintenance.
O	-Unit or aviation unit level can remove, replace, and use the Item.
F	-Direct support or aviation intermediate level can remove, replace, and use the Item.
H	-General support level can remove, replace, and use the Item.
L	-Specialized repair activity can remove, replace, and use the Item.
D	-Depot level can remove, replace, and use the Item

C-3. EXPLANATION OF COLUMNS (SECTIONS II AND III) (CONT).

(b) The maintenance code entered In the fourth position tells whether or not the Item is to be repaired and Identifies the lowest maintenance level with the capability to do complete repair (I e, perform all authorized repair functions). (NOTE: Some limited repair may be done on the Item at a lower level of maintenance, If authorized by the Maintenance Allocation Chart (MAC) and SMR codes I This position will contain one of the following maintenance codes

Code	Application/Explanation
O	-Unit or aviation unit is the lowest level that can do complete repair of the item
F	-Direct support or aviation Intermediate is the lowest level that can do complete repair of the Item
H	-General support is the lowest level that can do complete repair of the item .
L	-Specialized repair activity is the lowest level that can do complete repair of the item
D	-Depot is the lowest level that can do complete repair of the Item
Z	-Nonreparable. No repair Is authorized
B	-No repair is authorized. (No parts or special tools are authorized for the maintenance of a "B" coded item) However, the Item may be reconditioned by adjusting, lubricating, etc , at the user level

(3) Recoverability Code. Recoverability codes are assigned to Items to Indicate the disposition action on unserviceable items The recoverability code is entered In the fifth position of the SMR code as follows.

Recoverability Codes	Application/Explanation
Z	Nonreparable item unserviceable, condemn and dispose of the item at the level of maintenance shown In 3d position) of SMR code
O	Reparable Item. When uneconomically reparable condemn and dispose of the item at unit or aviation unit level
F	Reparable Item. When uneconomically reparable, condemn and dispose of the Item at the direct support or aviation level
H	Reparable Item. When uneconomically reparable, condemn and dispose of the Item at the .general support level

Recoverability Codes	Application/Explanation
D	-Reparable item. When beyond lower level repair capability, return to depot. Condemnation and disposal of item not authorized below depot level.
L	-Reparable item. Condemnation and disposal not authorized below Specialized Repair Activity (SRA).
A	-Item requires special handling or condemnation procedures because of specific reasons (e.g., precious metal content, high dollar value, critical material, or hazardous material). Refer to appropriate manuals/directives for specific instructions.

c. NSN (Column (3)). Indicates the National stock number (NSN) assigned to the item and which will be used for requisitioning.

d. CAGEC (Column (4)). The Contractor and Government Entity Code (CAGEC) is a 5-digit numeric code which is used to identify the manufacturer, distributor, or Government agency, etc., that supplies the Item.

e. PART NUMBER (Column (5)). Indicates the primary number used by the manufacturer (individual, company, firm, corporation, or Government activity), which controls the design and characteristics of the item by means of its engineering drawings, specifications standards, and inspection requirements to identify an item or range of items.

NOTE

When you use an NSN to requisition an item, the item you receive may have a different part number from the part ordered.

f. DESCRIPTION AND USABLE ON CODE (UOC) (Column (6)). This column includes the following Information:

(1) The Federal item name and, when required, a minimum description to identify the Item.

(2) The physical security classification of the item is indicated by the parenthetical entry which is a physical security classification abbreviation, e.g., Phy Sec C1 (C) Confidential, Phy Sec C1 (S) Secret, Phy Sec C1 (T) Top Secret.

(3) Items that are included in kits and sets are listed below the name of the kit or set.

(4) Spare/repair parts that make up an assembled item are listed immediately following the assembled item line entry.

(5) Part numbers for bulk materials are referenced in this column In the line item entry for the Item to be manufacture red/fabricated.

(6) When the item is not used with all serial numbers of the same model, the effective serial numbers are shown on the last line(s) of the description (before UOC).

(7) The usable on code, when applicable (see paragraph 5, Special Information).

C-3. EXPLANATION OF COLUMNS (SECTION II AND III) (CONT).

(8) In the Special Tools List section, the basis of issue (BOI) appears as the last line(s) in the entry for each special tool, special TMDE, and other special support equipment. When density of equipment supported exceeds density spread indicated in the basis of Issue, the total authorization is increased proportionately.

(9) The statement "END OF FIGURE" appears just below the last item description in Column 5 for a given figure In both Section II and Section III.

g. Oty (Column (7)) . The Qty (quantity per figure column) indicates the quantity of the item used in the breakout shown on the illustration figure, which is prepared for a functional group, subfunctional group, or an assembly. A "V" appearing In this column in lieu of a quantity indicates that the quantity is variable and the quantity may vary from application to application.

C-4. EXPLANATION OF COLUMNS (SECTION IV).

a. NATIONAL STOCK NUMBER (NSN) INDEX.

(1) STOCK NUMBER column. This column lists the NSN by National Item Identification Number (NIIN) sequence. The NIIN consists of the last nine digits of the NSN.

NSN
(i.e., 5305-01-674-1467). When using this column to locate an item, ignore the first 4 digits NIIN of the NSN. However, the complete NSN should be used when ordering items by stock number.

(2) FIG. column. This column lists the number of the figure where the item Is Identified/located. The figures are in numerical order in Section II and Section III.

(3) ITEM column. The item number identifies the item associated with the figure listed in the adjacent FIG. column. This item is also identified by the NSN listed on the same line.

b. PART NUMBER INDEX. Part numbers in this index are listed by part number in ascending alphanumeric sequence (i.e., vertical arrangement of letter and number combination which places the first letter or digit of each group In order A through Z, followed by the numbers 0 through 9 and each following letter or digit in like order).

(1) CAGEC column. The Contractor and Government Entity Code (CAGEC) is a 5-digit numeric code used to identify the manufacturer, distributor, or Government agency, etc., that supplies the item.

(2) PART NUMBER column. Indicates the primary number used by the manufacturer (individual, firm, corporation, or Government activity), which controls the design and characteristics of the item by means of Its engineering drawings, specifications standards, and inspection requirements to identify an Item or range of Items.

(3) STOCK NUMBER column. This column lists the NSN for the associated part number and manufacturer identified In the PART NUMBER and CAGEC columns to the left.

(4) FIG. column. This column lists the number of the figure where the item is identified/located in Sections II and III.

(5) ITEM column. The item number is that number assigned to the item as it appears in the figure referenced in the adjacent figure number column.

C-5. SPECIAL INFORMATION.

a. ASSEMBLY INSTRUCTION. Detailed assembly instructions for items source coded to be assembled from component spare/repair parts are found in Chapter 3. Items that make up the assembly are listed immediately following the assembly item entry or reference is made to an applicable figure.

b. ASSOCIATED PUBLICATIONS. The publications listed below pertain to the M16A2 rifle, M4/M4A1 carbine, and its components:

Publication	Short Title
TM 9-1005-319-10	M16A2 Rifle and M4/M4A1 Carbine

c. USABLE ON CODE. The usable on code appears in the lower left corner of the DESCRIPTION column heading, Usable on codes are shown as "UOC..." in DESCRIPTION column (justified left) on the first line applicable item description/nomenclature. Uncoded items are applicable to all models. Identification of the usable on codes used in the RPSTL are:

Code	Used On
AR8	M16A2 Rifle
AS1	M4 Carbine
AY6	M4A1 Carbine

C-6. HOW TO LOCATE REPAIR PARTS.

a. When National Stock Number or Part Number is Not Known.

(1) First. Using the table of contents, determine the assembly group or subassembly group to which the item belongs. This is necessary since figures are prepared for assembly groups and subassembly groups, and listings are divided into the same groups.

C-6. HOW TO LOCATE REPAIR PARTS (CONT).

(2) **Second.** Find the figure covering the assembly group or subassembly group to which the item belongs.

(3) **Third**. Identify the item on the figure and use the Figure and Item Number Index to find the NSN.

b. When National Stock Number or Part Number is Known.

(1) **First.** Using the National Stock Number or the Part Number Index, find the pertinent National Stock Number or Part Number. The NSN index is in National Item Identification Number (NIIN) sequence. (See C4.a.(1).) The pat numbers in the Part Number Index are listed in ascending alpha numeric sequence. (See C4.b.) Both indexes cross-reference you to the illustration/figure and item number of the item you are looking for.

(2) **Second.** Turn to the figure and item number, verify that the item is the one you're looking for, then locate the item number in the repair parts list for the figure.

C-7. ABBREVIATIONS.

Not applicable.

(1) ITEM NO QTY	(2) SMR CODE	(3) NSN	(4) CAGEC	(5) PART NUMBER	(6) DESCRIPTION AND USABLE	(7) ON CODE (UOC)
					GROUP 00 (M16A2) RIFLE 5.56MM 9349000,(M4) CARBINE 5.56MM 9390000, AND (M4A1) CARBINE 5.56MM 12972700 FIG. C-1. (M16A2) RIFLE, (M4) AND CM4A1) CARBINES	
1	PAOOO	1005000179546	19204	8448517	HANDLE ASSEMBLY,CHA CHARGING....................1	
2	AFFFF		19204	8448501	BOLT CARRIER ASSEMBLY (M16A2)....................1 UOC:AR8	
2A	AFFFF	19200	12011849	BOLT	AND BOLT CARRI ASSEMBLY (M4....................1 AND M4A1).................... UOC:AS1,AY6	
3	AFFFF	1005011879716	19200	9349050	UPPER RECEIVER AND BARREL ASSEMBLY ....1 (M16A2).................... UOC:AR8	
3A	AFFFF	19200	12972680	UPPER	RECEIVER AND BARREL ASSEMBLY....................1 (M4 AND M4A1).................... UOC:AS1,AY6	
4	PACZZ	1005009215004	19200	8448670	MAGAZINE,CARTRIDGE (30 ROUND)....................1	
5	PACZZ	1005012164510	19204	12624561	SLING,SMALL ARMS (M16A2)....................1 UOC:AR8	
5A	PACZZ	1005013689852	19200	12011996	SLING,SMALL ARMS (M4 AND M4A1)....................1 UOC:AS1,AY6	
6	XAFFA	19200	9349100	LOWER	RECEIVER AND BUTTSTOCK ....1 ASSEMBLY (M16A2).................... UOC:AR8	
6A	XAFFA		19200	9390011	LOWER RECEIVER AND BUTTSTOCK ....1 ASSEMBLY (4).................... UOC:AS1	
6B	XAFFA		19200	12972690	LOWER RECEIVER AND BUTTSTOCK ....1 ASSEMBLY (M4A1).................... UOC:AY6	
7	AFFFF		19200	12951011	CARRYING HANDLE ASS (M4 AND M4A1)....................1 UOC:AS1,AY6	

END OF FIGURE

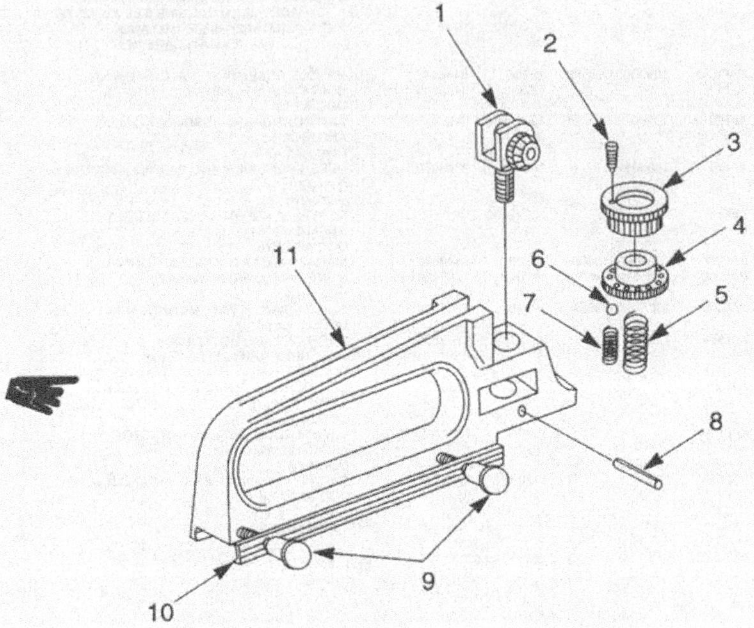

Figure C-1A. (M4 and M4A1) Carrying Handle Assembly 12951011.

(1) ITEM NO	(2) SMR CODE	(3) NSN	(4) CAGEC	(5) PART NUMBER	(6) DESCRIPTION AND USABLE ON CODE (UOC)	(7) QTY
					GROUP 00.1 (M4 AND M4A1) CARRYING HANDLE ASSEMBLY 12951011 FIG. C-1A. (M4 AND M4A1) CARRYING HANDLE ASSEMBLY	
1	AFFFF		19200	12951026	REAR SIGHT ASSY UOC:AS1,AY6	1
2	PAFZZ	5305011343622	19200	9349065	SCREW,INDEX	1
3	PAFZZ	1005013827089	19200	12951018	ELEVATING MECHANISM UOC:AS1 ,AY6	1
4	PAFZZ	5355013826801	19200	12951019	KNOB UOC:AS1,AY6	1
5	PAFZZ	5360011343710	19200	9349070	SPRING ,HELICAL,COMP	1
6	PAFZZ	3110001839175	96906	MS19060-4808	BALL,BEARING UOC:AS1,AY6	1
7	PAFZZ	5360013826802	19200	12951020	SPRING,HELICAL,COMP UOC:AS1,AY6	1
8	PAFZZ	5315008403812	96906	MS16562-121	PIN,SPRING	1
9	PAOZZ	5310013826793	19200	12951023	NUT,PLAIN,KNURLED UOC:AS1,AY6	2
10	PAOZZ	5340013823201	19200	12951017	BRACKET,MOUNTING UOC:AS1,AY6	1
11	PAFZZ	1005013827083	19200	12951021	HANDLE,GUN CARRYING UOC:AS1,AY6	1

END OF FIGURE

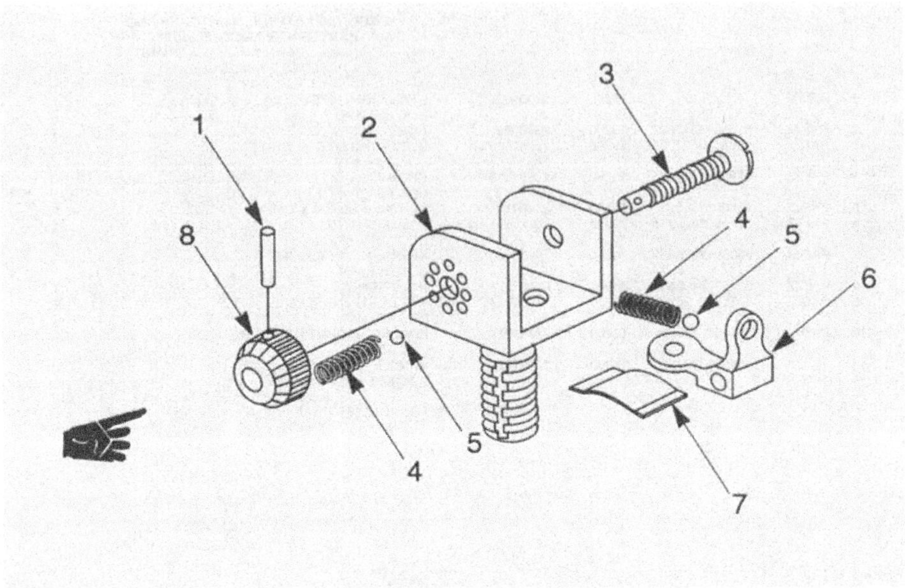

Figure C-1B. (M4 and M4A1) Rear Sight Assembly 12951026.

GROUP 00.101 (M4 AND M4A1) REAR SIGHT ASSEMBLY 12951026
FIG. C-18. (M4 AND M4A1) REAR SIGHT ASSEMBLY

(1) ITEM NO	(2) SMR CODE	(3) NSN	(4) CAGEC	(5) PART NUMBER	(6) DESCRIPTION AND USABLE ON CODE (UOC)	(7) QTY
1	PAFZZ	5315000586678	96906	MS16562-103	PIN,SPRING	1
2	PAFZZ	1005013827086	19200	12951028	BASE,REAR SIGHT UOC:AS1,AY6	1
3	PAFZZ	5305011441490	19200	9349076	SCREW,EXTERNALLY RE	1
4	PAFZZ	5360011481751	19200	9349069	SPRING,HELICAL,COMP UOC:AS1,AY6	2
5	PAFZZ	3110001839175	96906	MS19060-4808	BALL,BEARING UOC:AS1,AY6	2
6	PAFZZ	1005011353697	19200	9349075	APERTURE,SIGHT	1
7	PAFZZ	5360013816183	19200	12011987	SPRING,FLAT	1
8	PAFZZ	5355011343627	19200	9349077	KNOB,WINDAGE	1

END OF FIGURE

SECTION II

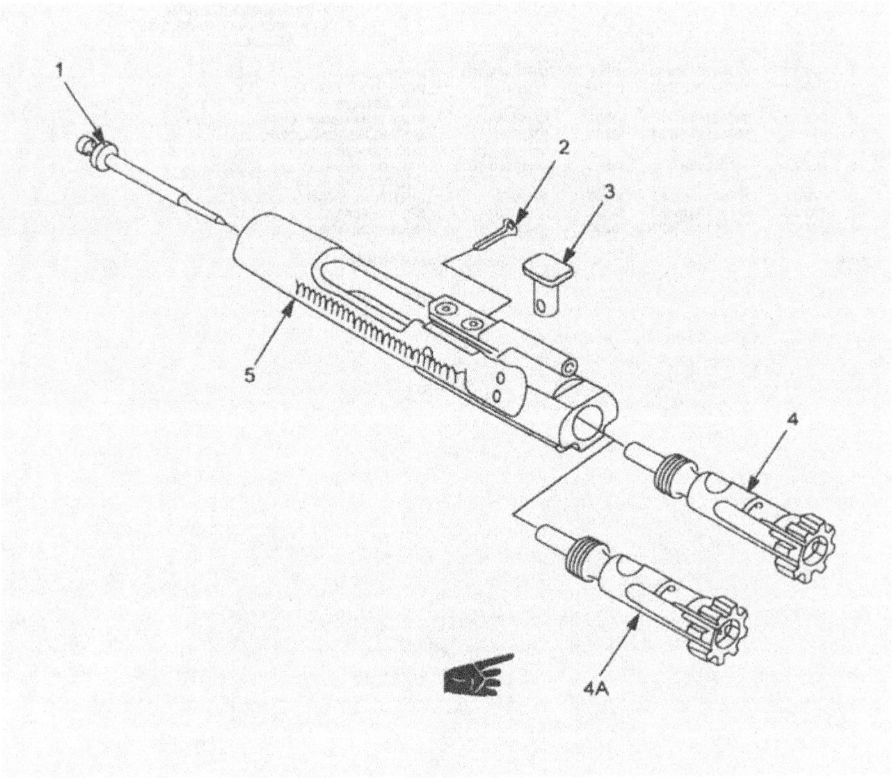

Figure C-2. Bolt Carrier Assembly (M16A2) 8448501 and (M4 and M4A1) 12011849.

(1) ITEM NO	(2) SMR CODE	(3) NSN	(4) CAGEC	(5) PART NUMBER	(6) DESCRIPTION AND USABLE ON CODE (UOC)	(7) QTY
					GROUP 01 BOLT CARRIER ASSEMBLY (M16A2) 8448501 AND (M4 AND M4A1) 12011849 FIG. C-2. BOLT CARRIER ASSEMBLY	
1	PAFZZ	1005000179547	19204	8448503	PIN,FIRING	1
2	PAOZZ	1005009991509	19204	8448504	PIN,FIRING PIN RETA RETAINING	1
3	PAOZZ	5315009927294	19204	8448502	PIN,GROOVED,HEADED BOLT CAM	1
4	PAFFF	1005009927285	19200	8448509	BOLT ASSEMBLY (M16A2) UOC:AR8	1
4A	PAFFF	1005014223770	19200	12972691	BOLT,BREECH ASSEMBLY (M4 AND M4A1) UOC:AS1,AY6	1
5	AFFFF		19204	8448505	KEY AND BOLT	1

END OF FIGURE

SECTION II

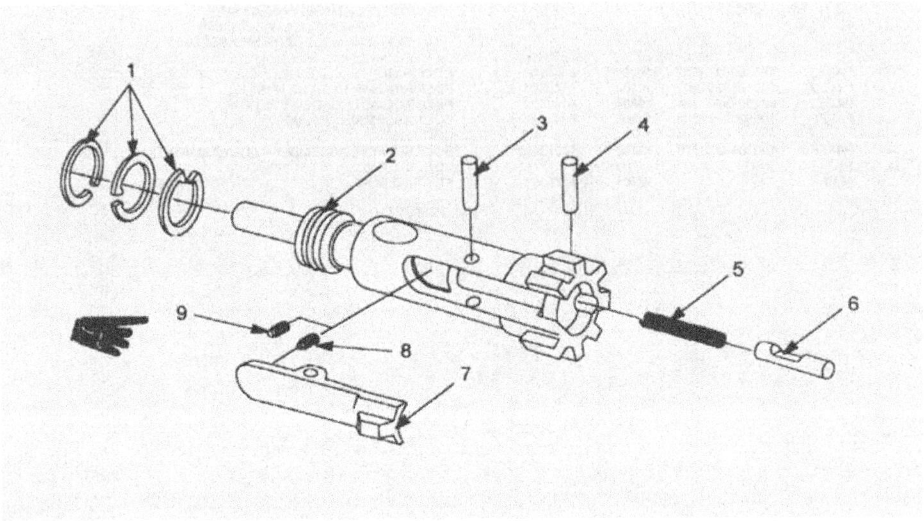

Figure C-3. Bolt Assembly (M16A2) 8448509 and (M4 and M4A1) 12972691.

(1) ITEM NO	(2) SMR CODE	(3) NSN	(4) CAGEC	(5) PART NUMBER	(6) DESCRIPTION AND USABLE ON CODE (UOC)	(7) QTY
					GROUP 0101 BOLT ASSEMBLY (M16A2) 8448509 AND (M4 AND M4A1) 12972691 FIG. C-3. BOLT ASSEMBLY	
1	PAFZZ	1005009927287	19204	8448511	RING,BOLT	3
2	XAFZZ		19204	8448510	BOLT	1
3	PAOZZ	1005009927290	19204	8448513	PIN,EXTRACTOR	1
4	PAOZZ	5315005975086	96906	MS16562-9B	PIN,SPRING EJECTOR	1
5	PAOZZ	5360009927292	19204	8448516	SPRING,HELICAL,COMP COMPRESSION, EJECTOR	1
6	PAOZZ	1005009927291	19204	8448515	EJECTOR,CARTRIDGE	1
7	PAOZZ	1005009927288	19204	8448512	EXTRACTOR,CARTRIDGE	1
8	PAOZZ	1005007603768	19200	8448755	SPRING ASSEMBLY,EXT EXTRACTOR (M16A2) UOC:AR8	1
9	PAOZZ	1005014245899	19200	12972692	SPRING ASSEMBLY,EXT EXTRACTOR (M4 AND M4A1)(BLACK) UOC:AS1 ,AY6	1

END OF FIGURE

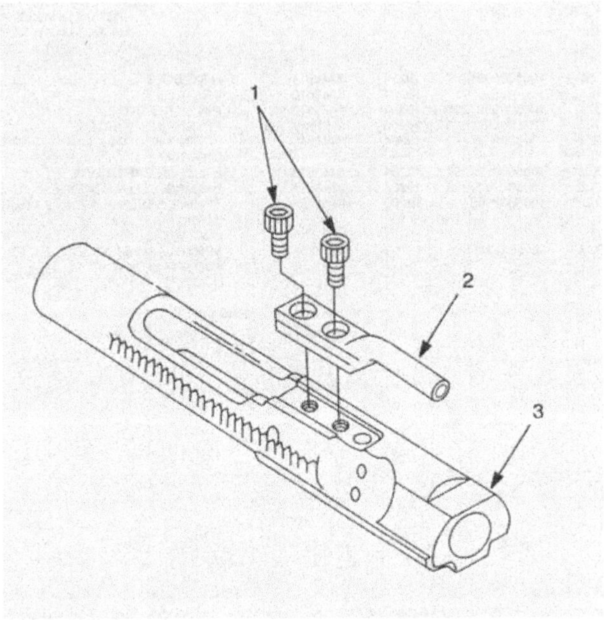

Figure C-4. Key and Bolt Carrier Assembly 8448505.

(1) ITEM NO	(2) SMR CODE	(3) NSN	(4) CAGEC	(5) PART NUMBER	(6) DESCRIPTION AND USABLE ON CODE (UOC)	(7) QTY
					GROUP 0102 KEY AND BOLT CARRIER ASSEMBLY 8448505 FIG. C-4. KEY AND BOLT CARRIER ASSEMBLY	
1	PAFZZ	5305009927284	19204	8448508	SCREW,CARRIER AND K KEY	2
2	PAFZZ	1005009927283	19200	8448506	KEY,BOLT CARRIER	1
3	PAFZZ		19200	8448507	CARRIER,BOLT	1

END OF FIGURE

SECTION II

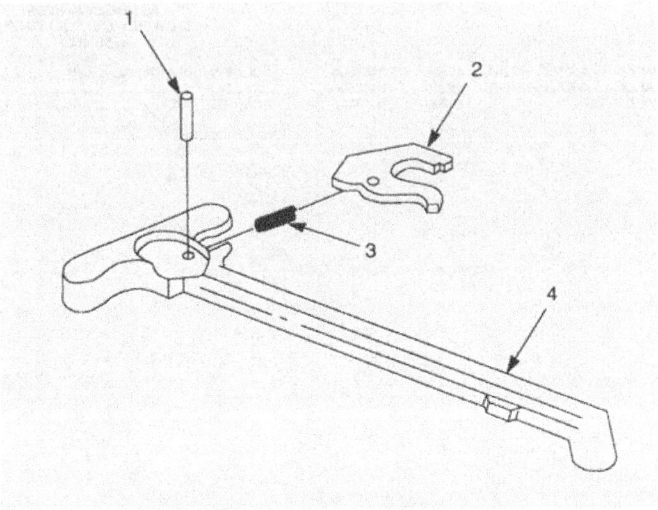

Figure C-5. Handle Assembly 8448517.

(1) ITEM NO	(2) SMR CODE	(3) NSN	(4) CAGEC	(5) PART NUMBER	(6) DESCRIPTION AND USABLE ON CODE (UOC)	(7) QTY
					GROUP 02 HANDLE ASSEMBLY 8448517 FIG. C-5. HANDLE ASSEMBLY	
1	PAOZZ	5315010489372	19204	8448521-2	PIN,SPRING CHARGING HANDLE......................................	1
2	PAOZZ	5340009990405	19200	8448519	LATCH,CHARGING HAND HANDLE....................................	1
3	PAOZZ	5360009990404	19204	8448520	SPRING,HELICAL,COMP COMPRESSION, ... CHARGING HANDLE	1
4	XAOZZ		19204	8448518	HANDLE ,CHARGING	1

END OF FIGURE

SECTION II

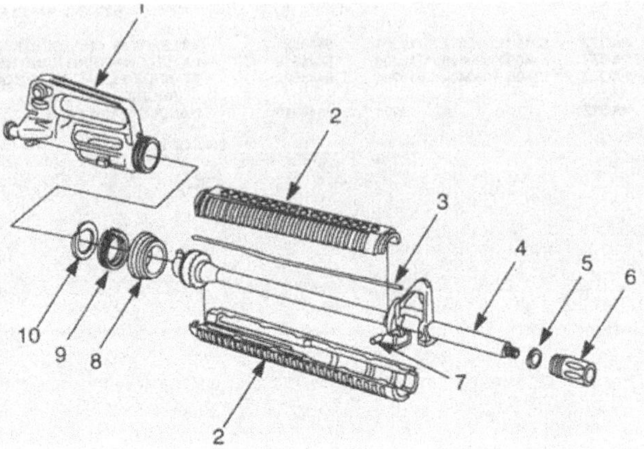

M16A2 Rifle

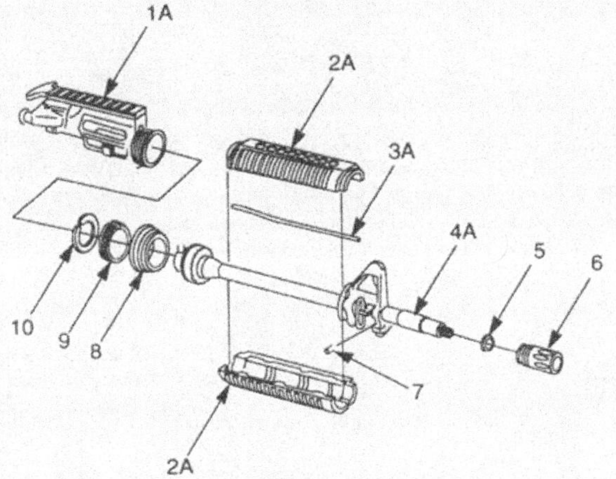

M4/M4A1 Carbine

Figure C-6. Upper Receiver and Barrel Assembly (M16A2) 9349050, and (M4 and M4A1) 12972680.

(1) ITEM NO	(2) SMR CODE	(3) NSN	(4) CAGEC	(5) PART NUMBER	(6) DESCRIPTION AND USABLE ON CODE (UOC)	(7) QTY
					GROUP 03 UPPER RECEIVER AND BARREL ASSEMBLY (M16A2) 9349050, (M4 AND M4A1] 12972680 FIG. C-6. UPPER RECEIVER AND BARREL ASSEMBLY	
1	AFFFF	19200	9349062	UPPER	RECEIVER ASSEMBLY (M16A2).. UOC:AR8	1
1A	AFFFF	19200	12972675	UPPER	RECEIVER ASSY (M4 AND M4A1)...................................... UOC:AS1,AY6	1
2	PAOZZ	1005011343629	19200	9349059	HANDGUARD ASSEMBLY (M16A2)................................... UOC:AR8	2
2A	PAOZZ	1005012342297	19200	9390003	GUARD,HAND,GUN (M4 AND M4A1)................................. UOC:AS1,AY6	2
3	PAFZZ	4710009781038	19200	8448567	TUBE,BENT,METALLIC (M16A2).. UOC:AR8	1
3A	PAFZZ	4710012338637	19200	9390016	TUBE,BENT,METALLIC (M4 AND M4A1) UOC:AS1,AY6	1
4	PAFFF	1005011467684	19200	9349124	BARREL ASSEMBLY (M16A2).. UOC:AR8	1
4A	PAFFF	1005012338529	19200	9390007	BARREL AND FRONT SI SIGHT ASSEMBLY, ... REPLACEMENT (M4 AND M4A1)... UOC:AS1,AY6	1
5	PAFZZ	5365011441496	19200	9349052	SHIM COMPENSATOR...	1
6	PAFZZ	1005011343633	19200	9349051	COMPENSATOR...	1
7	PAFZZ	5315000586044	96906	MS16562-106	PIN,SPRING ,GAS TUBE	1
8	PAFZZ	1005000878998	19204	8448712	RING,SLIP,HAND GUAR GUARD	1
9	PAFZZ	5360009781036	19204	8448555	SPRING,SLIP RING,HA HANDGUARD................................ UPPER RECEIVER...	1
10	PAFZZ	5325009990863	96906	M5S16626-3137	RING,RETAINING .... END OF FIGURE	1

SECTION II

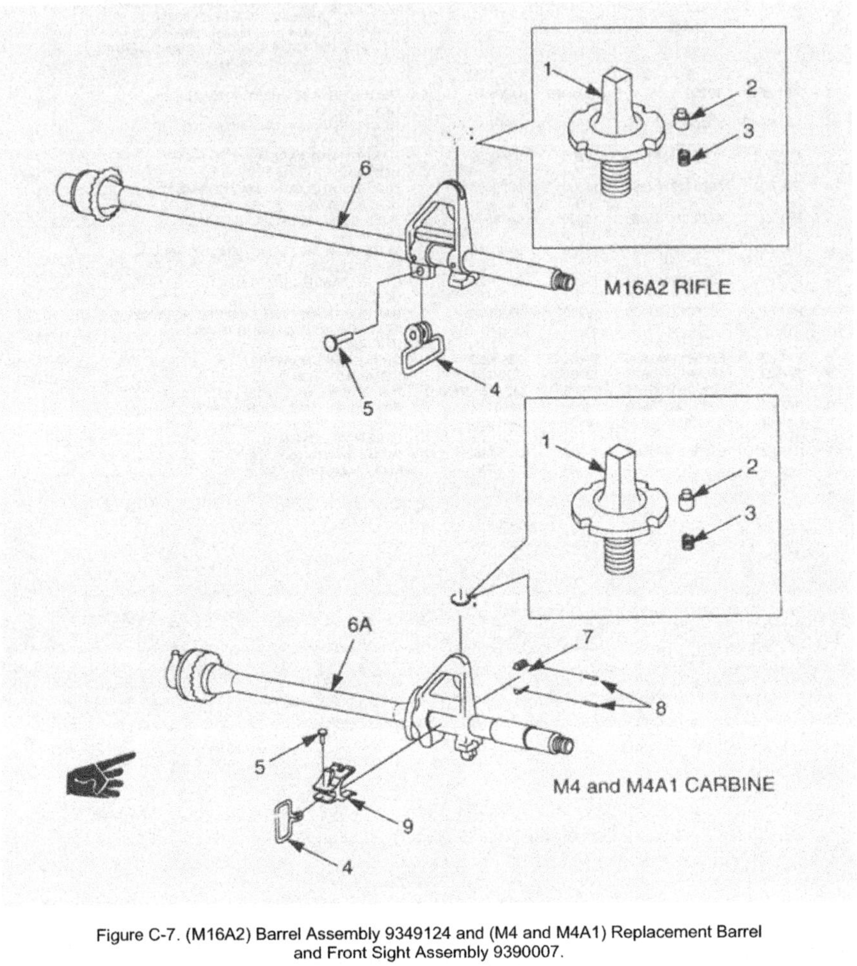

Figure C-7. (M16A2) Barrel Assembly 9349124 and (M4 and M4A1) Replacement Barrel and Front Sight Assembly 9390007.

(1) ITEM NO	(2) SMR CODE	(3) NSN	(4) CAGEC	(5) PART NUMBER	(6) DESCRIPTION AND USABLE ON CODE (UOC)	(7) QTY
					GROUP 0301 (M16A2) BARREL ASSEMBLY 9349124, AND (M4 AND M4A1) REPLACEMENT BARREL AND FRONT SIGHT ASSEMBLY 9390007 FIG. C-7. (M16A2) BARREL ASSEMBLY, AND (M4 AND M4A1) REPLACEMENT BARREL AND FRONT SIGHT ASSEMBLY	
1	PAOZZ	1005011343625	19200	9349056	POST,FRONT SIGHT	1
2	PAOZZ	5315009793930	19204	8448573	PIN,SHOULDER,HEADLE HEADLESS (DETENT, FRONT SIGHT)	1
3	PAOZZ	5360009793931	19204	8448574	SPRING,HELICAL,COMP ,COMPRESSION, FRONT SIGHT	1
4	PAOZZ	1005000179543	19204	8448571	SWIVEL,SLING,SMALL	1
5	PAOZZ	5320010637635	19204	8448697	RIVET,TUBULAR	1
6	XAFZZ		19200	9349054	BARREL AND BARREL E EXTENSION ASSEMBLY (M16A2) UOC:AR8	1
6A	XAFZZ		19200	9390009	BARREL AND BARREL EXTENSION ASSEMBLY (M4 AND M4A1) UOC:AS1,AY6	1
7	PAOZZ	5340012646530	19200	12598618	CLAMP,RIM CLENCHING (M4 AND M4A1) UOC:AS1,AY6	1
8	PAOZZ	5315006900544	96906	MS39086-93	PIN,SPRING (M4 AND M4A1) UOC:AS1,AY6	2
9	PAOZZ	1010012646517	19200	12598617	MOUNT,SWIVEL (M4 AND M4A1) UOC:AS1,AY6	1

END OF FIGURE

SECTION II

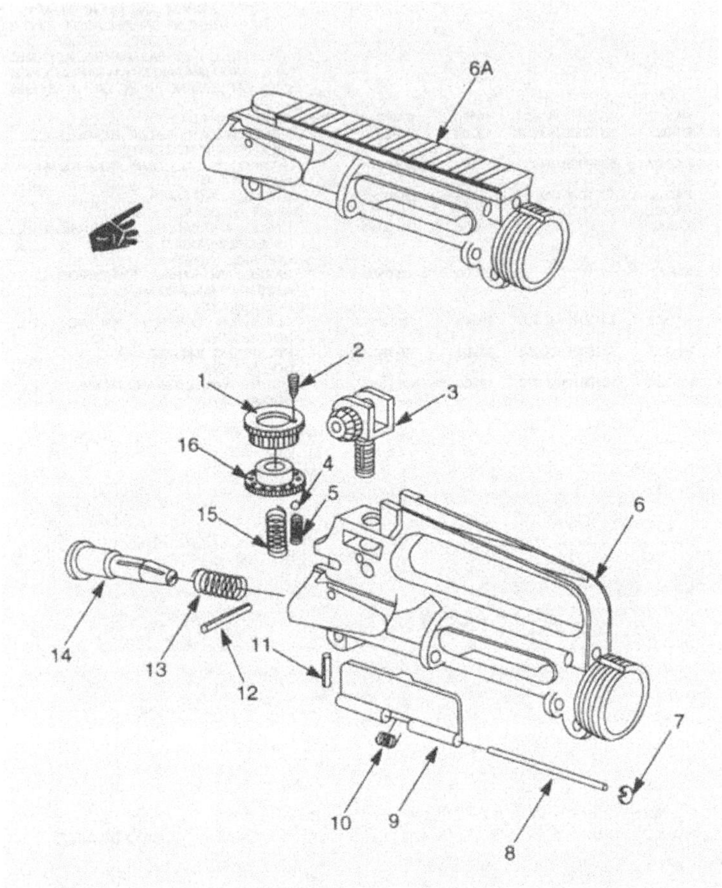

Figure C-8. Upper Receiver Assembly (M16A2) 9349062, and (M4 and M4A1) 12972675.

(1) ITEM NO	(2) SMR CODE	(3) NSN	(4) CAGEC	(5) PART NUMBER	(6) DESCRIPTION AND USABLE ON CODE (UOC)	(7) QTY
					GROUP 0302 UPPER RECEIVER ASSEMBLY (M16A2) 9349062, (M4 AND M4A1) 12972675 FIG. C-8. UPPER RECEIVER ASSEMBLY	
1	PAFZZ	1005011343621	19200	9349066	INDEX,ELEVATION (M16A2) UOC:AR8	1
2	PAFZZ	5305011343622	19200	9349065	SCREW,INDEX (M16A2)........	1
3	AFFFF	19200	9349072	REAR	SIGHT ASSEMBLY (M16A2)........ UOC:AR8	1
4	PAFZZ	3110001839175	96906	MS19060-4808	BALL,BEARING (M16A2) UOC:AR8	1
5	PAFZZ	5360011481751	19200	9349069	SPRING,HELICAL,COMP COMPRESSION, ... INDEX (M16A2)........ UOC:AR8	1
6	PAFZZ	1005011343701	19200	9349063	RECEIVER,UPPER (M16A2)........ UOC:AR8	1
6A	PAFZZ	1005013826795	19200	12972670	RECEIVER,CARTRIDGE (M4 AND M4A1)........ UOC:AS1,AY6	1
7	PAOZZ	5325009990864	96906	MS16632-3012	RING,RETAINING ,COVER........	1
8	PAOZZ	5315009781023	19204	8448533	PIN,GROOVED,HEADLES COVER	1
9	PAOZZ	1005009781022	19204	8448525	COVER,EJECTION PORT	1
10	PAOZZ	5360009781025	19204	8448532	SPRING,HELICAL,TORS ,COVER........	1
11	PAFZZ	5315008403812	96906	M5S16562-121	PIN,SPRING FORWARD ASSIST	1
12	PAFZZ	5315008403812	96906	M5S16562-121	PIN,SPRING (M16A2)	1
13	PAFZZ	5360000179541	19200	8448540	SPRING,HELICAL,COMP COMPRESSION .... FORWARD ASSIST	1
14	AFFFF		19204	9349086	FORWARD ASSIST ASSY ASSEMBLY	1
15	PAFZZ	5360011343710	19200	9349070	SPRING,HELICAL,COMP COMPRESSION, ... ELEVATION (M16A2)........	1
16	PAFZZ	5355011354972	19200	9349067	KNOB,ELEVATION (M16A2)........ UOC:AR8	1

END OF FIGURE

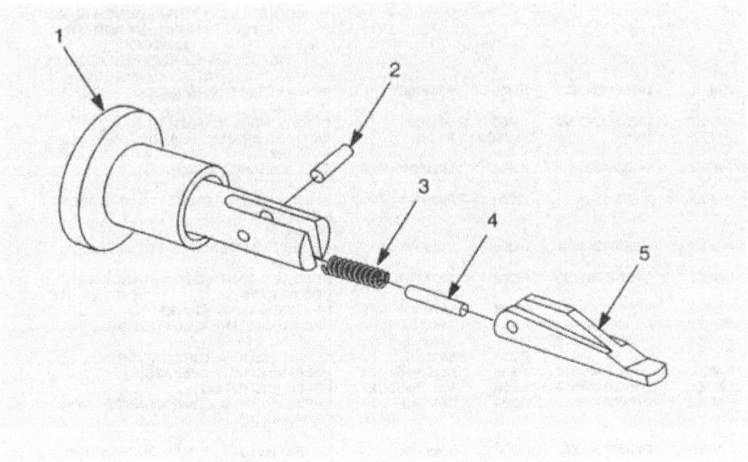

Figure C-9. Forward Assist Assembly 9349086.

(1) ITEM NO	(2) SMR CODE	(3) NSN	(4) CAGEC	(5) PART NUMBER	(6) DESCRIPTION AND USABLE ON CODE (UOC)	(7) QTY
					GROUP 030201 FORWARD ASSIST ASSEMBLY 9349086	
					FIG. C-9. FORWARD ASSIST ASSEMBLY	
1	PAFZZ	1005011441468	19200	9349085	PLUNGER ASSEMBLY	1
2	PAFZZ	5315010489372	19204	8448521-2	PIN,SPRING ,PAWL	1
3	PAFZZ	5360005238084	19200	8448542	SPRING,HELICAL,COMP COMPRESSION PAWL	1
4	PAFZZ	1005000179540	19204	8448544	DETENT,PAWL	1
5	PAFZZ	3040000179539	19204	8448543	PAWL,FORWARD ASSIST	1

END OF FIGURE

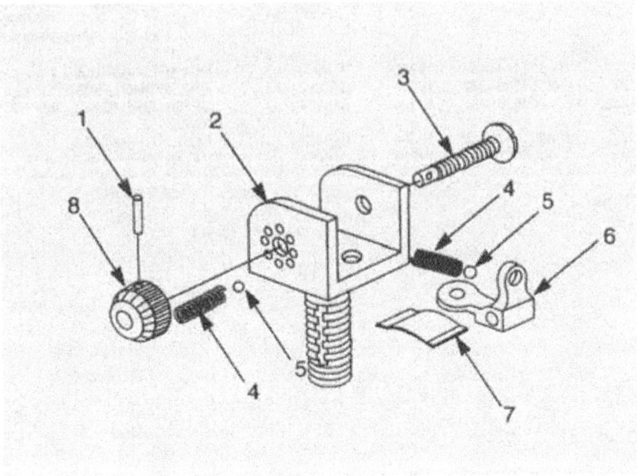

Figure C-10. (M16A2) Rear Sight Assembly 9349072.

(1) ITEM NO	(2) SMR CODE	(3) NSN	(4) CAGEC	(5) PART NUMBER	(6) DESCRIPTION AND USABLE ON CODE (UOC)	(7) QTY
					GROUP 030202 (M16A2) REAR SIGHT ASSEMBLY 9349072 FIG. C-10. REAR SIGHT ASSEMBLY	
1	PAFZZ	5315000586678	96906	MS16562-103	PIN,SPRING ,WINDAGE	1
2	PAFZZ	1005011343631	19200	9349074	BASE,REAR SIGHT UOC:ARB	1
3	PAFZZ	5305011441490	19200	9349076	SCREW,EXTERNALLY RE	1
4	PAFZZ	5360011481751	19200	9349069	SPRING,HELICAL,COMP ,REAR SIGHT UOC:AR8	2
5	PAFZZ	3110001839175	96906	M5S19060-4808	BALL,BEARING UOC:AR8	2
6	PAFZZ	1005011353697	19200	9349075	APERTURE,SIGHT	1
7	PAFZZ	5360013816183	19200	12011987	SPRING,FLAT REAR SIGHT	1
8	PAFZZ	5355011343627	19200	9349077	KNOB,WINDAGE	1

END OF FIGURE

SECTION II

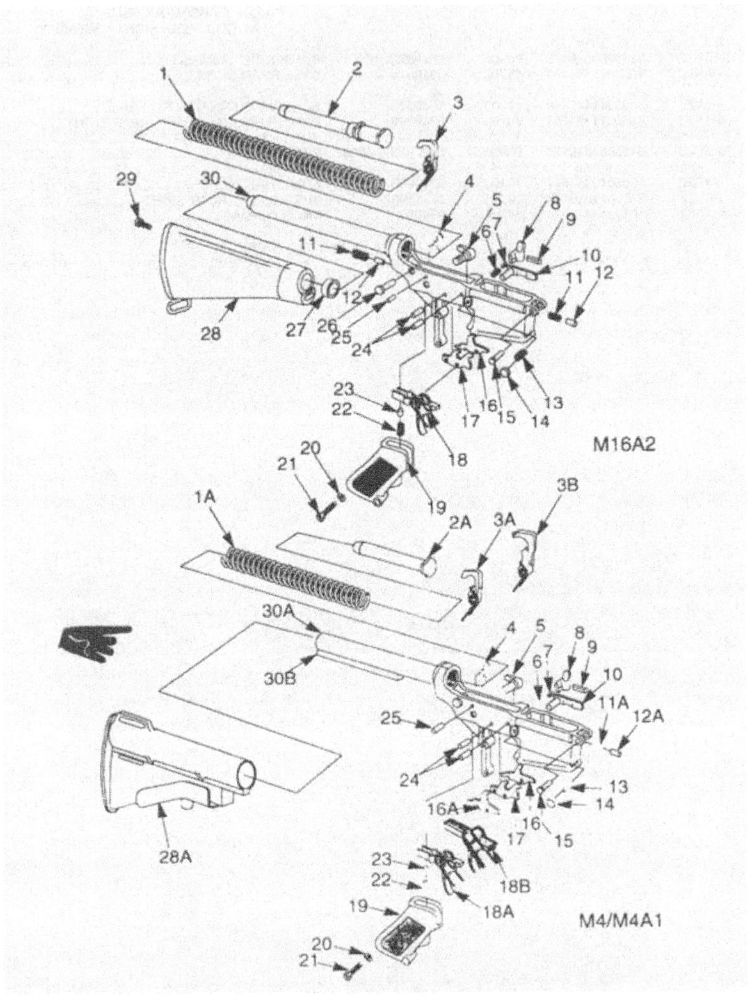

Figure C-11. Lower Receiver and Buttstock Assembly (MI 6A2) 9349100, (M4) 9390011 and (M4A1) 12972690.

(1) ITEM NO	(2) SMR CODE	(3) NSN	(4) CAGEC	(5) PART NUMBER	(6) DESCRIPTION AND USABLE ON CODE (UOC)	(7) QTY
					GROUP 04 LOWER RECEIVER AND BUTTSTOCK ASSEMBLY (M16A2) 9349100, (M4) 9390011, AND (M4A1) 12972690 FIG C-11 LOWER RECEIVER AND BUTTSTOCK ASSEMBLY	
1	PAOZZ	5360009926665	19204	8448629	SPRING, HELICAL, COMP ACTION (M16A2) UOC:AR8	1
1A	PAOZZ	5360012338617	19200	9390022	SPRING, HELICAL, COMP COMPRESSION (M4 AND M4A1) UOC:AS1, AY6	1
2	PAOZZ	1005009373078	19200	8448615	BUFFER ASSEMBLY (MI6A2) UOC:AR8	1
2A	PAOZZ	1005012313138	19200	9390023	BUFFER ASSEMBLY (M4 AND M4AI) UOC:AS1, AY6	1
3	AFFFF		19200	9349106	HAMMER ASSEMBLY (M16A2) UOC:AR8	1
3A	AFFFD		19200	9390032	HAMMER ASSEMBLY (M4) UOC:AS1	1
3B	AFFFF		19204	8448610	HAMMER ASSY (M4A1) UOC:AY6	1
4	PAFZZ	1005009926649	19200	8448595	SEAR	1
5	PAFZZ	5340012258339	19200	9381367	SELECTOR, FIRE CONTROL	1
6	PAFZZ	5360000562246	19204	8448633	SPRING, HELICAL, COMP COMPRESSION, BOLT CATCH	1
7	PAFZZ	1005000562247	19204	8448634	PLUNGER, BOLT CATCH	1
8	PAFZZ	1005000179548	19200	8448628	CATCH, BOLT	1
9	PAFZZ	5315008123312	96906119	MS16562-119	PIN, SPRING , BOLT CATCH	1
10	PAFZZ	1005000562201	19204	8448638	CATCH, MAGAZINE	1
11	PAOZZ	5360009926655	19204	8448586	SPRING, HELICAL, COMP COMPRESSION, TAKE DOWN/PIVOT PIN (M16A2) UOC:AR8	2
11A	PAOZZ	5360009926655	19204	8448586	SPRING, HELICAL, COMP COMPRESSION (M4 AND M4AI) UOC:AS1	1
12	PAOZZ	5315009926654	19204	8448585	PIN, STRAIGHT, HEADLE DETENT, TAKEDOWN PIN (M16A2) UOC:ARS	2
12A	PAOZZ	5315009926654	19204	8448585	PIN, STRAIGHT, HEADLE HEADLESS (M4 AND M4A1) UOC:AS1, AY6	1
13	PAFZZ	5360009927301	19204	8448637	SPRING, HELICAL, COMP COMPRESSION, MAGAZINE CATCH	1
14	PAFZZ	1005009927302	19204	8448636	BUTTON, MAGAZINE CAT CATCH	1
15	PAOZZ	5315000179537	19204	8448621	PIN, GROOVED, HEADED (PIVOT PIN)	1
16	PAFZZ	5340011457910	19200	9349114	LEVER, LOCK-RELEASE SEMI (M16A2 AND M4) UOC:AR8, AS1	1
16A	PAFZZ	1005009990406	19200	8448635	DISCONNECTOR (M4A1) UOC:AY6	1
17	PAFZZ	5340011441499	19200	9349113	LEVER, LOCK-RELEASE , BURST (MI6A2 AND M4)	1

(1) ITEM NO	(2) SMR CODE	(3) NSN	(4) CAGEC	(5) PART NUMBER	(6) DESCRIPTION AND USABLE ON CODE (UOC)	(7) QTY
					UOC:AR8, AS1	
18	AFFFF		19200	9349115	TRIGGER ASSEMBLY (M16A2) UOC:AR8	1
18A	AFFFF		19200	12972697	TRIGGER ASSEMBLY (M4) UOC:AS1	1
18B	AFFFF		19200	12972698	TRIGGER ASSEMBLY (M4A1) UOC:AY6	1
19	PAOZZ	1005011484805	19200	9349127	GRIP, RIFLE PLASTIC, BLACK	1
20	PAOZZ	5310005273634	96906	MS35335-61	WASHER, LOCK , RIFLE GRIP	1
21	PAOZZ	5305012681191	88044	AN501D416-18	SCREW, MACHINE , RIFLE GRIP	1
22	PAOZZ	5360009927292	19204	8448516	SPRING, HELICAL, COMP SAFETY	1
23	PAOZZ	1005009926667	19204	8448631	DETENT, SAFETY	1
24	PAFZZ	5315009927309	19204	8448609	PIN, GROOVED, HEADLES TRIGGER AND HAMMER	2
25	PAFZZ	5315009926650	19204	8448599	PIN, GROOVED, HEADLES , AUTOMATIC SEAR	1
26	PAFZZ	5315009926653	19204	8448584	PIN, GROOVED, HEADED (M16A2)	1
27	PAOZZ	5365012672169	19200	12597640	SPACER, STEPPED (M16A2) UOC:AR8	1
28	PAOOO	1005011354973	19200	9349119	BUTTSTOCK ASSEMBLY, (M16A2) UOC:AR8	1
28A	AOOOO		19200	9390012	BUTTSTOCK ASSY (M4 AND M4A1) UOC:AS1, AY6	1
29	PAOZZ	5305011478585	19200	9349128	SCREW, MACHINE , BUTTCAP (M16A2) UOC:AR8	1
30	XAFFA		19200	9349101	LOWER RECEIVER AND RECEIVER EXTENSION ASSEMBLY (M16A2) UOC:AR8	1
30A	XAFFA		99999	NPN	LOWER RECEIVER ASSY ASSEMBLY (M4) UOC:AS1	1
30B	XAFFA		99999	NPN	LOWER RECEIVER ASSY (M4AI) UOC:AY6	1
					END OF FIGURE	

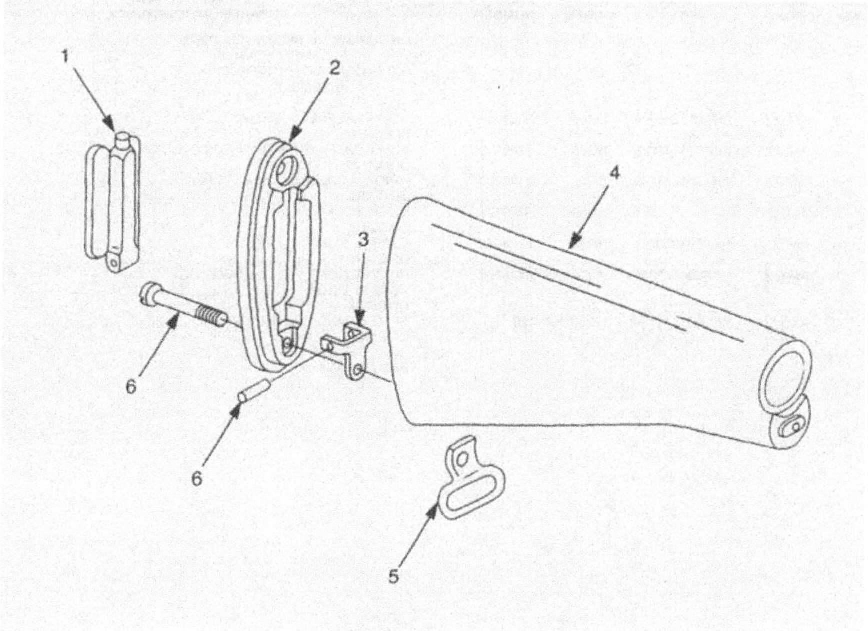

Figure C-12. (M16A2) Buttstock Assembly 9349119.

(1) ITEM NO	(2) SMR CODE	(3) NSN	(4) CAGEC	(5) PART NUMBER	(6) DESCRIPTION AND USABLE ON CODE (UOC)	(7) QTY
					GROUP 0401 (M16A2) BUTTSTOCK ASSEMBLY 9349119	
					FIG C-12 (M16A2) BUTTSTOCK ASSEMBLY	
1	PAOZZ	1005012288504	19200	9381380	DOOR ASSEMBLY, THUMB UOC:AR8	1
2	PAOZZ	1005011467685	19200	9349130	PLATE, BUTT, SHOULDER GUN STOCK UOC:AR8	1
3	PAOZZ	5340004633892	19200	8448653	HINGE, ACCESS DOOR BUTT PLATE UOC:AR8	1
4	XAOZZ		19200	9349121	BUTTSTOCK UOC:AR8	1
5	PAOZZ	1005004030964	19204	8448652	SWIVEL, SLING, SMALL UOC:AR8	1
6	PAOZZ	5315004633894	19204	8448655	PIN, STRAIGHT, HEADLE HEADLESS, ACCESS DOOR UOC:AR8	1
7	PAOZZ	5305011441494	19200 9349120		SCREW, MACHINE BUTT PLATE UOC:AR8	1

END OF FIGURE

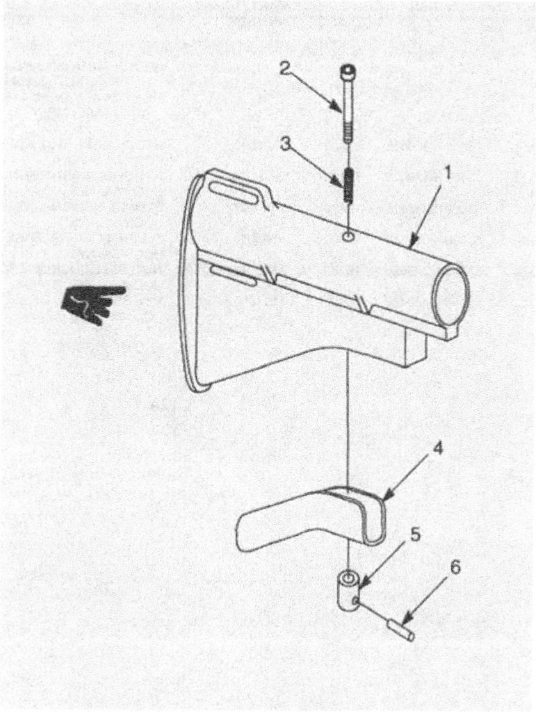

Figure C12A. (M4 and M4A1) Buttstock Assembly 9390012.

(1) ITEM NO	(2) SMR CODE	(3) NSN	(4) CAGEC	(5) PART NUMBER	(6) DESCRIPTION AND USABLE ON CODE (UOC)	(7) QTY
					GROUP 0401A (M4 AND M4A1) 9390012 BUTTSTOCK ASSEMBLY FIG C12A (M4 AND M4A1) BUTTSTOCK ASSEMBLY	
1	PAOZZ	1005012338636	19200	9390013	STOCK, GUN, SHOULDER UOC:AS1, AY6	1
2	PAOZZ	5315012338608	19200	9390025	PIN, SHOULDER, HEADLE UOC:AS1, AY6	1
3	PAOZZ	5360012338616	19200	9390027	SPRING, HELICAL, COMP COMPRESSION UOC:AS1, AY6	1
4	PAOZZ	5340012338638	19200	9390014	LEVER, LOCK-RELEASE UOC:AS1, AY6	1
5	PAOZZ	5310012338626	19200	9390026	NUT, SELF-LOCKING, RO UOC:AS1, AY6	1
6	PAOZZ	5315008439487	96906	MS16562-202	PIN, SPRING UOC:AS1, AY6	1

END OF FIGURE

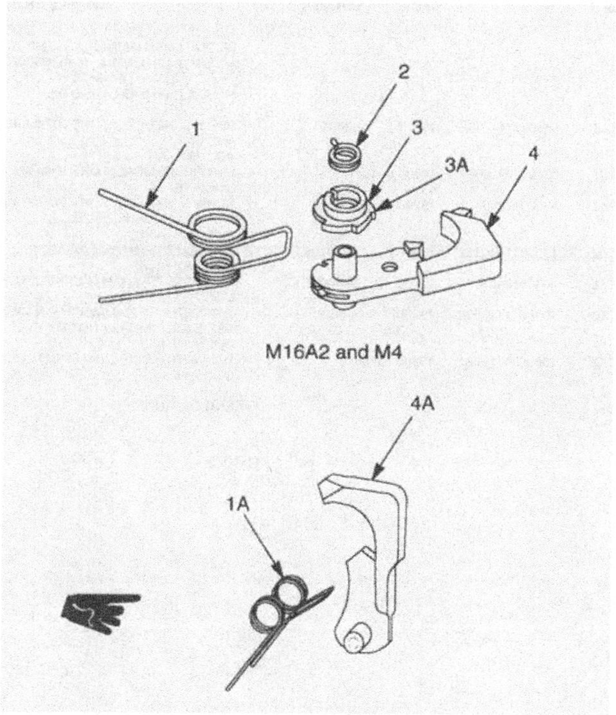

Figure C-13. Hammer Assembly(M16A2) 9349106, (M4) 9390032, and (M4A1) 8448610.

(1) ITEM NO	(2) SMR CODE	(3) NSN	(4) CAGEC	(5) PART NUMBER	(6) DESCRIPTION AND USABLE ON CODE (UOC)	(7) QTY
					GROUP 0402 HAMMER ASSEMBLY (M16A2) 9349106, (M4) 9390032, AND (M4A1) 8448610 FIG C-13 HAMMER ASSEMBLY	
1	PAFZZ	5360011441492	19200	9349107	SPRING, HELICAL, TORS HAMMER (MI6A2 AND M4) UOC:AR8, AS1	1
1A	PAFZZ	5360009926648	19204	8448611	SPRING, HELICAL, TORS (M4A1) UOC:AY6	1
2	PAFZZ	5360011365471	19200	9349109	SPRING, HELICAL, TORS BURST CAM (M16A2 AND M4) UOC:AR8, AS1	1
3	PAFZZ	1005011480172	19200	9349108	CAM, BURST (M16A2) (BLACK) UOC:AR8	1
3A	PAFZZ	3040012477969	19200	9390031	CAM, CONTROL (M4) (NICKEL/SHINY) UOC:AS1	1
4	PAFZZ	1005011343630	19200	9349110	HAMMER AND HAMMER P PIN RETAINER ASSEMBLY (MI6A2 AND M4) UOC:AR8, AS1	1
4A	PAFZZ	1005000179551	19200	8448612	HAMMER, FIRING, SMALL (M4A1) UOC:AY6	1
					END OF FIGURE	

SECTION II

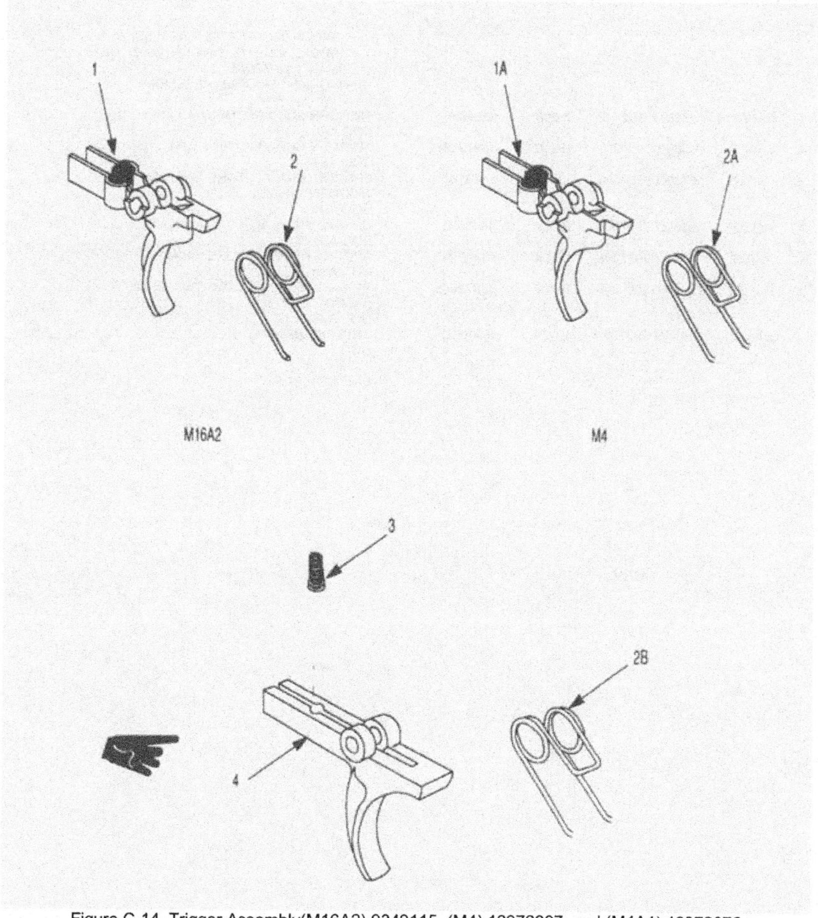

Figure C-14. Trigger Assembly(M16A2) 9349115, (M4) 12972697, and (M4A1) 12972676

(1) ITEM NO	(2) SMR CODE	(3) NSN	(4) CAGEC	(5) PART NUMBER	(6) DESCRIPTION AND USABLE ON CODE (UOC)	(7) QTY
					GROUP 0403 TRIGGER ASSEMBLY (M16A2) 9349115, (M4) 12972697, AND (M4A1) 12972676 FIG C-14 TRIGGER ASSEMBLY	
1	PAFFF	1005012192402	19200	9392518	TRIGGER SUB ASSEMBL (M16A2)...................................... UOC:AR8	1
1A	PAFFF	1005013954257	19200	12972696	TRIGGER SUBASSEMBLY (M4)... UOC:AS1	1
2	PAFZZ	5360009927308	19204	8448593	SPRING, HELICAL, TORS TORSION, ... TRIGGER (MI6A2)... UOC:AR8	1
2A	PAFZZ	5360009927308	19204	8448593	SPRING, HELICAL, TORS (M4)... UOC:ASI	1
2B	PAFZZ	5360009927308	19204	8448593	SPRING, HELICAL, TORS (M4A1).. UOC:AY6	1
3	PAFZZ	5360013960256	19200	12972695	SPRING, HELICAL, COMP DISCONNECT .... (M4A1)... UOC:AY6	1
4	PAFZZ	1005009927307	19204	8448592	TRIGGER (M4A1).. UOC:AY6	1

END OF FIGURE

SECTION II

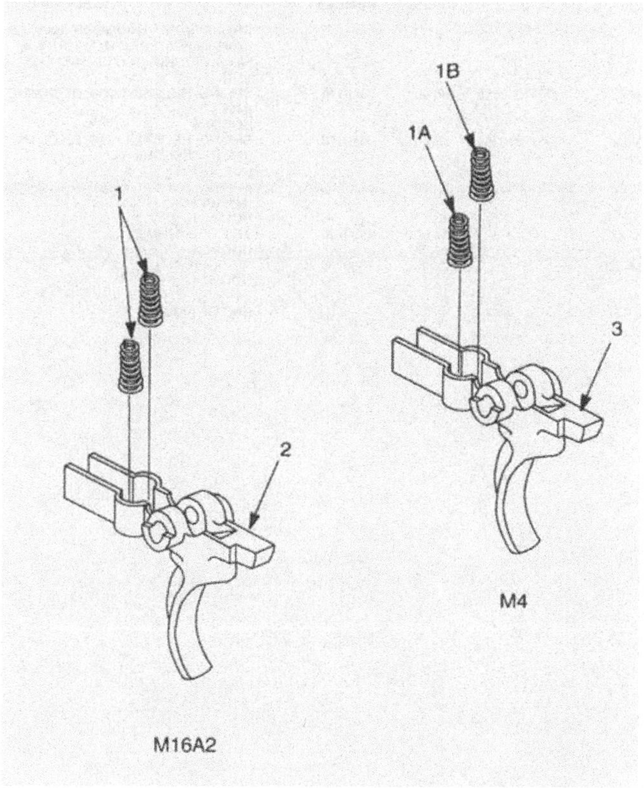

Figure C-15. Trigger Subassembly(M16A2) 9392518, (M4) 12972696.

(1) ITEM NO	(2) SMR CODE	(3) NSN	(4) CAGEC	(5) PART NUMBER	(6) DESCRIPTION AND USABLE ON CODE (UOC)	(7) QTY
					GROUP 040301 TRIGGER SUBASSEMBLY (M6A2) 9392518 AND (M4) 12972696 FIG C-15 TRIGGER SUBASSEMBLY	
1	PAFZZ	5360011350353	19200	9349116	SPRING, HELICAL, COMP DISCONNECT .... (M16A2).. UOC:AR8	2
1A	PAFZZ	5360011350353	19200	9349116	SPRING, HELICAL, COMP DISCONNECT .... (M4) (NICKEL/SHINY) UOC:AS1	1
1B	PAFZZ	5360013960256	19200	12972695	SPRING, HELICAL, COMP DISCONNECT .... (M4) (BLACK) UOC:AS1	1
2	XAFZZ		19200	9390736	TRIGGER (M16A2)... UOC:AR8	1
3	XAFZZ		19200	9390736	TRIGGER (M4) UOC:AS1	1

END OF FIGURE

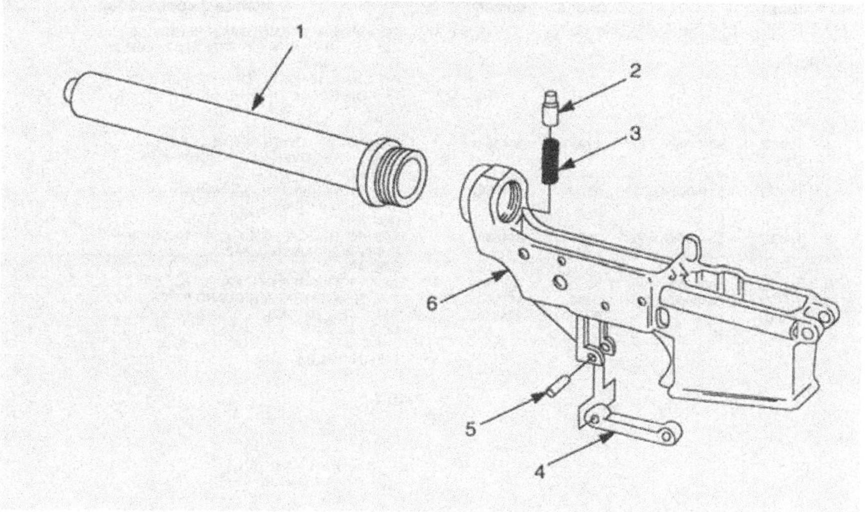

Figure C-16. (M16A2) Lower Receiver and Receiver extension Assembly 9349101.

(1) ITEM NO	(2) SMR CODE	(3) NSN	(4) CAGEC	(5) PART NUMBER	(6) DESCRIPTION AND USABLE ON CODE (UOC)	(7) QTY
					GROUP 0404 (M16A2) LOWER RECEIVER AND RECEIVER EXTENSION ASSEMBLY 9349101 FIG C-16 (M16A2) LOWER RECEIVER AND RECEIVER EXTENSION ASSEMBLY	
1	PAFZZ	5340009927297	19200	8448581	EXTENSION, LOWER REC RECEIVER .... (LOWER RECEIVER EXTENSION) (M1SA2)........................ UOC:AR8	1
2	PAFZZ	5315009926651	19204	8448582	PIN, SHOULDER, HEADLE BUFFER.................................... RETAINER (M16A2).. UOC:AR8	1
3	PAFZZ	5360009926652	19200	8448583	SPRING, HELICAL, COMP COMPRESSION, ... BUFFER RETAINER (NI6A2) UOC:AR8	1
4	PAFZZ	1005009927299	19204	8448587	GUARD, TRIGGER (M16A2)...	1
5	PAFZZ	5315000586081	96906	MS16562-129	PIN, SPRING TRIGGER GUARD (M16A2)............................	1
6	XAFDA		19200	9349102	RECEIVER, (M16A2).. UOC:AR8	1

END OF FIGURE

SECTION II

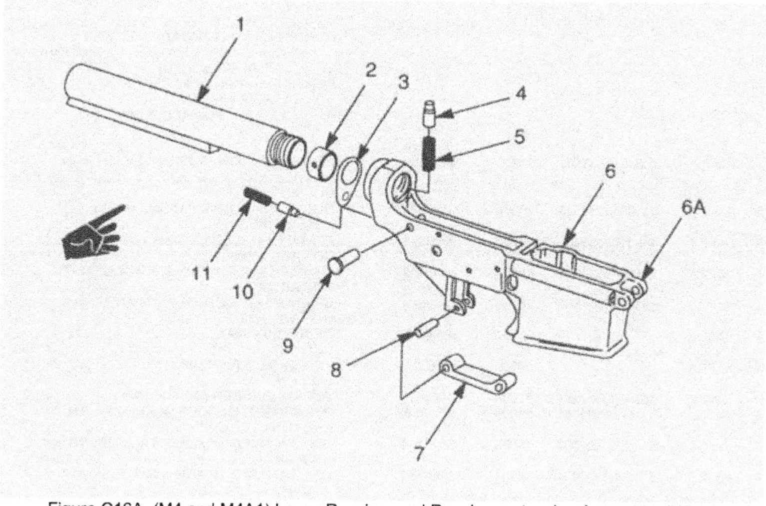

Figure C16A. (M4 and M4A1) Lower Receiver and Receiver extension Assembly, (NPN)

(1) ITEM NO	(2) SMR CODE	(3) NSN	(4) CAGEC	(5) PART NUMBER	(6) DESCRIPTION AND USABLE ON CODE (UOC)	(7) QTY
					GROUP 0404A (M4 AND M4A1) LOWER RECEIVER AND RECEIVER EXTENSION ASSEMBLY (NPN) FIG C16A (M4 AND M4A1) LOWER RECEIVER AND RECEIVER EXTENSION ASSEMBLY	
1	PAFZZ	1005012338531	19200	9390019	EXTENSION, LOWER RECEIVER (M4 AND M4A1).... UOC:AS1, AY6	1
2	PAFZZ	5310012338625	19200	9390020	NUT, PLAIN, ROUND (M4 AND M4A1) UOC:AS1, AY6	1
3	PAFZZ	105012338530	19200	9390021	PLATE, RECEIVER END (M4 AND M4A1).... UOC:AS1, AY6	1
4	PAFZZ	5315009926651	19204	8448582	PIN, SHOULDER, HEADLE (M4 AND M4A1).... UOC:AS1, AY6	1
5	PAFZZ	5360009926652	19200	8448583	SPRING, HELICAL, COMP (M4 AND M4A1).... UOC:AS1, AY6	1
6	XAFDA		19200	9390015	RECEIVER, (M4).... UOC:AS1	1
6A	XAFDA		19200	12972652	LOWER RECEIVER (M4A1).... UOC:AY6	1
7	PAFZZ	1005009927299	19204	8448587	GUARD, TRIGGER (M4 AND M4A1)....	1
8	PAFZZ	5315000586081	96906	MS16562-129	PIN, SPRING (TRIGGER GUARD PIN) (M4 AND M4A1)....	1
9	PAFZZ	5315009926653	19204	8448584	PIN, GROOVED, HEADED (TAKE DOWN) (M4 .... AND M4A1)....	1
10	PAFZZ	5315009926654	19204	8448585	PIN, STRAIGHT, HEADLE HEADLESS (M4 AND M4A1).... UOC:AS1, AY6	1
11	PAFZZ	5360009926655	19204	8448586	SPRING, HELICAL, COMP COMPRESSION (M4 AND M4A1).... UOC:AS1, AY6	1
					END OF FIGURE	

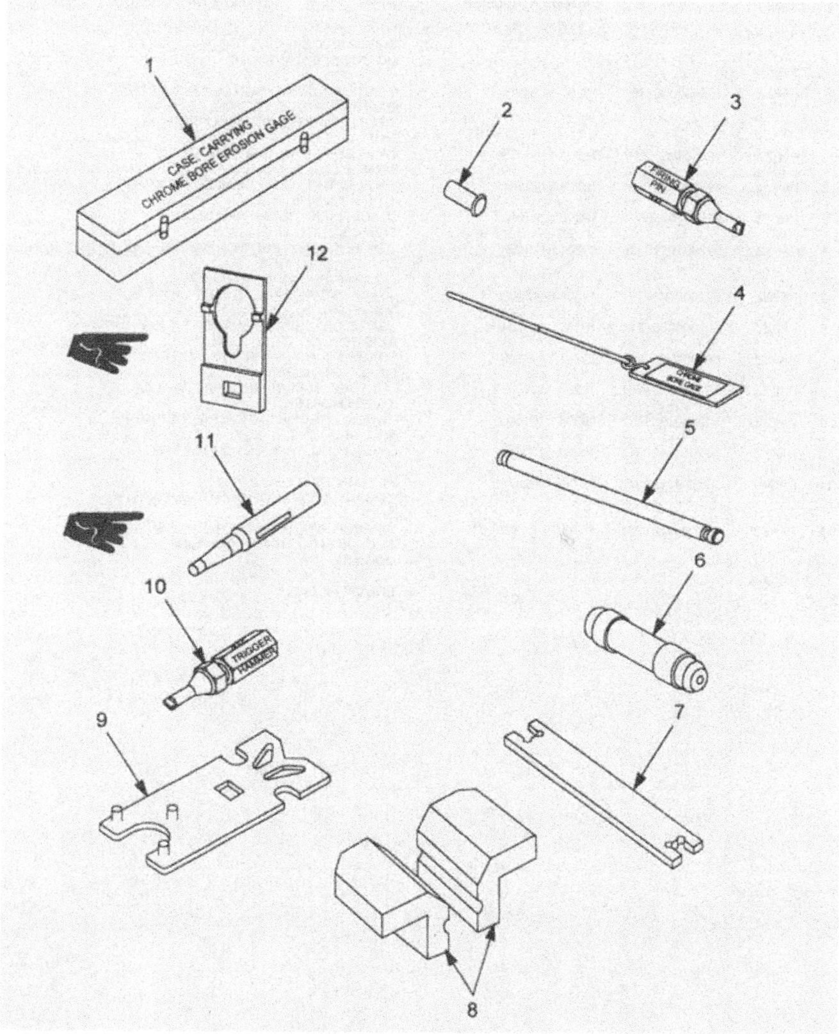

Figure C-17. Special Tools

(1) ITEM NO	(2) SMR CODE	(3) NSN	(4) CAGEC	(5) PART NUMBER	(6) DESCRIPTION AND USABLE ON CODE (UOC)	(7) QTY
					GROUP 9500 SPECIAL TOOLS FIG C-17 SPECIAL TOOLS	
	PEFZZ	1005000567106	19204	8426685	MAINTENANCE KIT, GUN DS/GS SUPPORT MAINTENANCE FOR 5.S6MM RIFLE, M16 RIFLE SERIES BOI: 2 PER SUPPORTING DSU/GSU	
1	PAFZZ	4933010355607	19204	12006359	CASE, BORE GAGE PART OF KIT P/N 8426685........	
2	PAFZZ	4933008007508	19204	8448201	..REFLECTOR TOOL, CHAMBER, PART OF KIT P/N 8426685	
3	PAFZZ	5220010755004	19200	12620101	..GAGE, PLUG, PLAIN PART OF KIT P/N 8426685........	
4	PAFZZ	5220010148183	19204	8448496	..GAGE, BARREL, EROSION BARREL EROSION PART OF KIT P/N 8426685(CHROME BARREL)........	
5	PAFZZ	5220002219391	19204	8448202	..GAGE, STRAIGHTNESS PART OF KIT P/N 8426685	
6	PAFZZ	5220000707814	19204	7799734	..GAGE, HEADSPACE PART OF KIT P/N 8426685........	
7	PAFZZ	6695000707815	19204	7799735	..GAGE, FIRING PIN PROTRUSION PART OF KIT P/N 8426685	
8	PAFZZ	4933000709151	19204	11010032	..FIXTURE, BARREL REMOVAL PART OF KIT P/N 8426685	
9	PAFZZ	5120000709152	19204	11010033	..WRENCH, COMBINATION PART OF KIT P/N 8426685	
10	PAFZZ	5220010439473	19204	12006472	GAGE, PLUG, TAPER CYLINDER PART OF KIT P/N 8426685	
11	PAOZZ	5315013100370	19200	12926769	KEY, MACHINE KEY, 2 PER ORGANIZATIONAL OR DIRECT SUPPORT SHOP	
12	PAFZZ	5120013246631	19200	9390035	WRENCH, SPANNER 2 PER DS/GS SUPPORT SHOP A/R FOR M4/M4A1 CARBINE UOC:AS1, AY6	
					END OF FIGURE	

CROSS-REFERENCE INDEXES

NATIONAL STOCK NUMBER INDEX

STOCK NUMBER	FIG	ITEM
5315-00-017-9537	C-11	15
3040-00-017-9539	C-9	5
1005-00-017-9540	C-9	4
5360-00-017-9541	C-8	13
1005-00-017-9543	C-7	4
1005-00-017-9546	C-1	1
1005-00-017-9547	C-2	1
1005-00-017-9548	C-11	8
1005-00-017-9551	C-13	4A
1005-00-056-2201	C-11	10
5360-00-056-2246	C-11	6
1005-00-056-2247	C-11	7
1005-00-056-7106	C-17	
5315-00-058-6044	C-6	7
5315-00-058-6081	C-16	5
	C16A	8
5315-00-058-6678	C-1B	1
	C-10	1
5220-00-070-7814	C-17	6
6695-00-070-7815	C-17	7
4933-00-070-9151	C-17	8
5120-00-070-9152	C-17	9
1005-00-087-8998	C-6	8
3110-00-183-9175	C-8	4
	C-1A	6
	C-1B	5
	C-10	5
5220-00-221-9391	C-17	5
1005-00-403-0964	C-12	5
5340-00-463-3892	C-12	3
5315-00-463-3894	C-12	6
5360-00-523-8084	C-9	3
5310-00-527-3634	C-11	20
5315-00-597-5086	C-3	4
5315-00-690-0544	C-7	8
1005-00-738-6213	C-4	3
1005-00-760-3768	C-3	8
4933-00-800-7508	C-17	2
5315-00-812-3312	C-11	9
5315-00-840-3812	C-8	11
	C-8	12
	C-1A	8
5315-00-843-9487	C12A	6
1005-00-921-5004	C-1	4
1005-00-937-3078	C-11	2
1005-00-978-1022	C-8	9
5315-00-978-1023	C-8	8
5360-00-978-1025	C-8	10
5360-00-978-1036	C-6	9
4710-00-978-1038	C-6	3
5315-00-979-3930	C-7	2
5360-00-979-3931	C-7	3
5360-00-992-6648	C-13	1A
1005-00-992-6649	C-11	4
5315-00-992-6650	C-11	25
5315-00-992-6651	C-16	2
	C16A	4
5360-00-992-6652	C-16	3
	C16A	5
5315-00-992-6653	C-11	26
	C16A	9
5315-00-992-6654	C-11	12
	C-11	12A
	C16A	10
5360-00-992-6655	C-11	11
	C-11	11A
	C16A	11
5360-00-992-6665	C-11	1
1005-00-992-6667	C-11	23
1005-00-992-7283	C-4	2
5305-00-992-7284	C-4	1
1005-00-992-7285	C-2	4
1005-00-992-7287	C-3	1
1005-00-992-7288	C-3	7
1005-00-992-7290	C-3	3
1005-00-992-7291	C-3	6
5360-00-992-7292	C-3	5
	C-11	22
5315-00-992-7294	C-2	3
5340-00-992-7297	C-16	1
1005-00-992-7299	C-16	4
	C16A	7
5360-00-992-7301	C-11	13
1005-00-992-7302	C-11	14
1005-00-992-7307	C-14	4
5360-00-992-7308	C-14	2
	C-14	2A
	C-14	2B
5315-00-992-7309	C-11	24
5360-00-999-0404	C-5	3
5340-00-999-0405	C-5	2
1005-00-999-0406	C-11	16A
5325-00-999-0863	C-6	10
5325-00-999-0864	C-8	7
1005-00-999-1509	'C-2	2
5220-01-014-8183	C-17	4
4933-01-035-5607	C-17	1
5220-01-043-9473	C-17	10
5315-01-048-9372	C-5	1
	C-9	2

CROSS-REFERENCE INDEXES

NATIONAL STOCK NUMBER INDEX

STOCK NUMBER	FIG	ITEM
5320-01-063-7635	C-7	5
5220-01-075-5004	C-17	3
1005-01-134-3621	C-8	1
5305-01-134-3622	C-8	2
	C-1A	2
1005-01-134-3625	C-7	1
5355-01-134-3627	C-1B	8
	C-10	8
1005-01-134-3629	C-6	2
1005-01-134-3630	C-13	4
1005-01-134-3631	C-10	2
1005-01-134-3633	C-6	6
1005-01-134-3701	C-8	6
5360-01-134-3710	C-8	15
	C-1A	5
5360-01-135-0353	C-15	1
	C-15	1A
1005-01-135-3697	C-1B	6
	C-10	6
5355-01-135-4972	C-8	16
1005-01-135-4973	C-11	28
5360-01-136-5471	C-13	2
1005-01-144-1468	C-9	1
5305-01-144-1490	C-1B	3
	C-10	3
5360-01-144-1492	C-13	1
5305-01-144-1494	C-12	7
5365-01-144-1496	C-6	5
5340-01-144-1499	C-11	17
5340-01-145-7910	C-11	16
1005-01-146-7684	C-6	4
1005-01-146-7685	C-12	2
5305-01-147-8585	C-11	29
1005-01-148-0172	C-13	3
5360-01-148-1751	C-8	5
	C-1B	4
	C-10	4
1005-01-148-4805	C-11	19
1005-01-216-4510	C-1	5
1005-01-219-2402	C-14	1
5340-01-225-8339	C-11	5
1005-01-228-8504	C-12	1
1005-01-231-3138	C-11	2A
1005-01-233-8529	C-6	4A
1005-01-233-8530	C16A	3
1005-01-233-8531	C16A	1
5315-01-233-8608	C12A	2
5360-01-233-8616	C12A	3
5360-01-233-8617	C-11	1A
5310-01-233-8625	C16A	2
5310-01-233-8626	C12A	5
1005-01-233-8636	C12A	1
4710-01-233-8637	C-6	3A
5340-01-233-8638	C12A	4
1005-01-234-2297	C-6	2A
3040-01-247-7969	C-13	3A
1010-01-264-6517	C-7	9
5340-01-264-6530	C-7	7
5365-01-267-2169	C-11	27
5305-01-268-1191	C-11	21
5315-01-310-0370	C-17	11
5120-01-324-6631	C-17	12
1005-01-368-9852	C-1	5A
5360-01-381-6183	C-1B	7
	C-10	7
5340-01-382-3201	C-1A	10
5310-01-382-6793	C-1A	9
1005-01-382-6795	C-8	6A
5355-01-382-6801	C-1A	4
5360-01-382-6802	C-1A	7
1005-01-382-7083	C-1A	11
1005-01-382-7086	C-1B	2
1005-01-382-7089	C-1A	3
1005-01-395-4257	C-14	1A
5360-01-396-0256	C-14	3
	C-15	1B
1005-01-422-3770	C-2	4A
1005-01-424-5899	C-3	9

CROSS-REFERENCE INDEXES

NATIONAL STOCK NUMBER INDEX

STOCK NUMBER	FIG	ITEM	STOCK NUMBER	FIG	ITEM
88044	AN5010416-18		5305-01-268-1191	C-11	21
96906	MS16562-103		5315-00-058-6678	C-1B	1
				C-10	1
96906	MS16562-106		5315-00-058-6044	C-6	7
96906	MS16562-119		5315-00-812-3312	C-11	9
96906	MS16562-121		5315-00-840-3812	C-8	11
				C-8	12
				C-1A	8
96906	MS16562-129		5315-00-058-6081	C-16	5
				C16A	8
96906	MS16562-202		5315-00-843-9487	C12A	6
96906	MS16562-98		5315-00-597-5086	C-3	4
96906	MS16626-3137		5325-00-999-0863	C-6	10
96906	MS16632-3012		5325-00-999-0864	C-8	7
96906	MS19060-4808		3110-00-183-9175	C-8	4
				C-1A	6
				C-1B	5
				C-10	5
96906	MS35335-61		5310-00-527-3634	C-11	20
96906	MS39086-93		5315-00-690-0544	C-7	8
99999	NPN			C-11	30A
				C-11	30B
19204	11010032		4933-00-070-9151	C-17	8
19204	11010033		5120-00-070-9152	C-17	9
19204	12006359		4933-01-035-5607	C-17	1
19204	12006472		5220-01-043-9473	C-17	10
19200	12011849			C-1	2A
19200	12011987		5360-01-381-6183	C-1B	7
				C-10	7
19200	12011996		1005-01-368-9852	C-1	5A
19200	12597640		5365-01-267-2169	C-11	27
19200	12598617		1010-01-264-6517	C-7	9
19200	12598618		5340-01-264-6530	C-7	7
19200	12620101		5220-01-075-5004	C-17	3
19204	12624561		1005-01-216-4510	C-1	5
19200	12926769		5315-01-310-0370	C-17	11
19200	12951011			C-1	7
19200	12951017		5340-01-382-3201	C-1A	10
19200	12951018		1005-01-382-7089	C-1A	3
19200	12951019		5355-01-382-6801	C-1A	4
19200	12951020		5360-01-382-6802	C-1A	7
19200	12951021		1005-01-382-7083	C-1A	11
19200	12951023		5310-01-382-6793	C-1A	9
19200	12951026			C-1A	1
19200	12951028		1005-01-382-7086	C-1B	2
19200	12972652			C16A	6A
19200	12972670		1005-01-382-6795	C-8	6A
19200	12972675			C-6	1A
19200	12972680			C-1	3A
19200	12972690			C-1	6B
19200	12972691		1005-01-422-3770	C-2	4A

CROSS-REFERENCE INDEXES

PART NUMBER INDEX

CAGEC	PART NUMBER	STOCK NUMBER	FIG	ITEM
19200	12972692	1005-01-424-5899	C-3	9
19200	12972695	5360-01-396-0256	C-14	3
			C-15	1B
19200	12972696	1005-01-395-4257	C-14	1A
19200	12972697		C-11	18A
19200	12972698		C-11	18B
19204	7799734	5220-00-070-7814	C-17	6
19204	7799735	6695-00-070-7815	C-17	7
19204	8426685	1005-00-056-7106	C-17	
19204	8448201	4933-00-800-7508	C-17	2
19204	8448202	5220-00-221-9391	C-17	5
19204	8448496	5220-01-014-8183	C-17	4
19204	8448501		C-1	2
19204	8448502	5315-00-992-7294	C-2	3
19204	8448503	1005-00-017-9547	C-2	1
19204	8448504	1005-00-999-1509	C-2	2
19204	8448505		C-2	5
19200	8448506	1005-00-992-7283	C-4	2
19200	8448507	1005-00-738-6213	C-4	3
19204	8448508	5305-00-992-7284	C-4	1
19200	8448509	1005-00-992-7285	C-2	4
19204	8448510		C-3	2
19204	8448511	1005-00-992-7287	C-3	1
19204	8448512	1005-00-992-7288	C-3	7
19204	8448513	1005-00-992-7290	C-3	3
19204	8448515	1005-00-992-7291	C-3	6
19204	8448516	5360-00-992-7292	C-3	5
			C-11	22
19204	8448517	1005-00-017-9546	C-1	1
19204	8448518		C-5	4
19200	8448519	5340-00-999-0405	C-5	2
19204	8448520	5360-00-999-0404	C-5	3
19204	8448521-2	5315-01-048-9372	C-5	1
			C-9	2
19204	8448525	1005-00-978-1022	C-8	9
19204	8448532	5360-00-978-1025	C-8	10
19204	8448533	5315-00-978-1023	C-8	8
19200	8448540	5360-00-017-9541	C-8	13
19200	8448542	5360-00-523-8084	C-9	3
19204	8448543	3040-00-017-9539	C-9	5
19204	8448544	1005-00-017-9540	C-9	4
19204	8448555	5360-00-978-1036	C-6	9
19200	8448567	4710-00-978-1038	C-6	3
19204	8448571	1005-00-017-9543	C-7	4
19204	8448573	5315-00-979-3930	C-7	2
19204	8448574	5360-00-979-3931	C-7	3
19200	8448581	5340-00-992-7297	C-16	1
19204	8448582	5315-00-992-6651	C-16	2
			C16A	4
19200	8448583	5360-00-992-6652	C-16	3
			C16A	5

CROSS-REFERENCE INDEXES

PART NUMBER INDEX.

CAGEC	PART NUMBER	STOCK NUMBER	FIG	ITEM
19204	8448584	5315-00-992-6653	C-11	26
			C16A	9
19204	8448585	5315-00-992-6654	C-11	12
			C-11	12A
			C16A	10
19204	8448586	5360-00-992-6655	C-11	11
			C-11	11A
			C16A	11
19204	8448587	1005-00-992-7299	C-16	4
			C16A	7
19204	8448592	1005-00-992-7307	C-14	4
19204	8448593	5360-00-992-7308	C-14	2
			C-14	2A
			C-14	2B
19200	8448595	1005-00-992-6649	C-11	4
19204	8448599	5315-00-992-6650	C-11	25
19204	8448609	5315-00-992-7309	C-11	24
19204	8448610		C-11	3B
19204	8448611	5360-00-992-6648	C-13	1A
19200	8448612	1005-00-017-9551	C-13	4A
19200	8448615	1005-00-937-3078	C-11	2
19204	8448621	5315-00-017-9537	C-11	15
19200	8448628	1005-00-017-9548	C-11	8
19204	8448629	5360-00-992-6665	C-11	1
19204	8448631	1005-00-992-6667	C-11	23
19204	8448633	5360-00-056-2246	C-11	6
19204	8448634	1005-00-056-2247	C-11	7
19200	8448635	1005-00-999-0406	C-1l	11A
19204	8448636	1005-00-992-7302	C-11	14
19204	8448637	5360-00-992-7301	C-11	13
19204	8448638	1005-00-056-2201	C-11	10
19204	8448652	1005-00-403-0964	C-12	5
19200	8448653	5340-00-463-3892	C-12	3
19204	8448655	5315-00-463-3894	C-12	6
19200	8448670	1005-00-921-5004	C-1	4
19204	8448697	5320-01-063-7635	C-7	5
19204	8448712	1005-00-087-8998	C-6	8
19200	8448755	1005-00-760-3768	C-3	8
19200	9349050		C-1	3
19200	9349051	1005-01-134-3633	C-6	6
19200	9349052	5365-01-144-1496	C-6	5
19200	9349054		C-7	6
19200	9349056	1005-01-134-3625	C-7	1
19200	9349059	1005-01-134-3629	C-6	2
19200	9349062		C-6	1
19200	9349063	1005-01-134-3701	C-8	6
19200	9349065	5305-01-134-3622	C-8	2
			C-1A	2
19200	9349066	1005-01-134-3621	C-8	1
19200	9349067	5355-01-135-4972	C-8	16
19200	9349069	5360-01-148-1751	C-8	5

CROSS-REFERENCE INDEXES

PART NUMBER INDEX

CAGEC	PART NUMBER	STOCK NUMBER	FIG	ITEM
19200	9349069	5360-01-148-1751	C-1B	4
			C-10	4
19200	9349070	5360-01-134-3710	C-8	15
			C-1A	5
19200	9349072		C-8	3
19200	9349074	1005-01-134-3631	C-10	2
19200	9349075	1005-01-135-3697	C-1B	6
			C-10	6
19200	9349076	5305-01-144-1490	C-1B	3
			C-10	3
19200	9349077	5355-01-134-3627	C-1B	8
			C-10	8
19200	9349085	1005-01-144-1468	C-9	1
19204	9349086		C-8	14
19200	9349100		C-1	6
19200	9349101		C-11	30
19200	9349102		C-16	6
19200	9349106		C-11	3
19200	9349107	5360-01-144-1492	C-13	1
19200	9349108	1005-01-148-0172	C-13	3
19200	9349109	5360-01-136-5471	C-13	2
19200	9349110	1005-01-134-3630	C-13	4
19200	9349113	5340-01-144-1499	C-11	17
19200	9349114	5340-01-145-7910	C-11	16
19200	9349115		C-11	18
19200	9349116	5360-01-135-0353	C-15	1
			C-15	1A
19200	9349119	1005-01-135-4973	C-11	28
19200	9349120	5305-01-144-1494	C-12	7
19200	9349121		C-12	4
19200	9349124	1005-01-146-7684	C-6	4
19200	9349127	1005-01-148-4805	C-11	19
19200	9349128	5305-01-147-8585	C-11	29
19200	9349130	1005-01-146-7685	C-12	2
19200	9381367	5340-01-225-8339	C-11	5
19200	9381380	1005-01-228-8504	C-12	1
19200	9390003	1005-01-234-2297	C-6	2A
19200	9390007	1005-01-233-8529	C-6	4A
19200	9390009		C-7	6A
19200	9390011		C-1	6A
19200	9390012		C-11	28A
19200	9390013	1005-01-233-8636	C12A	1
19200	9390014	5340-01-233-8638	C12A	4
19200	9390015		C16A	6
19200	9390016	4710-01-233-8637	C-6	3A
19200	9390019	1005-01-233-8531	C16A	1
19200	9390020	5310-01-233-8625	C16A	2
19200	9390021	1005-01-233-8530	C16A	3
19200	9390022	5360-01-233-8617	C-11	1A
19200	9390023	1005-01-231-3138	C-11	2A
19200	9390025	5315-01-233-8608	C12A	2

CROSS-REFERENCE INDEXES

PART NUMBER INDEX

CAGEC	PART NUMBER	STOCK NUMBER	FIG	ITEM
19200	9390026	5310-01-233-8626	C12A	5
19200	9390027	5360-01-233-8616	C12A	3
19200	9390031	3040-01-247-7969	C-13	3A
19200	9390032		C-11	3A
19200	9390035	5120-01-324-6631	C-17	12
19200	9390736		C-15	2
			C-15	3
19200	9392518	1005-01-219-2402	C-14	1

APPENDIX D
EXPENDABLE/DURABLE SUPPLIES AND MATERIALS LIST

Section I. INTRODUCTION

D-1. SCOPE. This appendix lists expendable durable supplies and materials you will need to operate and maintain the M16A2 Rifle This listing is for informational purposes only and Is not authority to requisition the listed Items. These items are authorized to you by CTA 50-970, Expendable, Durable Items (Except Medical, Class V, Repair Parts, and Heraldic Items), or CTA 8-100, Army Medical Department Expendable/Durable Items.

D-2. EXPLANATION OF COLUMNS.

a. Column (1)-Item Number. This number is assigned to the entry in the listing and is referenced In the narrative instructions to identify the material (e.g., "Use cloth, abrasive, crocus, Item 12, app D").

b. Column (2)-Level. This column identifies the lowest level of maintenance that requires the listed Item.

C - Operator/Crew Maintenance
O - Unit Maintenance
F - Direct Support Maintenance

c. Column (3)-National Stock Number. This is the National stock number assigned to the Item, use It to request or requisition the Item.

d. Column (4)-Description. Indicates the Federal Item name and, if required, a description to identify the item. The last line for each Item indicates the Commercial and Government Entity Code (CAGEC) in parentheses followed by the part number.

e. Column (5)-Unit of Measure (U/M). Indicates the measure used in performing the actual maintenance function. This measure is expressed by a two-character alphabetical abbreviation (e g., ea, In , pr). If the unit of measure differs from the unit of issue, requisition the lowest unit of issue that will satisfy your requirements.

Section II. EXPENDABLE/DURABLE SUPPLIES AND MATERIALS LIST

(1) ITEM NUMBER	(2) LEVEL	(3) NATIONAL STOCK NUMBER	(4) DESCRIPTION PART NUMBER AND CAGE	(5) UNIT OF MEAS
1	F	8040 00-944 7292	ADHESIVE, I<IT, (81348) MMM A 1754	KT
2	0	8020 00 244 0153	BRUSH, ARTIST'S metal ferrule, flat chisel ed(i(e 7 16 w, 1 1/8, exposed bristle (813481 H B 241	EA
3	F	1005-00-716 2702	BRUSH, CLEANING, SMALL ARMS (192051 7162702	EA
4	C	1005-00-903 1296	BRUSH, CLEANING, SMALL ARMS bore (19204) 11686340	EA
5	C	1005-00 999-1435	BRUSH, CLEANING, SMALL ARMS chamber (19204) 8432358	EA
6	C	1005 00-444-6602	BRUSH, CLEANING, SMALL ARMS tooth 1192041 8448462	EA
7	O	7920-00 205-2401	BRUSHCLEANING, TOOLS AND PARTS (813491 MILS43871	EA
8	O	6850-00-965-2332	CARBON REMOVING COMPOUND (813481 P-C-111	GL
9			CLEANER, LUBRICANT AND PRESERVATIVE (27412)	
	O	9150-01-079-6124	CLP- 4 oz (118 30 ml) bottle	EA
	O	9150-01-054 6453	CLP- 1 pt (0 47 1l bottle	EA
	O	9150-01 053-6688	CLP 7 gal (26 50 11 bottle	EA
10	C	9150 01-102 1473	CLEANER, LUBRICANT AND PRESERVATIVE (81349i MIL L 63460 1/2 oz (14 79 ml) bottle	EA
11	C	9920-00-292 9946	CLEANER, TOBACCO PIPE cotton turf, wire core (89855) DILLSPIPE cleaner (6 per pkg)	
			CLEANING COMPOUND, RIFLE BORE small arms bore cleaner solution (RBC) 1813491 MIL C 372	
	C	6850-00-224-6656	2 oz 159.15 mil) bottle	OZ
	O	6850-00-224-6657	8 oz (236 59 ml) can	CN
	O	6850-00-224-6663	1 gal (3.79 l) can	CN

EXPENDABLE/DURABLE SUPPLIES AND MATERIALS LIST (CONT)

(1) ITEM NUMBER	(2) LEVEL	(3) NATIONAL STOCK NUMBER	(4) DESCRIPTION PART NUMBER AND CAGE	(5) UNIT OF MEAS
13	0	5350-00-221-0872	CLOTH, ABRASIVE: (58536) A-A-1206	SH
14			Deleted See Item 23.2	
15	F		DICHLOROMETHANE, TECHNICAL: (81346) ASTM D 4701	
		6810-00-244-0290	5 gal (18.93 1) pail	CN
		6810-00-616-9188	600 lb (272.16 kg) drum	DR
16	0	6850-00-281-1985	DRY CLEANING SOLVENT: (58536) A-A-711 1 gal(3.79 l1) can	GL
17	0	8010-00-297-0560	ENAMEL: olive drab no 3407 (81348) TT-E-527 1 gal (3.791) can	GL
18	0		GLOVES, CHEMICAL AND OIL PROTECTIVE: (81348) ZZ-G-381	
		8415-00-823-7455	Size 9	PR
		8415-00-823-7456	Size10	PR
		8415-00-823-7457	Size 11	PR
19	F	9150-00-754-2595	GREASE, MOLYBDENUM DISULFIDE: (81349) MIL-G-21164	LB
20	C	1005-01-113-0321	HANDLE SECTION, CLEANING ROD, SMALL ARMS: (19204)8436776	EA
21	0	9150-01-260-2534	LUBRICANT, SOLID FILM: (81349) MIL-L-23398 16 OZ (473.18 ml) spray can	OZ
22	C	9150-00-292-9689	LUBRICATING OIL, WEAPONS: (LAW) (81349) MIL-L-14107 1 qt (0.95 1) can	QT

EXPENDABLE/DURABLE SUPPLIES AND MATERIALS LIST (CONT)

(1) ITEM NUMBER	(2) LEVEL	(3) NATIONAL STOCK NUMBER	(4) DESCRIPTION PART NUMBER AND CAGE	(5) UNIT OF MEAS
23			LUBRICATING OIL, WEAPONS: (LSA), semifluid (81349) MIL-L-46000	
	C	9150-00-935-6597	2 oz (59.15 ml) plastic btl	OZ
	C	9150-00-889-3522	4 oz (118.30 ml) plastic btl	OZ
	O	9150-00-687-4241	1 qt (0.95 I1) can	CN
	O	9150-00-753-4686	1 gal(3.79 I1) can	CN
23.1	0	5340-01-230-3181	MOUNTING BRACKET (M4/M4A1 ONLY) (19200)12556036	EA
23.2	0	8010-00-087-0102	PAINT, ENAMEL, SEMIGLOSS: paint for blank firing attachment (M1 5A2) (81348) TT-E-529 1 qt can (RED - M16A2)	EA
23.3	0	8010-01-031-1274	PAINT, ENAMEL, SEMIGLOSS: paint for blank firing attachment (M23) (81348) TT-E-529 1 pt can (YELLOW - M4/M4A1)	EA
24	0	4940-00-795-3595	PAN, WASH: (94453) 1211	EA
25	F	6850-00-826-0981	PENETRANT KIT: (81349) MIL-I-25135	KT
26	F	8135-01-019-1691	POLYETHYLENE SHEET: (84744) PE88-80-2	EA
26.1	0	1005-01-394-7677	PROTECTOR, RAIL (M4/M4A1 ONLY) (19200)12972676	EA
27	C	7920-00-205-1711	RAG, WIPING: (58536) A-A-531 50 lb (22.68 kg) bdl	LB
28	C	1005-00-050-6357	ROD SECTION, CLEANING SMALL ARMS: (19204) 8436775 (3 REQUIRED)	EA
29	F	8030-00-670-8553	SEALING COMPOUND (16059) DEVCONF	KT

EXPENDABLE/DURABLE SUPPLIES AND MATERIALS LIST (CONT)

(1) ITEM NUMBER	(2) LEVEL	(3) NATIONAL STOCK NUMBER	(4) DESCRIPTION PART NUMBER AND CAGE	(5) UNIT OF MEAS
30	C	1005-00-937-2250	SWAB HOLDER SECTION, CLEANING ROD, SMALL ARMS: (19204) 11686327	EA
31	C	1005-00-912-4248	SWAB, SMALL ARMS: (19204)11686408	SH
32	C	6920-01-395-2949	TARGET, 25 METER ZEROING, M16A2: (19200)12012024	BX

APPENDIX E

ILLUSTRATED LIST OF MANUFACTURED ITEMS

E-1. INTRODUCTION.

a. This appendix includes complete instructions for making items authorized to be manufactured or fabricated at unit or direct support maintenance.

b. A part number in alphanumeric order is provided for cross-referencing the part number of the item to be manufactured to the figure which covers fabrication criteria.

c. All bulk matenals needed for manufacture of an item are listed by part number or specification number In a tabular list on the illustration.

E-2. MANUFACTURED ITEMS PART NUMBER INDEX.

INDEX

Item	Figure Number
Front sight detent depressor	E-1
Front sight post removal and installation tool	E-2
Pivot pin removal tool	E-3
Pivot pin installation tool	E-4
Slave pin	E-5
Adapter bar for M12 arms rack	E-6
Modified needle nose pliers	E-7

E-3. MANUFACTURED ITEMS ILLUSTRATIONS - FORMAT AND CONTENT.

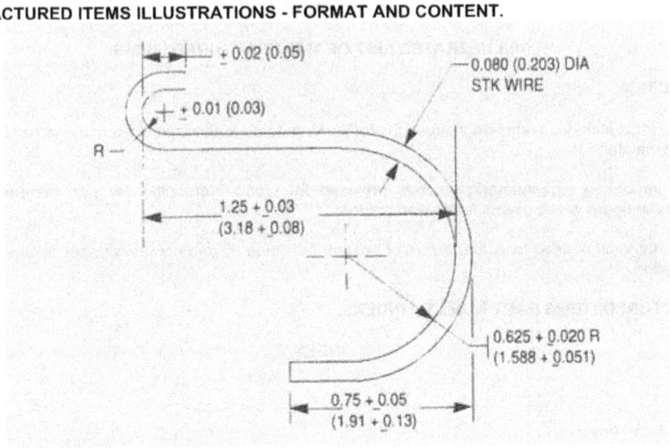

FABRICATE FROM 0.08 IN. MUSIC WIRE OR EQUIVALENT.
FINISH: NO. 5.3.1.2 OR 5.3.2.2 OF MIL-STD 17.

NOTE: ALL DIMENSIONS SHOWN ARE IN INCHES WITH THE METRIC CONVERSION TO CENTIMETERS IN PARENTHESIS.

Figure E-1. Front sight detent depressor.

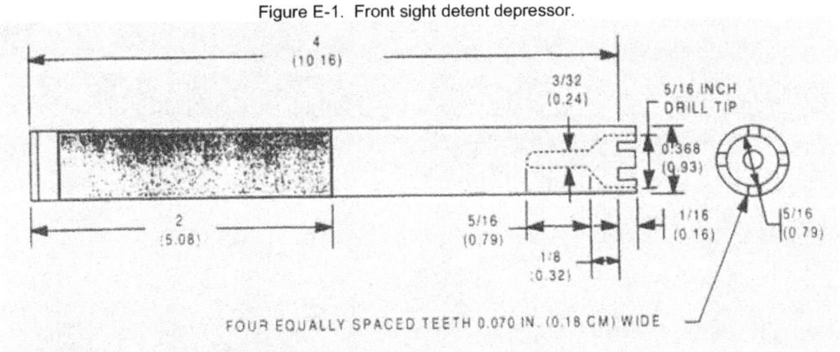

FABRICATE FROM 0.375 INCH ROUND METAL BAR,
ASTM A686. FSCM 81346, GRADE C. CLASS W2-09.
NSN 9510-00-640-4407 OR EQUIVALENT.

NOTES:
1. ALL DIMENSIONS SHOWN ARE IN INCHES WITH THE METRIC CONVERSION TO CENTIMETERS IN PARENTHESIS.
2. TEETH MUST BE HAND FILED TO FIT FRONT SIGHT POST.

Figure E-2. Front sight post removal and Installation tool.

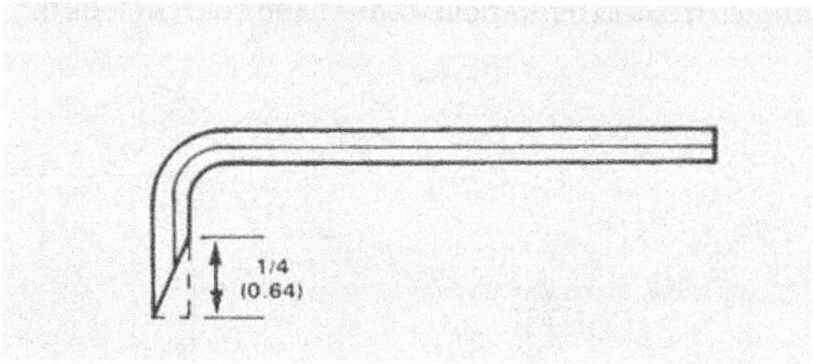

FABRICATE FROM 1'16 IN. SOCKET HEAD SCREW KEY
NSN 5120-00-198-5398 OR EQUIVALENT.

NOTE: ALL DIMENSIONS SHOWN ARE IN INCHES WITH THE METRIC CONVERSION TO CENTIMETERS IN PARENTHESES

Figure E-3. Pivot pin removal tool.

E-3. MANUFACTURED ITEMS ILLUSTRATIONS-FORMAT AND CONTENT (CONT).

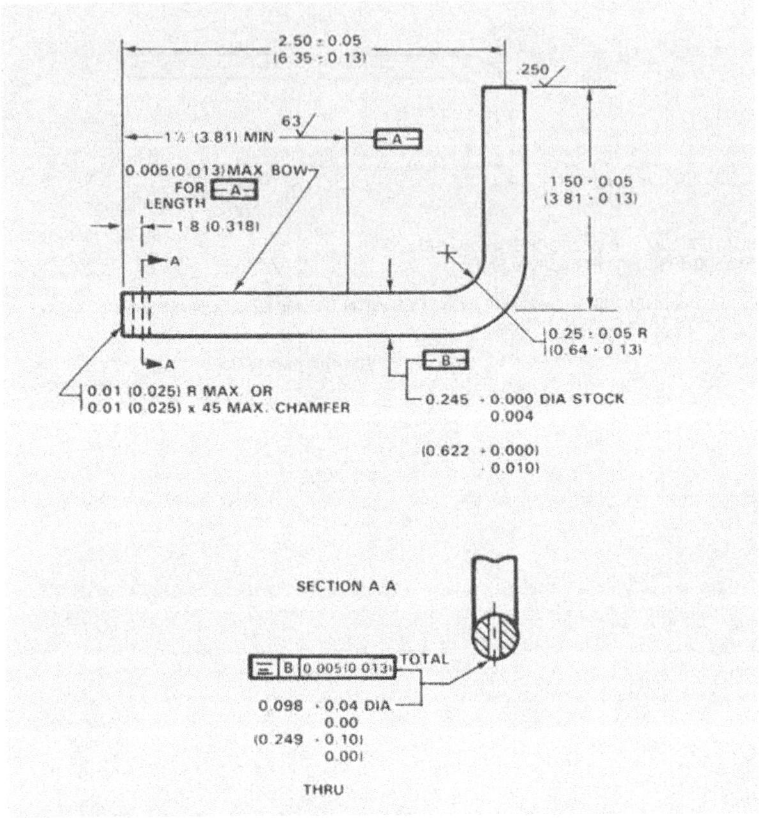

FABRICATE FROM 0 245 IN STEEL AISI 1095 OR EQUIVALENT
HARDEN AND TEMPER TO RC-57-61 FOR LENGTH -A-
FINISH 5 3 1.2 OR 5 3.2.2 OF MIL-STD-171

NOTE: ALL DIMENSIONS SHOWN ARE IN INCHES WITH THE METRIC CONVERSION TO CENTIMETERS IN PARENTHESES

Figure E-4. Pivot pin installation tool.

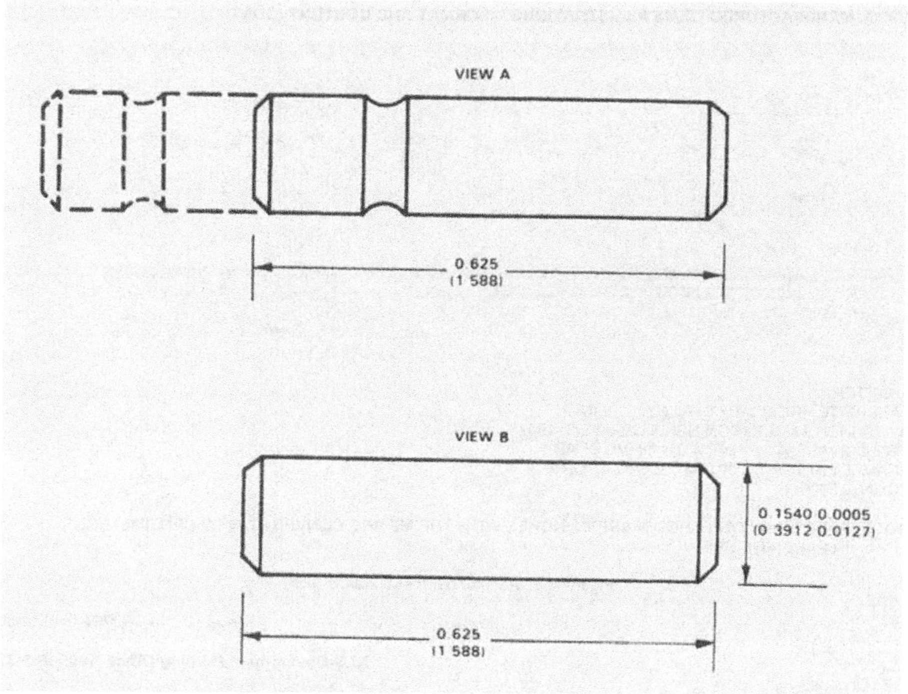

FABRICATE FROM OLD TRIGGER PIN (VIEW AI P,/N 8448609 OR FABRICATE SLAVE PIN (VIEW B) FROM MATERIAL BLOCK, WIRE, STEEL ALLOY, GRADE 4140, ASTM-A547 OR EQUIVALENT

NOTE: ALL DIMENSIONS SHOWN ARE IN INCHES WITH THE METRIC CONVERSION TO CENTIMETERS IN PARENTHESES

Figure E-5. Slave pin.

E-3. MANUFACTURED ITEMS ILLUSTRATIONS -- FORMAT AND CONTENT (CONT).

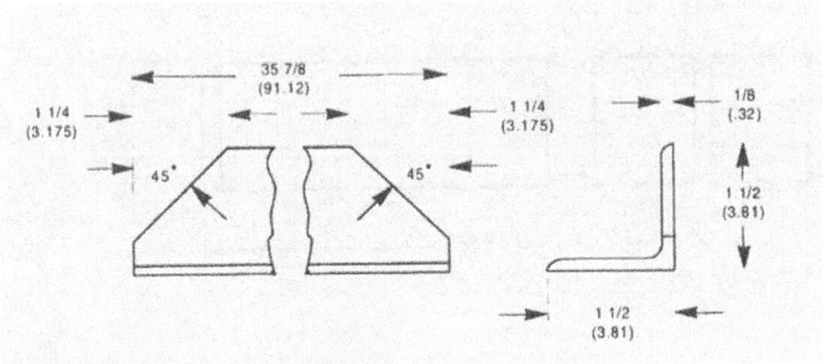

ADAPTER
FABRICATE FROM 1 1/2 INCH BY 1 1/2 INCH
BY 1/8 INCH ANGLE IRON NSN 9520-00-277-4902
OR EQUIVALENT. PAINT WITH OLIVE DRAB
ENAMEL PAINT, NSN 8010-01-350-5249, OR
EQUIVALENT.

NOTE: ALL DIMENSIONS SHOWN ARE IN INCHES WITH THE METRIC CONVERSION TO CENTIMETERS IN PARENTHESES.

Figure E-6. Adapter bar for M12 arms rack.

PIN: 060514-004

***U S Government Printing Office 1995- 648-378**

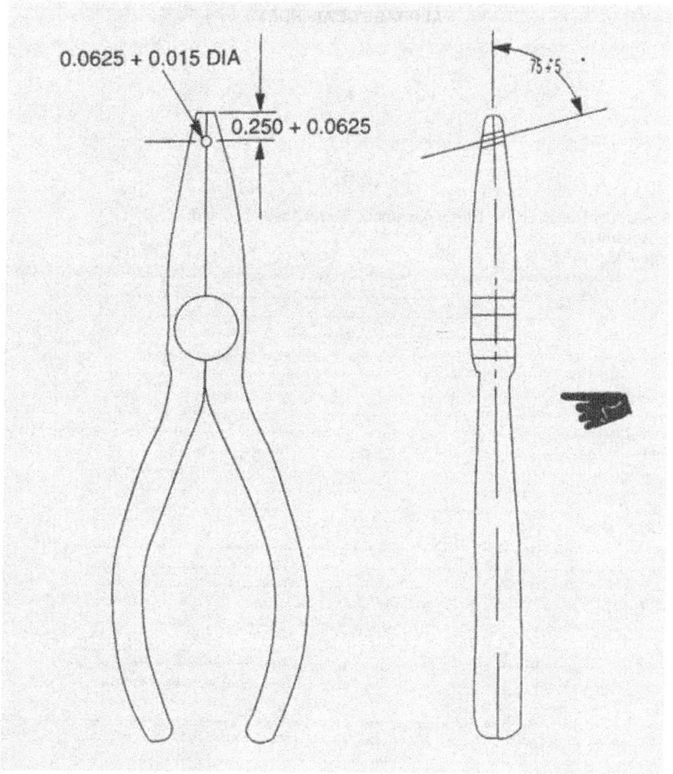

FABRICATE FROM NEEDLE NOSE PLIERS,
NSN 5120-00-268-3579 OR EQUIVALENT.

NOTE: ALL DIMENSIONS ARE IN INCHES.

Figure E-7. Modified needle nose pliers.

***U. S. GOVERNMENT PRINTING OFFICE 1997 - 545-039/4 077**

ALPHABETICAL INDEX

Subject	Page
A	
Auxiliary Equipment	4-1
B	
Barrel Assembly (See Upper Receiver and Barrel Assembly, Barrel Assembly, and Upper Receiver Assembly)	
Blank Firing Attachment M 15A12	
Cleaning	4-11
Inspection	4-11
Installation	4-9
Removal	4-10
Repainting	4-11
Replacement	4-11
Bolt Assembly	
Cleaning	2-40
Disassembly	2-39, 3-21
Inspection Repair	2-40, 3-21
Lubrication	2-41
Reassembly	2-41 3 24
Test	3-24
Bolt Carrier Assembly	
Cleaning	2-36, 3-16
Disassembly	2-35, 3-16
Inspection	3-17
Inspection Repair	2-36
Lubrication	2-37
Reassembly	2-37, 3-20
Repair	3-20
Test	3-20
Buttstock Assembly	
Cleaning	2-65
Disassembly	2-65
Inspection	2-66, 3-66
Lubrication	2-67
Reassembly	2-67
Repair	2-67, 3-67
C	
Charging Handle Assembly	
Cleaning	2-44
Disassembly	2-43
Inspection Repair	2-44
Lubrication	2-44
Reassembly	2-44
Common Tools and Equipment	2-1, 3-1
Corrosion Prevention and Control ,CPCI	1-2

ALPHABETICAL INDEX (CONT)

Subject	Page
D	
Destruction of Army Materiel to Prevent Enemy Use	1-1
E	
Equipment Characteristics, Capabilities, and Features	1-3
Equipment Data	1-5
Equipment Description and Data	1-3
Expendable Durable Supplies and Materials List	D-1
F	
Forward Assist Assembly	
Disassembly	3-59
Inspection	3-60
Lubrication	3-61
Reassembly	3-61
Repair	3-60
Functional Theory of Three-Round Burst Control (See M16A2 Rifle Final Inspection)	
G	
Gauging Requirements (See M16A2 Rifle Annual Gauging Requirements)	
General Information	1-1
General Inspection Criteria	3-97
H	
Hammer Assembly	
Disassembly	3-73
Inspection Repair	3-74
Reassembly	3-74
How to Use This Manual	iii
I	
Illustrated List of Manufactured Items	E 1
Initial Setup	2 33
K	
Key and Bolt Carrier Assembly	
Disassembly	3-26
Reassembly	3-27
Repair	3-26

ALPHABETICAL INDEX (CONT)

Subject	Page
L	
Location and Description of Major Components	1-4
Lock Plate	
Inspection	4-5
Installation	4-2
Removal	4-4
Lower Receiver and Buttstock Assembly	
Cleaning	2-59
Disassembly	2-57, 3-62
Inspection	6-0, 3-66
Lubrication	2-61
Reassembly	2-62, 3-70
Repair	2-61, 3-67
Test	3-68
Lower Receiver and Receiver Extension Assembly	
Disassembly	3-77
Inspection	3-79
Reassembly	3-80
Repair/Modify	3-79
Lubrication General	2-33
M	
Maintenance Allocation Chart	B-1
Maintenance Forms, Records, and Reports	1-1
Maintenance Procedures	2-33, 3-15
Major Components of M16A2 Rifle (See M16A2 Rifle)	
Mechanical Zero Procedures	2-54, 3-58
M15A12 Blank Firing Attachment (See Blank Firing Attachment M15A12)	
M16A2 Rifle	
Disassembly	2-34, 3-15
Inspection	2-69
Reassembly	2-68, 3-83
Stowage	2-70
Test	3-83
M16A2 Rifle Annual Gauging Requirements	3-94
M16A2 Rifle Final Inspection	
Final Inspection	3-85
Functional Theory of Three-Round Burst Control	3-88
Test	3-88
O	
Official Nomenclature, Names, and Designations	1-1
P	
Preembarkation Inspection of Materiel in Units Alerted for Overseas Movement	3-95
Preparation for Storage or Shipment	1-1, 2-70
Preventive Maintenance Checks and Services	2-3
Principles of Operation-	1-6

ALPHABETICAL INDEX (CONT)

Subject	Page
R	
Rear Sight Assembly (See Upper Receiver Assembly and Rear Sight Assembly)	
References	A-1
Repair Parts	2-1, 3-1
Repair Parts and Special Tools List	C-1
Repair Parts List	C-1 1
Repair Parts, Special Tools, TMDE, and Support Equipment	2-1, 3-1
Reporting Equipment Improvement Recommendations (EIR)	1-2
Reporting Errors and Recommending Improvements	
S	
Service Upon Receipt	2-2
Special Tools List	C-17-1
Special Tools, TMDE, and Support Equipment	2-1
Symptom Index	2-21, 3-2
T	
Top Sling Adapter	
Inspection	4-8
Installation	4-6
Removal	4-7
Trigger Assembly and Trigger Subassembly	
Disassembly	3-75
Inspection Repair	3-76
Reassembly	3-76
Trigger Subassembly (See Trigger Assembly and Trigger Subassembly)	
Troubleshooting	2-21, 3-2
U	
Unit Level Auxiliary Equipment Repair	4-1
Upper Receiver and Barrel Assembly	
Disassembly	3-30
Inspection Cleaning	3-33
Reassembly	3-39
Repair	3-35
Test	3-45
Upper Receiver and Barrel Assembly, Rifle Barrel Assembly, and Upper Receiver Assembly	
Cleaning	2-48
Disassembly	2-45
Inspection	2-49
Lubrication	2-52
Reassembly	2-52
Repair	2-52

Subject **Page**

Upper Receiver Assembly (See Upper Receiver and Barrel Assembly, Barrel Assembly, and Upper Receiver Assembly or Upper Receiver Assembly and Rear Sight Assembly)
Upper Receiver Assembly and Rear Sight Assembly
Disassembly 3-48
Inspection 3-51
Lubrication 3-53
Reassembly 3-54
Repair 3-53

W

Warnings .... a

Made in United States
Orlando, FL
05 November 2026

90229003R00272